101 Advances in Polymer Science

Polymer Compositions Stabilizers/Curing

With contributions by
G. R. Barshtein, M. L. Fridman,
S. G. Kulichikhin, A. Y. Malkin
J. Pospišil, O. Y. Sabsai,
R. P. Singh, S. Sivaram, J. P. Terent'eva

With 59 Figures and 19 Tables

Springer-Verlag
Berlin Heidelberg GmbH

ISBN 978-3-662-15005-4   ISBN 978-3-540-46462-4 (eBook)
DOI 10.1007/978-3-540-46462-4

Library of Congress Catalog Card Number 61-642

Originally published by Springer-Verlag Berlin Heidelberg New York in 1991
Softcover reprint of the hardcover 1st edition 1991

Typesetting: Th. Müntzer, Bad Langensalza

02/3020-543210 — Printed on acid-free paper

Editors

Table of Contents

Compositions with Mineralorganic Fillers

Grigorii Removitch Barshtein
Vtoraja Ulitsa Bebelya 26, apt. 139, 103220 Moscow, USSR

Otto Yulievitch Sabsai
Nizhnyaa Pervomaiskaya ul. 24, apt. 218, 105554 Moscow, USSR

The review contains results of recent studies of the technological properties of compositions filled with natural mineralorganic products in comparison with pure non-filled polymers. We will provide results of studying isothermal and non-isothermal flows, phase transitions kinetics, and the influence of the thermophysical properties of the compositions on their behaviour during processing. We will discuss the methods for forecasting technological properties of these processing regimes.

Advances in Polymer Science 101
© Springer-Verlag Berlin Heidelberg 1991

List of Symbols

ε_c^*	— critical reversible deformation
τ_c	— critical shear stress
K_τ	— empirical coefficient (Ex 15)
T_g	— glass transition temperature
C_1 and C_2	— Williams-Lendall-Ferry constants
ΔT_g	— changing of glass transition temperature
$\varepsilon_0(t)$	— crystallization degree on the time moment
z	— growth rate
n	— Kholmogorov's-Avramis index
A, K_y	— constant (Ex 21)
T_m^0	— equilibrium melting temperature
E	— activation energy
h	— specific crystallization heat
α	— thermal linear expansion coefficients
λ	— thermal conductivity coefficient
L	— length of filling the "snail" spiral form
ϱ	— density
F.C.	— flow curve
LDPE	— low density polyethylene
HDPE	— high density polyethylene
PS	— polystyrene
CEVA	— copolymer ethylene with vitylacetate
PE	— polyethylene
c	— organic component weight concentration in the filler
φ	— volume fraction
φ_{max}	— maximum volume concentrations
η	— viscosity
η_0	— Newtonian viscosity
γ	— shear rate on the wall
γ_{eff}	— apparent shear rate
τ	— shear stress
$a_T(\eta)$	— "Temperature" displacements of curves flow along
$a_T(\tau)$	— the viscosity and shear stress axles
$a_\psi(\eta)$	— "Concentration" displacements of curves flow
$a_\psi(\tau)$	— along the viscosity and shear stress axles
K_τ	— empiristic coefficient (Ex 3)
P_c	— losses of pressure on the composition flow in the channel
P_{en}	— entrance pressure losses
P_{ex}	— exite pressure losses
P_R	— pressure in the reservoire
S	— constante (Ex 7)
m	— constante (Ex 7)
b	— piezocoeffients of viscosity
μ	— constante (Ex 11)
$\varkappa$	— constante (Ex 10)
Q	— volume rate
Q_c	— critical volume rate
σ^*, ε^*	— correspondingly tension and reversible deformation

1 Introduction

The review of commercial polymer fillers [1], published not long ago, is very concise. It consists of short abstracts, each containing the filler name, alist of polymers it is introduced into and companies which manufacture the compositions. This simple enumeration is 40 pages long and includes 186 references. The review [1] demonstrates the evergrowing interest of leading companies in the production and utilization of filled polymers. Use of filled polymers instead of conventional unfilled polymers is a very effective method of economizing on the polymer in production of non-constructive items like decorative panels for interiors, plinths, furniture fittings, etc. Until the present decade fillers were introduced into polymers in comparatively limited quantities (up to 10% of total volume) to improve durability and hardness of the finished product, as well as for other purposes. Development and commercial production and utilization of high-power mixers, dosimeters and pelletizers became profitable due to increases in the world prices for oil and gas as the basic raw materials for polymer synthesis − filling is a means of saving raw materials. Developments in this field and of this character have been carried out in the USSR for rather a long time [2].

For filling to be profitable and the composition to have satisfactory physical, mechanical and technological properties, the filler should be up to the requirements, briefly outlined in Table 1.

The data in the above Table are surely disputable and far from complete. It does not include specific requirements for fillers and is of a general character. Though, this Table allows us to define the minimum range of problems we have to answer, when assessing the reasons for using a new thermoplast filler. These requirements make a good basis for formulating advantages and disadvantages of two filler classes, namely: organic and non-organic, according to the existing classification.

From the literature analyses we see that organic fillers have thermal expansion coefficients (TEC) values close to those of thermoplasts [7], they have low abrasiveness, their density is similar to that of thermoplasts [6, 8] but as a rule, their grindig is difficult [9], they are combustible and explosive, not thermostable and not always cheap [9]. Mineral fillers are comparatively easy to grind, their cost is low, they are thermostable and non-combustible, but their TEC's are an order lower than those of thermoplasts. As a rule, their density 2−3 times higher than the thermoplasts' [6]. Their abrasiveness is too high.

One of the possible ways of overcoming the fillers' drawbacks is utilization of organomineral products, combining advantages of the fillers of both types [10].

Table 1. General requirements of fillers

Requirements of fillers	Results	Results of inadequate fulfilling of requirements
filler density should be similar to polymer density	composition density is maintained at the basic polymer density level	high filler density leads to dramatic increase of composition density and, consequently to heavier finished products
filler and polymer thermal expansion coefficients should be comparable	low thermal tensions in the finished product	difference between filler and polymer thermal coefficients leads to development of thermal tensions and, consequently, reduces durability of the finished product [3, 4]
filler should not considerably increase abrasiveness of composition	equipment depreciation is insignificant	abrasiveness increase leads to significant depreciation of processing equipment
introduction of filler should not lead to dramatic increase in composition viscosity	maintains the capacity of processing equipment of the basic polymer level	reduction of equipment capacity and appearance of undesired rheologic effects (e.g. unsteady flow), [5]
filler particle sizes and their distribution should correspond to the type of the finished proproduct	optimally balanced complex of physical, mechanical and rheological properties [6]	some physical, mechanical of technological properties of the composition do not correspond to the requirements of the finished product
filler should be thermally stable at processing temperatures	properties of the finished product are as required	the finished product will not be solid, it's physical and mechanical properties will be unadequate
combustion and explosion safety	no need for explosion-safe equipment	need for additional construction improvements
filler stock should be sufficient for commercial applications	high economic efficiency of the material	limited economic efficiency of the filler utilization

2 Mineralorganic Fillers

Mineralorganic fillers can be produced by mixing mineral fillers, like ash from heating and power plants, chalk, glass spheres, etc., with organic fillers such as saw-dust, ground nut shells, starch, etc. These immense possibilities have not been exploited yet.

There are also natural mineralorganic fillers, namely plant and animal residues of different geologic age: coals, shales, sapropels. The stocks of these materials amount to trillions of tonnes. In particular, sapropels (products of geologic evolution of lake silts) are excavated during dredging of lakes and used as a fertilizer to enrich poor soils. Sapropels have started to be used as polymer fillers very recently.

Combustible shales and kerogenes (products of their processing [12, 13]) have been used more widely. Kerogenes are produced on a commercial scale. They are marked per weight content of organic component. For example, Kerogene 70 contains 70% organic matter, in this case, kerogene itself. In Estonia, USSR, they are carrying out broad studies in the physics, chemistry and problems of the utilization of combustible shales. Interest in these problems in other countries is also significant [14]. The properties of kerogenes as polymer fillers are well known [13]. In Table 2 characteristics of Kerogene 70 as compared to mineral and organic fillers can be seen. Table 2 demonstrates that the Kerogene 70 commercial product can be used as a polymer filler.

The organic matter of coals, shales and sapropels is a multipolymer in its chemical nature. This multipolymer consists of multifunctional components, connected by buffer bridges into a non-soluble gel. (For more detail see [14].) In [13] it is proved that at t = 70 °C, kerogene transfers from a glassy to a rubbery state, demonstrating a viscous-elastic instead of a purely elastic reaction to applied tensions. This means that on pelletizing and processing of mineralorganic filled compositions they behave like round rubbers [5].

Thus researchers' interest in production, processing and utilization of polymers, filled with natural mineraloganic products is easily understandable. The review contains results of recent studies of the technological properties of these compositions in comparison with pure non-filled polymers and mineral products. We will provide results of studying isothermal and non-isothermal flows, phase transitions Kinetics, influence of thermophysical properties of the compositions on their behaviour during processing. We will also discuss methods of forecasting technological properties of these processing regimes.

Table 2. Comparative characteristics of organic, mineral fillers and Kerogene 70

Characteristic Filler	Organic	Mineral	Kerogene 70
Density kg/m^3 × 10^{-3}	0.8–1.0	2.2–4.8	1.35
Thermal linear expansion coefficient degree^{-1} × 10^5	15.0	1.0	10.8
Dispersiveness (particles size), μm	100–500	0.5–40	max. 120
Thermal stability	non thermo-stable	thermostable	thermostable up to 180 °C
Combustibility	combustible	non-combustible	combustible

3 Isothermal Flows

3.1 Superposition of the Flow's Curves

Rheologic behaviour of kerogene-filled thermoplasts is thoroughly examined in [13, 16–21]. As in [12] these compositions were marked "keroplasts" with the addition of the abbreviated name of the corresponding basic thermoplast. The authors [13], [16–21] varied the organic component content in the filler from 0% to 90% (of the gross weight) and total concentration of fillers in keroplasts from 0% to 40% (of volume). The fillers were produced by two methods:

1. by enriching combustible shales (rock);

2. by mixing the concentrate of the shales organic component, containing 90% (gross weight) of organic component, i.e. Keroene 90, with dispersed mineral fillers, i.e. shale ash, chalk and tufa. The basic polymers were: LDPE — brands 15303-003, 10802-020; HDPE — brands 277-73, 20906-040; block polystyrene PSM 115 v/s; ethylene co-polymer with vinyl acetate CEVA 113-075. Studies of these compositions basically resulted in the following:

1. Keroplast rheologic properties do not depend on the composition production method and type of kerogene mineral component.

2. Keroplast rheologic properties do not depend on the organic component content in the filler, but only on the polymer matrix type and total filler concentration.

Let us discuss the results of studies [13, 16–21], obtained through studying isothermal flows of keroplasts. In compliance with the above-mentioned facts these results can be applied to the description of the rheologic behaviour of compositional polymer materials with various disperse inert fillers. At displacement speeds corresponding to the speeds realized under the conditions of processing thermoplastic compositions, the Newton flow area was obtained on the flow curves (FC) of sevilene-based keroplasts but not with other keroplasts (polyethylene and polystyrene-based).

It has been experimentally demonstrated that the principles of temperature and concentration superposition of flow curves are applicable to melted keroplasts. In other words, in the fixed matrix FC compositions with different filler content, φ and T measured at different temperatures, can be interpolated through a flat and parallel displacement along the coordinate axes (in double logarithmic coordinates).

Let us study the flow curve in the form $\lg \eta = f(\lg \tau)$, where τ is shear stress on the channel wall, η is viscosity. The displacement vector along which the FC's are matched can be presented as consisting of 4 components:

1. $a_\tau(\eta)$ — "temperature" displacement along the viscosity axis. Existence of the superposition itself shows that it is a dependence of the basic polymer's Newton viscosity on the temperature, described by the Williams-Landell-Ferry or Arrenius equations (e.g. see [22]).

2. $a_\varphi(\eta)$ — "concentration" displacement along the viscosity axis. This is nothing but a dependence of the composition's Newton viscosity on the filler concentration. Experimental studies [10] have demonstrated that T. Chong's [23] equation

produces the best possible similarity to experiment for kerogene and other disperse fillers:

$$\eta_0(\varphi) = \eta(0) \cdot \left(1 + 0.75 \frac{\varphi}{\varphi_{max} - \varphi}\right), \tag{1}$$

where φ_{max} is the maximum volume concentration (filling) in the solid phase. φ_{max} value depends on the fraction composition of fillers [24]. If a filler is composed of large particles by 75% and of small ones by 25%, than, depending on the ratio of their diameters Δ, φ_{max} values will change from 0.605 at $\Delta = 1$ to 0.756 at $\Delta = 0.138$ [23, 24]. V. V. Moshev [23] in his study uses the algorithm for calculating φ_{max} value for the case of arbitrary distribution by the size of the filler's spherical particles. Dependence of $\eta_0(\varphi)$ becomes significant at $\varphi \geq 50\%$ (of volume) [23, 24]. In general, it is recommended [23] to describe the dependence of the suspensions' viscosity of filler concentration in the given coordinates φ/φ_{max}. In experiments with keroplasts, where φ values did not exceed 40% (of volume), the authors [13, 16–21] neglected dependence of $\eta_0(\varphi)$ on the filler granulometric composition, on the assumption that real commercial fillers have, "on the average", similar φ'_{max} and dependence $\eta(\varphi)$ on φ_{max} is insignificant up to concentration values $\varphi \leq 40\%$ (of volume) (according to [24]). φ_{max} value for keroplasts was 0.6 [21, 25].

3. $a_T(\tau)$ — "temperature" displacement along the stress axis, in a sense, a dependence of a melted elastic module on the temperature:

$$a_T(\tau) = T/T_D. \tag{2}$$

4. $a_\varphi(\tau)$ — "concentration" displacement along the stress axis. It was experimentally proved in [21, 25] that

$$\lg a(\tau) = K_\tau \cdot \varphi, \tag{3}$$

where K_τ is an empirical coefficient, constant for the specific basic polymer sample. To define K_τ, the FC of a pure polymer and that of a composition on its basis with any concentration of any disperse filler, should be measured, K_τ values for different polymers are shown in Table 3.

After having calculated composition FC, it is easy to determine losses of P_c pressure on this composition flow in the channel at specified volume consumption rate Q. The FC should be reconstructed into a dependence of displacement efficient speed $\dot{\gamma}_{eff}$ of τ:

$$\lg \dot{\gamma}_{ef} = \lg \dot{\gamma} - \lg (3 + d \lg \dot{\gamma}/d \lg \tau), \tag{4a}$$

$$\lg \dot{\gamma}_{ef} = \lg \dot{\gamma} - \lg (4 + 2 d \lg \dot{\gamma}/d \lg \tau). \tag{4b}$$

Table 3. K_τ coefficients values for different
thermoplasts

Thermoplast	K_τ
LDPE 15303-003	0.60
HDPE 277-73	0.42
HDPE 209006-40	0.47
PS-115	0.21
CEVA 113-0.75	0.75

Equation (4a) is applied for a round channel while (4b) is for a flat one. In both (4a) and (4b) $\dot{\gamma}$ is displacement speed on the channel wall. Both equations are approximate. It is assumed that $d \lg \dot{\gamma}/d \lg \tau = d \lg \dot{\gamma}_{ef}/d \lg \tau$, which is true with the accuracy up to $d^2 \lg \dot{\gamma}_{ef}/d(\lg \tau)^2$. For more exact correlations see [38]. The P_c value dependence of Q is calculated through multiplying τ and $\dot{\gamma}_{ef}$ by the corresponding channel form coefficients.

If the channel has an arbitrary (not round) section, the calculation can be carried out for an equivalent round channel with a radius $r = 2S/p$, where S is the channel section area, P is the section perimeter. (On applications of this method see a recent review [26].)

3.2 Pressure Losses at the Channel Entrance

Complete pressure losses in the composition flow are composed of 3 components:

$$P_R = P_c + P_{ent} + P_{ex}.$$

(5)

As applied to the channels of great relative length (length L ratio to diameter D), pressure losses at the channel outlet and in the flow in the channel are much bigger than pressure losses at the channel outlet [27], and

$$P_R = P_c + P_{ent}.$$

(6)

Correlation (6) provides for determining the P_{ent} value through extrapolation of P_R dependence of L/D to a zero-length capillar. It is stated in [28–31] that increase of the filler concentration leads to increasing P_{ent} value. To predict the $P_{ent}(\varphi)$ dependence, the authors [32] used the equation, formerly applied for pure polymers in [33]:

$$P_{ent} = S \cdot \tau^m.$$

(7)

S and m of (7) are constant. After having analysed literature data [28–31] and results of their own studies [13, 26–21], the authors [32] came to the following conclusions:

1. The m value does not depend on the filler type and concentration and is constant for a specific basic polymer sample.

2. The S value depends on the channel form and size as well as the outlet configuration. For the compositions with disperse fillers

$$S_\varphi = S_0(1 - \varphi). \tag{8}$$

3.3 Piezoeffects in a Melted Filled Polymer Flow

Utilization of superposition methods (Sect. 3.1) and correlations (7) and (8) (Sect. 3.2) allows us to calculate the value of complete pressure losses $P_R(\varphi)$ in the composition flow only when the dependence of viscosity and pressure losses at the channel inlet of hydrostatic pressure is so small that it can be neglected in calculations. It is not always so. For example, this dependence is rather significant at a pressure value exceeding 30 MPa during polysterene and polysulfone moulding, and it cannot be neglected [33]. In the case of laminar regimes of melted non-filled polymer flow

$$\eta(\dot{\gamma}, P) = \eta(\dot{\gamma}, 0) \cdot \exp [b(\dot{\gamma}) \cdot P] \tag{9}$$

(see [34–37])

$$P_{ent}(\dot{\gamma}, P) = P_{ent}(\dot{\gamma}, 0) \cdot \exp [\varkappa(\dot{\gamma}) \cdot P] \tag{10}$$

(see [33]).

Piezocoefficients of viscosity-b and losses at the point of entry-$\varkappa$ are weak functions of $\dot{\gamma}$ and [33]:

$$\varkappa(\dot{\gamma}) = \mu \cdot b(\dot{\gamma}), \tag{11}$$

where the constant μ is independent of shear rate $\dot{\gamma}$. Applicability of Eqs. (9)–(11) to the descriptions of melted kerogene 70 filled PS as well as of shales-ash filled PS flows is checked in [38]. When viscosity depends on hydrostatic pressure, the pressure itself non-linearly drops along the channel length [36] equation (9). Therefore, in this case, the equation for calculating the pressure gradient by pressure overfalls, generally used in rheologic studies:

$$\frac{dP}{dz} = -\frac{P_2 - P_1}{L_2 - L} = -\frac{\Delta P}{\Delta L} \tag{12}$$

is not correct. For correct determination of dP/dz (and, consequently, $\tau = (R/2L)(dP/dz)$, the authors of [38] studied the PS keroplast flow through a

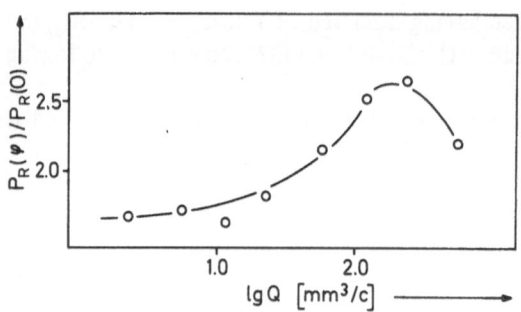

Fig. 1. Dependence of $P_p(\varphi)/P_p(0)$ value on PS Keroplast melt volume consumption (T = 433 K; L/D = 40/2; φ = 0.4 (vol.) share) [10]

long flat capillary ($200 \times 9.9 \times 1.9$ mm) equipped with 4 pressure sensors, placed along its length. The fifth sensor was fixed in the viscosimeter reservoir to allow assesment of $P_{bx}(P)$. (For the assay method see Ref. [38].)

It is experimentally proved in [38] that b, $\varkappa$, μ do not depend on the filler type and concentration. In the temperature interval from 413 to 483 K and at volume consumption of 4.52 to 9.02×10^3 mm^3/s, the b coefficient drops with the increase in volume consumption and/or the temperature from 1.2×10^{-3} to 0.8×10^{-3} (MPa)$^{-1}$. In the case of polystyrene, the parameter $\mu = 1.7 \pm 0.1$.

The results obtained allow us to derive a non-obvious equation for calculating $P_R(\varphi, Q)$ with consideration of the piezoeffects on the basis of $P_R(0, Q)$ dependence:

$$\ln (P_R(\varphi) + b^{-1} \ln (1 - K(\tau(0), 0) \cdot (\eta(\dot{\gamma})_{rel}) L \cdot b))$$

$$= \ln (P_p(0) + b^{-1} \ln (1 - K \cdot (\tau(0)\, 0) \cdot L \cdot b)) + b\mu \ln (P_R(\varphi)$$

$$- P_R(0)) + \mu \ln (\eta(\dot{\gamma})_{rel}) + \ln (1 - \varphi) \ln (\eta(\dot{\gamma})_{rel})$$

$$= 2.3 \cdot \varphi[K_\tau + (K_\eta - K_\tau)/(d \lg \gamma/d \lg \tau) , \tag{13}$$

$$K = \tau/(dP/dz) .$$

As seen from Fig. 1, theory and practice coincide. The methods and equations, given in Sect. 3.1–3.3 allow us to calculate the head and consumption characteristics of the melted thermoplast, filled with a disperse inert filler, in particular, a mineralorganic one, at various pre-set temperatures and concentrations by using a single known head and consumption basic polymer characteristic, measured at any fixed temperature. The authors of this review have developed appropriate algorithms and programs for carrying out these calculations.

3.4 Conditions of Transfer to an Irregular Flow Regime

The laminar character of the flow is violated in thermoplast and thermoplast-based compositions flow throught the channels at high volume consumption rates. In extrusion formation processes this leads to undesired distortion of the extrudate form. The problem of determining the critical volume consumption rates Q_c is very important for a technologist. The Q_c value is the upper limit of equipment capacity, close to which (but lower!) the optimum technological processing regimes are realized.

Depending on the type, molecular weight, molecule and weight thermoplast distribution, as well as the filler type and concentration, two phenomena, different in principle, may occur.

In some cases, for example, the flow of monodisperse homopolymers derangement phenomenon (dramatic increase of volume consumption rates within a very narrow pressure interval [39, 40]) is observed. G. V. Vinogradov et al. proved that at the shear rate characteristic of the derangement, the melted material is transformed into a highly elastic state [39, 40]. Simultaneously, polymer adhesion to the surface of molding equipment reduces sharply. The melted material starts to slide along the channel walls [41]. This transfer from the regime of flowing to the regime of polymer "plug" sliding in the channel is easily fixed by developing a significant electric potential at the "dry" friction region on the "polymer – channel wall " interface. This was demonstrated by G. V. Vinogradov et al. in [41]. As was proved in [42, 43], the assumption that the inlet has a great influence on the melted material movement at volume consumption rates corresponding to derangement regimes is evidently not true. This effect obviously fades in short capillaries.

The derangement phenomenon is not observed in polydisperse homopolymer flow. The head and consumption characteristic lacks the usual vertical part in the Q on P dependence. Though, when the Q value exceeds the critical Q_c value, i.e. $Q > Q_c$, pressure oscillations take place. Extrudates will have a distorted form and a rough not smooth surface. This effect is named "the melted material disintegration phenomenon". It is assumed [27, 44] that such a change in the flow character is associated with the accumulation of elastic energy at the channel inlet when in excess of a certain "critical" value [27], leading to the "melted material disintegration". The authors [44] have measured the dependence of the sound velocity in the melted polymer on the volume consumption rate. When Q_c was reached, sound velocity sharply dropped, demonstrating formation of cavities – gaps in the flow continuity. In this case the Q_c value does not depend on the channel length and its cross-sectional area [27]. It is a proof of the disintegration of the melted material occurring at the channel inlet.

When mineral rock fillers are introduced into the melted polymer, the Q_c value is reduced in all cases. In some, even when the basic polymer did not demonstrate derangement in its flow, it occurred [28–30, 45, 46] when filling reached about 30% by volume. In other cases with a filler introduction, derangement of melted materials was observed at $Q > Q_c$ [27, 47].

In all cases of the flow of melted mineralorganic filled polymers studied, transition to irregular flow regimes was not accompanied by a dramatic growth of consumption rates in narrow pressure intervals [20]. At Q_c, the pressure started to fluctuate. The head and consumption characteristic being smooth, the average pressure around which oscillation occurred, was taken as the pressure value. The oscillation amplitude became less along the length of the channel, reaching-zero at the outlet.

Therefore, it can be stated that melted keroplasts disintegrate in the flow before transfer into a highly elastic state. To explain this better we would like to discuss the studies in Refs. [48–50]. It is known that expansion deformation is the principle type of deformation the the polymer is exposed to at the channel inlet. In [48] using as an example uncharged rubbers and in [49] using as the example high impact polystyrene, modified by a triple-block styrene and diene copolymer, the condition for melted polymer disintegration on deformation with single-axis expansion was formulated. Independent of the law (background) of deformation this condition can be described:

$$\sigma^*/(\varepsilon^* - \varepsilon^*) = G = \text{Constant}. \tag{14}$$

In Eq. (14) σ^* and ε^* are, correspondingly, tension and reversible deformation at disintegration, ε_c^* is a "critical" reversible deformation, the lower margin for disintegration. Thus, the melted material can flow without rupture indefinitely. At reversible deformations higher than ε_c^*, melted polymers can disintegrate by the mechanism of cross-linked rubbers. Durability (time before desintegration) of such sytems depends on the degree of tension. Filling with hard additives reduces polymers ability to undergo large reversible deformations and it can be assumed that ε_c^* will reduce with the increase of the filler concentration. This was convicingly demonstrated in Ref. [50]. When shale ash was introduced into 1.2-polybutadiene, ε_c^* linearly diminished along with the growth of φ. When $\varphi \geq 25\%$, $e_c^* = 0$. This means that melted highly filled thermoplasts can desintegrate at any reversible deformation. So, in this case Q_c (or in the flow curve φ_c) is only determined by the existence of the deformed medium. Correlation between the critical deformation speed at which the composition is transformed into a highly elastic state and the melted material durability determines the character of transition to the irregular flow regime on reaching Q_c.

Although it is not the main subject of this review, we would like to note that the above-mentioned ideas can serve as a working hypothesis for composing a theory of deformational disintegration for polymers and polymer-based compositions. The essence of the phenomenon discovered and studied by Academician N. S. Enikolonov and his colleagues lies in the fact that under certain conditions (temperature, shear rate and hydrostatic pressure are varied) when extruding compositions through nozzles, a polymer will disintegrate and come out of the nozzle in powder form. Evidently, high hydrostatic pressure is essential for carrying out the process, to ensure polymer adhesion to the channel wall, a high displacement speed is also needed to provide for disintegration of the melted material, as well as sufficient time for the composition to disintegrate.

Now, let us go back to the methods of forecasting conditions for obtaining critical regimes for keroplast flow. As demonstrated in [20], critical tension τ_c of melted filled polymer can be calculated as a regular point on the flow curve, i.e.

$$\lg (\tau_c(\varphi)) = \lg \tau_c(0) + K_\tau \cdot \varphi, \tag{15}$$

where K_τ is an empirical coefficient, constant for a specific basic polymer sample, as discussed in Sect. 3.1. It is worth mentioning another time that for PS keroplasts we should introduce corrections on their dependence of hydrostatic pressure when calculating the flow curves, inlet pressure losses and $\tau_c(\varphi)$.

3.5 Specific Effects in an Isothermal Flow

When keroplasts are pressed through small diameter (D = 0.5 mm) capillaries, as in the flow of mineral-material filled compositions [46, 51, 52, 55] channel "blocking" occurs.

The blocking has not been observed when pressing compositions based polymerization-filled polymers, norplasts [46]. This effect develops in the following way. At a specific volume consumption rate the flow of the melted material from the channel ceases, and pressure in the viscosimeter reservoir starts to increase dramatically. When it reaches a certain level (pressure value level is fortuitous and could change several times) a batch of material is pushed out. The resulting extrudate has a considerably lower (by 2–5 times) filler concentration, as compared to the basic composition. As in Refs. [51, 52], a "package" of the carrier accumulates at the channel inlet and the polymer filters through it. Existence of the blocking effect in natural mineralorganic-filled thermoplast compositions requires caution in the choice of a moulding tool for processing these media. The nozzle diameters for keroplast extrusion and the diameter of the moulding form channels should not be less than 1 mm.

To understand the channel blocking effect in keroplast flows we have to analyse the microstructure of the natural mineralorganic fillers more thoroughly. These geological formations consists of a "puff-pastry micropie", in which organic areas about 40 microns in size alternate with non-organic. When coals, shales or sapropels are ground by mills or even disintegrators, the multipolymer (the organic component in the filler) hardly disintegrates at all. This means that natural mineralorganic substances are large particles dispersion fillers, which is the reason for the observed blocking effects.

Another effect to be discussed in this section is described in [19] and is an abnormal change of thermal dependence of the viscosity of Kerogene 70 filled polystyrene. Let us discuss the results [19] in more detail.

It is known [53] that introduction of fillers can alter the glass transition temperature T_g of the composition as compared to the basic polymer T_g. Variation of the composition viscosity can be described by the equation [54]:

$$\frac{\eta_0(\Delta T_g)}{\eta_0(0)} = \exp\left[\frac{C_1 \cdot C_2 \cdot \Delta T_g}{\theta(\theta - \Delta T_g + 2C_2) + C_2(C_2 - \Delta T_g)}\right]. \tag{16}$$

In Eq. (16) $\eta_0(0)$, $\eta_0(\Delta T_g)$ is viscosity at $\Delta T_g = 0$ and ΔT_g; C_1 and C_2 are Williams-Lendall-Ferry constants; $\theta = T - T_g$ where T is the experiment temperature. It follows from Eq. (16) that viscosity will change the more, the less θ is and the more ΔT_g is. Therefore, even if the T_g value for PE – filler – type compositions is displaced, it should not be observed even at higher θ values, because T_g PE = 150 K, and the processing temperature is circa 473 K. This explains the fact why abnormal changes of viscosity thermal dependence (VTD) is not observed in PE Keroplasts [19]. Polystyrene processing temperatures are close to its T_g. Because of low θ for PS — hard filler systems we can expect significant VTD as compared to the basic polymer. The authors [19] observed such an effect only with the introduction of commercial Kerogene 70 into PS. Addition of shales ash or other kerogenes does not result in an alteration of the composition glasse — transition temperature as compared to the basic polymer T_g. Commercial Kerogene-70-filled polystyrene T_g's were 368, 364, 361, 361 K at concentrations 0%, 20%, 30%, 40% (by volume) correspondingly. Viscosity values forecasted by these methods in consideration of the glass transition temperature displacement by Eq. (16) agreed well with experimentally measured viscosity values. It is important to note that commercial Kerogene 70 reduces the composition η_0 as compared to the basic polymer. This leads to a significant decrease in the efficient viscosity value as compared to the same for PS + mineral filler materials with an equal volume of content. For example, at 413 K, 30% (vol.) Kerogene 70 PS composition has the same head and consumption characteristics as the basic polystyrene [19].

This significant alteration of T_g is explained by the fact that commercial Kerogene 70 is delivered with an additive (10–15% green oil), to improve fire safety during transportation. We can assume that the green oil plays the role of a polystyrene "surfactant plasticizer".

This method of reducing viscosity by coating fillers with plasticizers is widely known [56, 57]. In this sense, mineralorganic fillers preserve all the advantages of purely organic ones. The organic particle surface easily retains organic plasticizers and modifiers in general. Expensive surface coatings capable of sticking to the mineral fillers are not required for surface modification of such fillers.

4 Crystallization Kinetics

Introduction of fillers into polymers changes the latters' crystallization Kinetics. Experimental data and theoretical ideas on this matter are summarized in [58]. The authors [58] write on p 39: "The concentration of heterogenous centers and speed of crystallization in them increase when high filler content compositions are developed. Simultaneously, homogenous centers number and polymer crystallization speed in the volume decrease. Limitation of molecular mobility, loss of a degree of freedom under the influence of a high energy-surface and the influence of long-range forces slows down macromolecule restructuring from the disordered into the ordered state. It reduces crystallization speed, inhibits or completely supresses this process. Existence of loose layers with low density macromolecule packing and expansion of free space promotes higher molecule mobility and crystallization speed. Simultaneously the impact of all the enumerated factors results in a complex dependence of the crystallization process in the presence of a solid surface. Surface energy influence the crystallization process: high-energy surfaces slow down crystallization, while low-energy surfaces have no impact on crystallization. Therefore, depending on the interactions at the border of phases distribution and the solid phase content, the crystallization process is considerably changed".

Before we start to describe thoroughly the kinetics of mineralorganic-filled thermoplasts we would note that in Ref. [13], with the example of PE keroplasts, it was proved that kerogens only very slightly influence polymer crystallization kinetics, and, consequently can be classified as low-energy surface fillers.

The authors of Ref. [13] studied crystallization kinetics of kerogene-filled HDPE, brand 15803-02.

In Table 4 see the values of speed growth "z" and degree index "n" (the constant of Kholmogorov-Abvami equation [64, 65]):

$$\varepsilon_0(t) = \exp(-zt^n), \tag{17}$$

where $\varepsilon_0(t)$ is crystallization degree on the time moment t from the start of isothermic crystallization process.

As can be seen from Table 4, introduction of kerogene into polyethylene does not lead to significant change in isothermic crystallization kinetics.

It is very important to understand non-isothermic kinetics to analyse the composition's technological properties, especially at very high speeds of cooling, which occur in thermoplasts and thermoplast-based compositions processing. The method for describing and forecasting non-isothermic crystallization kinetics is proposed in Ref. [59]. Alternative methods for describing kinetics are in the studies cited in Ref. [59] (see also Refs. [66, 67]). The essence of the method in Ref. [59] is as follows.

As it is known (see, for example, Refs. [60, 61]), in many cases experimentally measured $\ln \varepsilon_0(t)$ functions depend on the principle of temperature and time superposition (TTS). In other words, the dependences $\ln(-\ln \varepsilon_0(t))$ on $\ln t$, obtained at different constant temperatures can be matched by displacement along the $\ln t$ axis.

Table 4. Kholmogorov-Avrami, i.e. Equation (17), constants for PE-based keroplasts, the filler is Kerogene 70

Crystallization temperature (K)	Filler concentration (% vol)	Speed growth constant z $c^{-1} \times 10^5$	Degree index n of equation (17)
371.7	0	3.50	1.96
371.7	20	4.1	2.12
371.7	30	4.5	2.16
371.7	40	4.0	2.18
373.7	0	0.92	2.01
373.7	20	0.10	2.11
373.7	30	0.10	2.20
373.7	40	0.11	2.20

When the TTS principle is carried out, the non-isothermic crystallization kinetics can be described by replacing argument t in the dependence of non-crystallinity degree of time during isothermic crystallization. The replacement equation is

$$t' = \int_0^t \frac{dy}{a_{T_0}(T(y))},\tag{18}$$

where $a_{T_0}(T)$ is the factor of the dependence, describing isothermic crystallization at temperature T, displaced to the dependence at temperature T_0. $T(y)$ is the law of temperature changing in non-isothermic crystallization.

The values $a_{T_0}(T)$ can be calculated by Hopkin's equation [62] the same as when describing non-isothermic viscous and elastic behaviour [62, 63]:

$$a_{T_0}(T) = \frac{d(\ln \varepsilon(t))}{dt} \bigg/ \frac{d(\ln \varepsilon_0(t))}{dt},\tag{19}$$

where $\varepsilon_0(t)$ and $\varepsilon(t)$ are experimentally measured values of ε on the moment t from the start of crystallization in isothermic and arbitrary chosen regimes of temperature changing respectively. Let us note that $a_{T_0}(T)$ is an invariant function to the law of temperature changing, and Eq. (19) completely solves the task of numerical definition of $a_{T_0}(T)$. It is sufficient to experimentally determine only two dependences $\varepsilon_0(t)$ and $\varepsilon(t)$. Equation (19) allows us to calculate the dependence of non-crystallinity degree of t at any other (interesting for the researcher) law of temperature changing according to the already measured function $\varepsilon_0(t)$.

It is of interest to describe the function $a_{T_0}(t)$ analytically. This was done in [59] on the assumption that isothermic crystallization kinetics of thermoplasts and thermoplast-based compositions is described by the Kholmogorov-Avrami Equation (17). Let us write down Eq. (17) in a different way:

$$\ln \varepsilon_0(t) = -\ln 2 \left(\frac{t}{\tau}\right)^n,\tag{20}$$

where τ is a semi-period of isothermic crystallization. Then, $a_{T_0}(T)$ definition will take the form of the definition of temperature dependence. The last one is:

$$\ln \tau(T) = \frac{E}{RT} + \frac{K_y T_{ml}^0}{T_{ml}^0 - T} - 2.3A_n + \frac{1}{n} \ln (\ln 2).\tag{21}$$

In Eq. (21) ΔE is the activation energy of transport process, calculated by Eq. (22):

$$\Delta E = -2.3 \cdot C_1 \cdot C_2 \frac{RT^2}{(C_2 + T - T_c)^2},\tag{22}$$

Table 5. Dependence of Eq. (21) constants on the filler content in LDPE, the filler is Kerogene 90

Filler content (% vol)	K_y	$-2.3A_n + \dfrac{1}{n}\ln(\ln 2)$.
0	−60.7	8.7
20	−55.1	8.0
30	−52.0	5.0
40	−43.0	3.2

where C_1 and C_2 are the constants of the Williams-Landell-Ferry Equation. The ΔE value does not depend on the filler concentration φ, because of the principle of temperature and time superposition, existing for keroplasts.

In Eq. (21) R is a gas constant; T_{ml}^0 is the equilibrium melting temperature; K_y — the parameter, reflecting nucleation and spherolite growth characteristics; A_n is the preexponential factor of the expression, describing nucleation. The methods of experimental determination and calculation of "K_y", "A_n" and "n" are described in [59].

As shown in Table 5, K_y and A_n change rather insignificantly on introduction of kerogene into polyethylene. We can consider, with accuracy sufficient for engineering practice, that the crystallization kinetics of PE keroplasts do not depend on the content of filler in polyethylene. In Fig. 2 we demonstrate the good correlation of theoretical calculations with the experiment.

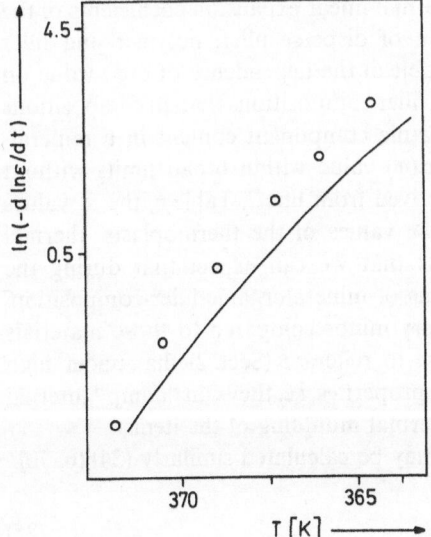

Fig. 2. Keroplast crystallization kinetics ($\varphi = 0.3$ (vol.) share) at cooling speed 20.0 °C/min. Points-experimental values; curve — calculations results [13]

The specific crystallization heat h of PE keroplasts can be calculated by the additive equation [13]:

$$h(\varphi) = h(0) (1 - \varphi).\tag{23}$$

5 Thermophysical Properties

The materials in Sect. 3 and 4 demonstrate the following principle idea: rheological properties and crystallization kinetics of thermoplasts, filled with inert disperse fillers, including mineralorganic, do not depend on the filler type and the organic component content.

Through the processes of thermoplasts and thermoplast-based compositions are substantially non-isothermic. Therefore, variation of thermophysical properties is an important method of target regulation of filled thermoplasts technological properties. Utilization of mineralorganic materials as fillers for thermoplasts has great potential in this sphere.

In Table 6, we can see the average thermophysical properties of kerogenes as compared to the values of the same characteristics of thermoplasts. Thermophysical properties of processed compositions in the area of phase transitions are of prime importance. In Table 6, we can see the average values of the corresponding thermal co-efficients at 100–150 °C. The data for polymers are cited from [69, 70]. Assuming that there are no local thermal tensions in the melt, we can calculate the composition's thermal linear expansion coefficient by the additive equation [69, 70]:

$$\alpha(\varphi) = \alpha_f \cdot \varphi + \alpha_p(1 - \varphi),\tag{24}$$

where $\alpha(\varphi)$, α_p, α_f are the composition thermal linear expansion coefficients of the composition, contains φ volume fraction of disperse filler, polymer and filler correspondingly. Therefore (see line 2, Table 6) the dependence of $\alpha(\varphi)$ value on the (φ) value is expressed better in mineral filler compositions than in compositions with a organic additive. Varying the organic component content in a mineralorganic filler, it is possible to alter the $\alpha(\varphi)$ value within broad limits without changing the general filler content. As derived from line 2, Table 6, the α_f values of mineralorganic fillers are similar to the values of the thermoplasts' thermal linear expansion coefficients. This means that we can expect that during the structure formation in production of items of mineralorganic-filler-composition, thermal tensions in these items will be very minor compared to those materials with 100% mineral fillers. We would like to reiterate (Sect. 2) that under high temperatures, kerogenes posess "rubbery" properties, i.e. they can "damp" internal thermal tensions, arising under non-isothermal moulding of the items.

The filled thermoplast's heat capacity may be calculated similarly (24) [6, 70]:

$$C(\varphi) = C_f \cdot \varphi + C_p(1 - \varphi),\tag{25}$$

where $C(\varphi)$ C_p and C_f are heat capacities of the composition, basic polymer and filler correspondingly. Estimates using Eq. (25) utilizing the data of line 4, Table 6, demonstrate that heat capacity only slightly changes dependence on the filler concentration and type.

There are a lot of correlations, proposed in the literature, allowing one to calculate the composition thermal conductivity coefficient $\lambda(\varphi)$ by the known values λ_p and λ_f [72]. The authors of Ref. [72] studied thermoplastic polymer compositions containing four types of filler. The volume concentration varied within the interval of 0.1 to 0.4 volume shares of the filler, the value of λ_f/λ_p ratio varied from 3 through to 354. Resulting from the comparison of the results of calculations by different equations with experimental data, it was found that the following expression is the best possible [72]:

$$\lambda(\varphi) = \lambda_p \frac{(1 + 3\varphi)}{(1 - \varphi) + 3/m}, \tag{26}$$

where

$$m = \frac{\lambda_f}{\lambda_p} - 1. \tag{27}$$

We should also mention the study [73] in which the authors divide the total heat conductivity of the compositional material by three components — by the cluster of intercontacting filler particles, by the disperse medium and by the border "medium-filler". Such an approach is more "physical". The authors forecast followed [75] the inflection of a curve $\lambda(\varphi)$ of φ in the area $0.1 < \varphi < 0.2$ in the terms of the flow theory [74].

We would remind that the value of the critical volume concentration of inert filler continuous cluster formation is 16%.

Table 6. Comparison of thermophysical properties of Kerogenes and thermoplasts at the temperatures 100–150 °C

NN	Material	Kerogene-90	Kerogene-70	Kerogene-35 basic rock-ku-kersyte-shale	Thermoplasts polyethylene, polystyrene
1.	density $(kg/m^3 \times 10^{-3})$	1.1	1.3	2.5	1.0
2.	thermal expansion coefficient $K^{-1} \times 10^4$	1.0	1.1	0.1	1.5
3.	thermal conductivity coefficient $W/(m\,K) \times 10^2$	9.3	9.8	29.1	12.0
4.	heat capacity $KJ/kg\,K$	1.6	1.5	0.8	1.5

The authors of Ref. [19] used Eqs. (26) and (27) for estimating $\lambda(\varphi)$ of keroplasts. Basic data for calculations see line 3, Table 6. As in Ref. [72], introduction of a mineral filler into a polymer (in the case of [19] — shale ash) greatly (by 2–3 times) increases the value of the composition heat conductivity as compared to λ_p. Replacement of the composition 40% (vol.) shale ash for 40% (vol.) Kerogene 70 leads to $\lambda(\varphi)$ value reduction by more than 50%.

6 Non-Isothermic Flows

Let us discuss the influence of changing the complex of thermophysical properties of the mineralorganic-filled-composition melts on the flow character of these media in the field of variable temperatures. This problem was studied in [18, 19, 32].

The case, when the channel wall temperature along the flow axle decreases, but stays higher than the temperature of the composition phase transfer was realized in the keroplasts flow in a flat slit channel. To create a temperature gradient along the flow axis, only the viscosimeter reservoir was heated. The zones of heating the flat slit capillary were switched off. After two hours of heating, the realized temperature gradient was measured. The temperature at the channel outlet was always higher than the temperature of the composition phase transfer. It changed from 453 K at the channel inlet down to 413 K at the outlet. Pressure was taken by four sensors placed along the channel axis.

In the case of flows of polyethylene-based keroplast melts, the pressure profile along the channel length stayed linear. Calculations demonstrated that this is obviously explained by a relatively minor dependence of the composition viscosity on the temperature. The η value changes along the channel length by not more than two times. In a PE-keroplast flow in the low temperature gradient channels, pressure loss values are practically the same as in the isothermic flows. They can be calculated in compliance with the methods explained in Sect. 3.

The situation observed in PS-Keroplast melts non-isothermic flow is different. In these compositions, viscosity depends substantially on the temperature, the η value decreases by two orders, when the temperature changes from 453 K to 413 K. In Fig. 3, pressure profiles along the channel length are shown. It is clear

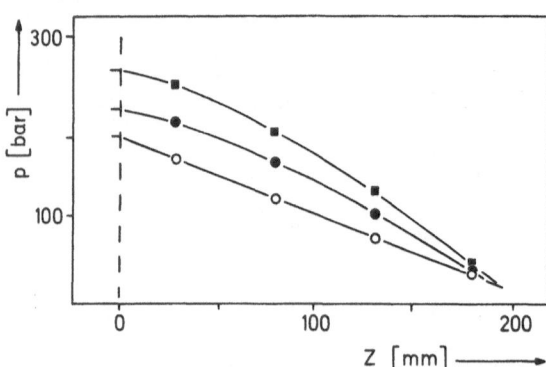

Fig. 3. Pressure profile inside the flat slit channel in PS Keroplast melt flow (contains 0.3 (vol.) shares of a filler). Volume consumption is 1.3 mm³ per s. Filler-Kerogene 90 (○), shale ash (□). Light marks-isothermic regime (T = 453 K), dark-non-isothermic (reservoir temperature 453 K) [10]

that the level of pressures under which non-isothermic PS-Keroplast consumption is realized, is higher for compositions, containing shale ash, than Kerogene 90. The explanation is that ash-containing compositions have a heat conductivity coefficient value much greater than Kerogene-90-containing compositions (see Sect. 5). Heat transport from the former toward the channel walls is much faster, than with Kerogene-90-filled PS keroplasts. These results prove that varying the organic component content in the filler is an effective means of target-regulating hydrodynamics of non-isothermic flows of mineralorganic material-filled thermoplast melts.

Under great volume consumption rates of melts in the same experiments, an interesting effect was observed: the extrudate "became narrowed" at the channel slit size 1.9×9.9 mm. The extrudate was 2 mm thick but only $5.5 + 0.5$ mm wide. Reduction of the extrudate cross-section means: transfer to a "multilayer" flow. It can be assumed that this phenomenon is caused by the existence of temperature cross-gradients along the channel section. This leads to a non-linear distribution of speed gradients and tensions over the channel section and along the flow axis, that in some places conditions for the melts-cohesion disintegration exist (see Sect. 4.4).

Now let us discuss the case of non-isothermic flows when the channel walls temperature is lower than the composition phase transfer temperature. This is a common situation for filling moulding forms. In [18, 19, 32] this is realized by filling the "snail" spiral form. Before filling, the form was kept at room temperature. After filling, the form was taken apart. The length of the melt L was measured.

To separately study the influence of rheological properties on the length of filling, the authors of the quoted works initially studied kerogene-90-filled PS keroplasts. As demonstrated in Table 6, in this case, thermophysical characteristics of the composition are similar to those of basic polystyrene. The result in this cas was simple, as expected:

$$L_{rel} \approx (\eta(\dot{\gamma}))_{rel}^{-1}. \tag{28}$$

In Eq. (28) L_{rel} is length ratio of filling the "snail" by the filled composition and the basic polymer respectively. η_{rel} is the ratio of respective viscosities. The viscosity values are chosen under similar shear rate $\dot{\gamma}$ values, equal to the rated volume consumption realized during filling the form.

Increasing the mineral component content in the filler (φ = const) leads to the growth of the melt heat conductivity coefficient. As a result, L_{rel} decreases along with decreasing the "C" organic component content in the filler. Processing of experimental data showed that in this case Equation (28) will change to:

$$L_{rel}^{-1} = \eta(\dot{\gamma})_{rel} \times \left(1 + \varphi \left(1 - \frac{C}{100} \right) \right). \tag{29}$$

In (29) φ is the filler volume share, c — organic component weight concentration in the filler (%).

Experimentally observed L_{rel} values, measured for crystallizing polymer (poly-ethylene) compositions are somewhat higher than L_{rel} as calculated by (29). It is possible to explain this fact by the reduced heat release at PE-keroplast crystallization as compared to the basic polymer (see Eq. (23)). In all cases, the difference between experimental and calculated by (29) L_{rel} values never exceeded 15%. Therefore, Eq. (29) can be recommended for engineering calculation of the length of filling the moulding form of any thermoplast.

7 Production of Mineralorganic-Material-Filled Thermoplastic Foam Items

The theoretical density of mineralorganic-material-filled thermoplast compositions can be calculated by the additiveness law:

$$\varrho(\varphi) = \varrho_f \cdot (1 - \varphi) + \varrho_f \cdot \varphi \,. \tag{30}$$

Equation (30) demonstrates that increasing the filler density ϱ_f (reduction of the organic component content) leads to a significant increase of the composition density $\varrho(\varphi)$.

We see the readiness of the organic part of mineralorganic fillers for limited thermodestruction as a useful characteristic. According to the data in [12, 17], starting from 443 K, kerogenes desintegrate with a minor weight loss (less than 1%) forming volatile products. Limited thermodestruction is observed up to 503 K. Afterwards, disintegration becomes more intensive. The thermodestruction data on sapropels (lake silts) are similar [11, 76]. But in their case, destruction starts at lower temperatures. Release of volatile products is more intensive. Mineral-organic fillers posess "limited nonthermostability" within a significant range of temperatures. This interval coincides with the thermal area processing of majority of commercial thermoplasts like polyethylene, polystyrene, polypropylene. This property allows the technologist to produce foam (foam-filled) products.

Foaming degree of mineralorganic material-filled thermoplast compositions is subject to a number of factors [17].

1) organic component content "C" in the filler; naturally, the higher "C" is under the fixed φ value, the higher the degree of foaming;
2) products can be formed by a chosen method, maximum density will be achieved on composition processing by pressure moulding, minimum − by extrusion;
3) processing temperature regime; temperature increase will lead to higher foaming degree in the products; this is especially important when processing by extrusion.

Low speed and low intensive kerogene destruction within the temperature range of 443 K to 503 K allow one to produce both monolyth and foam items [17] trough variations on pelletizing and keroplasts processing regimes. As demonstrated in Table 7, to decrease the foaming degree of samples produced by extrusion, it is necessary to pelletize the composition together with its degassing at temperatures 10° to 20° higher than the extrusion temperaturs of the same

Table 7. HDPE 209-01-based Keroplast density, filler-Kerogene 70, 20% (vol. content) at different regimes of pelleting and extrusion

Processing temperature K	Samples density ($\varrho \cdot 10^{-3}$ kg/m³) at pelleting temperature, K				
	443	453	463	473	483
443	0.970	–	0.970	–	–
453	0.950	0.970	0.970	0.970	–
463	0.930	0.950	0.960	0.970	–
473	0.910	0.930	0.950	0.960	0.940
483	0.810	0.910	0.930	0.940	0.940

composition during processing. During pelletizing the major portion of gaseous products is removed during the melt degassing. In technological regimes, pelletizing and processing are specially chosen (see Table 7), the foaming degree of keroplast extrudate items can reach 25–30%. It is not so bad if we keep in mind that keroplasts are filled thermoplastic polymers.

Alteration of the temperature regime of keroplasts pelleting has practically no influence on the density of the items, produced by pressure moulding. The maximum foaming degree of such samples is 3%. Foam moulded items can be produced with sapropel-filled thermoplasts [11].

The numerous examples (moulding regimes, concentrations, etc.) see [11].

An important and useful property, found during studying the possibilities of using sapropels as foaming agents is that the structure of the items, produced from sapropel-based compositions is an integral character (see Fig. 4). As clear from Fig. 4, the integral structure of sapropel-containing products is better than

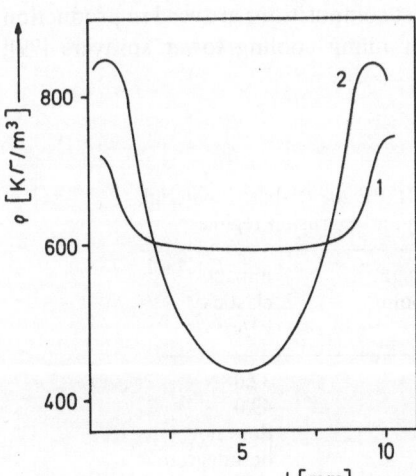

Fig. 4. Distribution of material density over the sample thickness (for high density polyethylene). Gas formers: (1) azodicarbonamide (1 part by weight) sapropel (2) (5 parts by weight). (From data in [11])

of that, containing conventional foaming agents, for example, azodicarbonamide (ChHZ-21). In the first case, we observe a smooth transfer from practically unfoamed surface unfoamed layer to a significantly foamed nucleus, in the 2nd case this transfer is uneven. The border of this transfer is a source of additional internal tensions, which leads to great deterioration in the product's operational qualities.

8 Conclusions

The operational mechanical qualities of mineralorganic-materials-filled thermoplast compositions are similar to mineral-materials-filled compositions [17]. We can see in Table 8 that relative extention at the yield point and impact elasticity in keroplasts are somewhat higher than in shale ash-filled polyethylene. Durability of these materials is similar.

In comparison with mineral fillers, mineralorganic ones have a number of advantages.

Abrasiveness of keroplasts melts is only 2–3 times higher than in the basic Keroplasts [17]. Introduction of mineral fillers into a keroplast increases the abrasiveness by dozens of times [77].

Anisotropy of mechanical properties at the increase of filler concentration [70, 78] rises significantly in the items produced from mineral media-filled thermoplasts. When a mineral filler is replaced for a mineralorganic, the effect of mechanical anisotropy increase fades considerably [18]. At the same time φ values, durability of "cold" seal moulded products, made of kerogenes is higher than in products of mineral materials-filled compositions [18].

A rather high level of operational properties of thermoplasts compositions with a mineralorganic filler and their substantial technological advantages make utilization of mineralorganic media for polymer fillers profitable.

Mineralorganic material-filled thermoplast compositions are used in production of linoleum floor covering [12]; sheets for filling cooling tower sprayers [79]; technical casings, furniture fittings, etc.

Table 8. Physical and mechanical properties of HDPE 209-01-based material, filled with Kerogene 70, under different pelleting and extrusion regimes

filler	yield point n/mm²	relative extension at the yield point %	impact elasticity kJ/m²
shale ash	20.0	13.0	32.6
Kerogene 70	20.5	18.0	49.0
zero filler	21.3	20.2	does not desintegrate

Under isothermic flow conditions, head and consumption characteristics of mineralorganic filler-containing thermoplasts melts coincide with head and consumption characteristic of mineral-material-filled compositions (at similar volume concentrations). These characteristics can be calculated by the methods, explained in Sect. 3 of the review.

9 References

1. Seymour RB (1985) In: Developments in plastics technology, vol 2. Elsevier, London, p 219
2. Enikopolov RS, Wol'fsson SA (1978) Plast Massy 1: 39
3. Wettergren VI, Dzyubenko LS i dr (1979) V knige: Struktura i svoistva polimernykh materialov (in Russian). Riga p 31
4. Wang TT, Kwei KI (1969) J Polymer Sci, A 15: 889
5. Popov VL, Sabssai OYu, Lobkova ML, Fridman ML (1980) V knige: Reologiya v pererabotke polimerov (in Russian). NPO "Plastik", Moskva: 24
6. Handbook of Fillers and Reinforcements for Plastics (1978) Eds Katz HS, Milewski JW Van-Nostrand-Reinhold Co
7. Seymour RB (1976) Plastic Engineering 8: 29
8. Teplofizicheskie i reologicheskie svoistva polimerov (1977) Naukova dumka, Kiev
9. Lightsey GR, Hines AL, Arnold PV, Sinha VK (1975) Plastic Engineering 5: 40
10. Barshtein GR, Sabssai OYu (1988) Tekhnologicheskie svoistva termoplastov c mineralorganicheskimi napolnitelyami (in Russian). NIITEKhIM, Moskva
11. Avtorskoye svidetel'stvo SSSR N 1351951 ot 15. 07. 1987
12. Yakovlev VI, Proskuryakov BYa (1977) Struktura, svoistva i primeneniye malozol'nogo kontzentrata goryachego slantza (in Russian). Leningr goc universitet, Leningrad
13. Barshtein GR, Sabssai OYa (1982) V knige: Pererabotka napolnennykh kompozitzionnykh materialov (in Russian). NPO "Plastik" Moskva: 29
14. Goryuchie slantzy (1980) Nedra Leningrad
15. Barshtein GR (1982) V knige: Tezisy dokladov Bsesoyuznoi nauchno-tekhnicheskoi konferentzii po vysokonapolnennym kompozitzionnym materialam (in Russian). Moskva: 65
16. Barshtein GR, Sabssai OYa, Fridman ML (1984) Plast Massy 1: 61
17. Barshtein GR, Sabssai OYu, Popov VL (1984) V knige: Tekhnicheskie doklady Vsesoyuznogo simpoziuma po reologii (in Russian). Volgograd: 86
18. Barshtein GR, Sabssai OYu, Fridman ML, Lapitzki YuA (1985) Plast Massy 3: 53
19. Barshtein GR, Sabssai OYu (1986) V knige: Tezisy dokladov I Vsesoyuznoi nauchnotekhnicheskoi konferentzii po reologii i optimizatzii protzessov pererabotki polimernykh materialov. Ustinov: 81
20. Barshtein GR, Sabssai OYu (1986) V knige: Tezisy dokladov I Vsesoyuznoi nauchnotekhnicheskoi konferentzii po reologii i optimizatzii protzessov pererabotki polimernykh marerialov. Ustinov: 82
21. Vinogradov GV, Malkin AYa (1977b) Reologiya polimerov. Khimiya, Moskva
22. Chong IS, Cristiansen EB, Baer AD (1971) J of Appl Polymer Sci, 8: 2007
23. Moshev VV (1977) V knige: Reologiya (polimery i neft') (in Russian). Novosibirsk: 53
24. Barshtein GR, Sabssai OYu (1985) Tezisy dokladov III Vsesoyuznogo simpoziuma: Teoriya mekhanicheskoi pererabotki polimerov (in Russian). Perm': 12
25. Tim S (1985) In: Developments in plastics technology, vol 2. Elsevier, London: 297
26. Han CD (1976) Rheology in Polymer Processing. Academic Press, NY
27. Popov VL, Sabssai OYu, Lobkova ML, Fridman ML (1980) V knige: Reologiya v pererabotke (in Russian). NPO "Plastik", Moskva: 24

28. Crowson RS, Folkes MI, Bright PF (1980) Polymer Eng Sci, 14: 925
29. Popov VL, Fridman ML, Abramov VV (1981) Reologicheskie i tekhnologicheskie svoistva napolnennykh polymernykh materialov (in Russian). Pererabotka plastmass. NIITEKhIM, Moskva
30. Oyanagi Y, Yamaguchi Y, Okada I, Tagami K (1972) Res Reports Kogakuin Univ, 32: 81
31. Barshtein GR, Sabssai OYu, Akutin MS (1980) Khimiya i khimicheskaya tekhnologiya
32. Kalinchev EL, Sakovtzeva MB (1983) Svoistva i perestroika termoplastov (in Russian). Khimiya, Leningrad
33. Aibinder SB, Tyunin EL, Tzirule KI (1981) Svoistva polimerov v razlichnykh napryazhennykh sostoyaniyakh (in Russian). Khimiya, Moskva
34. Semenov V (1968) Adv Polymer Sci, 5: 387
35. Penwell RC, Porter RS, Middelman S (1971) J Polymer Sci A-2, 4: 731
36. Westover RF (1961) SPE Transactions, 1: 14
37. Barshtein GR, Sabssai OYu, Popov VL, Fridman ML, Enikolonyan NS (1985) Dokl Akad Nauk SSSR, 2: 271
38. Vinogradov GV (1971) Vysokomolekulyarnye soedineniya, 2: 294
39. Vinogradov GV (1974) V knige: Khimiya i tekhnologiya vysokomolekulyarnykh soedinenii. Moskva, 5: 130
40. Vinogradov GV, Protasov VP, Dreval VE (1984) Rheolog Acta, 1: 46
41. Paskhin ED, Polyak YuF, Fridman ML (1985) Dokl Akad Nauk SSSR, 2: 392
42. Paskin ED (1978) Rheol Acta, 6: 663
43. Heronen M, Savolainen A (1984) Rheol Acta, 6: 461
44. Clegg DW, Collger AA, Griffin NC (1980) Polymer, 12: 1360
45. Fridman ML, Popov VL, Lobkova ML, Sabssai OYu, Gevorgyan MA (1980) Dokl Akad Nauk SSSR, 5: 1185
46. Zoltor AM (1970) Modern Plat 12: 63
47. Borisenkova Ek, Sabssay OYu, Kurbanaliev MK, Dzeval VE, Vinogradov GV (1978) Polymer, 19: 1473
48. Chalaya NM, Sabssai OYu, Abramov VV, Vinogradov GV, Akutin MS (1982) V knige: Pererabotka napolnennykh kompozitzionnykh materialov. NPO "Plastik", Moskva: 80
49. Optov BA, Borisenkova EK, Sabssai OYu (1988) Kolloid zhurnal, 6: 1033
50. Malkin AYa, Epple GV, Gritzuk AI (1972) Kolloid zhurnal, 4: 550
51. Fridman ML, Yarlykov BV i dr (1976) Plast Massy, 1: 54
52. Lipatov YuC (1977) Fizicheskaya khimiya napolnennykh polimerov (in Russian). Khimiya, Moskva
53. Droste D, Dibenedetto A (1969) J Appl Polymer Sci, 10: 2149
54. Enikolopian NS, Fridman ML, Popov WL, Sralnow IO (1986) J Appl Polymer Sci, 7: 6107
55. Patent Yaponii N 50-18897 ot 11. 03. 1971, kl 25(1) F212.2, opubl 2. 07. 1975
56. Poluyanovich VB, Yatzenko VV I dr (1977) V knige: Khimiya i khimicheskaya tekhnologiya, vypusk 12 (in Russian). "Vyshaya shkola", Moskva: 91
57. Simonov-Emel'yanov ID, Kuleznev VN (1987) Osnovy sozdaniya kompozitzionnykh materialov (in Russian). Mosk institut tonkoi khimicheskoi tekhnologii, Moskva
58. Sabssai OYu, barshtein GR, Fridman ML, Godovskiy YuK (1985) Vysokomolekulyarnye soedineniya, 8: 1697
59. Nakamura K, Katayama K, Watanabe T, Amano T (1972) J Appl Polymer Sci, 5: 1077
60. Borokhovskiy VA, Gasparyan KA, Mirzoev RG, Sevast'yanov LK, Baranov VG (1975) Vysokomolekulyarnye soedineniya B 1: 35
61. Hopkins I (1958) J Polymer Sci 4: 631
62. Malkin AYa, Sabssai OYu, Gromakovskaya NE (1974) Mekhanika polimerov, 4: 945
63. Kolmogorov AN (1937) Izvestiya Akademii Nauk SSSR, seriya Matematicheskaya, 3: 355
64. Avrami M (1939) J Chem Phys 12: 1103; (1940) 2: 212; (1941) 2: 177
65. Begishev VP, Kilin IA, Malkin AYa (1982) Vysokomolekulyarnye soedineniya, B 9: 656
66. Malkin AYa (1976) J Polymer Sci 5: 315
67. Godovsky YuK, Slonimsky IG (1974) J Polymer Sci 2: 1053

68. Van Krevelen DW (1972) Properties of polymers correlations with chemical structure. Elsevier, London
69. Godovsky YuK (1982) Teplofizika polimerov Khimiya, Moskva
70. Richardson MO (1977) Polymer Engineering Composites Elsevier, London
71. Sundstrom DW, Lee VD (1972) J Appl Polymer Sci, 12: 3159
72. Zarichenok YuP, Novikov VV (1978) Inzhenerno-fizichesky zhurnal: 648
73. Stauffer D (1985) Introduction to percolation theory, London: 237
74. Godovsky YuK, Kiguradze OR, Lolua DG (1986) Vysokomolekulyarnye soedineniya, V 8: 564
75. Kashcheeva NI, Mitrofanov AD, Okunev PA, Panov YuT (1985) V knige: Tezisy dokladov III Vsesoyuznogo simpoziuma: Teoriya mekhanicheskoi pererabotki polimernykh materialov (in Russian) Perm': 85
76. Stambursky EA (1982) V knige: Vysokonapolnennye kompozitzionnye polimernye materialy, razvitie ikh proizvodstva n primemenie v narodnom khozyaistve (in Russian) NIITEKhIM, Moskva: 79
77. Shchupak EN, Tochin VA, Tumanov VV (1982) V knige: Tezisy dokladov Vsesoyuznoi nauchno-tekhnicheskoi konferentzii: Vysokomolekulyarnye materialy, razvitie ikh proizvodstva (in Russian). Moskva: 17
78. Nikolaeva NE, Sabssai OYu, Fridman ML (1985) Reologicheskie svoistva gazonapolnennykh polimerov i ikh proyavlenie pri formirovanii izdeliya. NIITEKhIM, Moskva: 44

Editor N. S. Enikolopyan
Received March 7, 1990

Compositions Based on Aminoresins

J. P. Terent'eva
B. Spasskaya ul. 10, kv. 108, 129010 Moscow, USSR

M. L. Fridman
Ul. Chernishevskogo 41, kv. 70, 103062 Moscow, USSR

The results of studies and development of composite materials with specified properties based on amino resins and their modification products are considered. The fields of aminoplast applications and as well as methods of estimating aminoplasts' technological properties are analysed. Emphasis is put on rheological methods of evaluating the technological characteristics of aminoplasts. This is based mainly on the various modifications of rotational and capillary instruments, as well as applying chemical and physical methods for the estimation of the technological properties. Generalization are made on the basis of the authors' work on evaluating the technological characteristics of aminoplasts in non-steady state of their processing with varying temperature and time. New methods for evaluating rheological properties of aminoplasts based on the analysis of elastic-deformation grinding when aminoplasts are setting are discussed. Various methods of aminoplasts processing including pressing, injection moulding and extrusion, as well as performance characteristics of compositions based on amino resins are given.

Advances in Polymer Sciences 101

1 Aminoplasts. General and their Modifications

1.1 Preparing Compositions Based on Aminoresins

In the production of composite plastics based on aminoresins those with the greatest popularity are resins based on carbamide, melamine and their mixtures. These resins are solidifying binders with the highest production volumes.

Carbamide- and melamine-formaldehyde resins are synthesized in aqueous solutions by using, as a rule, 1.5–2 mol of formaldehyde to 1 mol of carbamide or 2.9–3.0 mol od dormaldehyde to 1 mol of melamine in a weakly-alkaline medium. When using carbamide and melamine mixtures their ratio must be 1 : 1 or 2 : 1.

Combining aminoformaldehyde resins with the fillers in the production of composite materials is achieved by impregnating the filler with the aqueous solutions of resins with a low degree of condensation with subsequent drying and prehardening.

Hardening of aminoformaldehyde resins takes place as a result of the interaction of methylol groups with one another and with the active hydrogen atoms of the amino groups. The reaction is accompanied by the release of water and formaldehyde. To speed up the process of resin hardening, 0.5–1.0% of hardening catalysts are introduced, i.e. organic acids (oxalic, phthalic, etc.), their derivatives, and inorganic salts (zinc sulfate, ammonium tetrachloride) [1].

The optimum temperature of polycondensation of urea with formaldehyde is 30–40 °C.

The results of investigations carried out by a group of chemists from Washington University [2] give grounds for assuming that urea-formaldehyde resins (UFR) are, in effect, linear polymers and are colloidal dispersions. Solidification of the urea-formaldehyde resin proceeds due to the agglomeration of colloidal particles. The said particles are stabilized by formaldehyde molecules which are liberated in the process of agglomeration. The linearity of the structure of urea-formaldehyde resins (UFR) is confirmed by gel-permeating chromatography (GPC), scanning electron microscopy and by measuring the viscosity of the UFR solution on hardening.

1.2 Fillers

In preparing compositions use is made of organic and mineral fillers. The most important organic filler is bleached wood pulp. The pulp fiber length depends on the type of wood: thus, pulp produced from beech has 1 mm-long fibers, whilst that obtained from conifers –3 mm-long fibers.

Pulp-containing compositions are characterized by good mechanical and dielectric properties, high heat resistance and an outstanding coloring property.

Woodmeal is rarely used as a filler for aminoplasts. This filler is preferred in the case of phenol-formaldehyde resins. Woodmeal is produced, as a rule, by crushing sawdust or wood wastes. Woodmeal particles must be crushed to such an extend that they are capable of being easily introduced into the resin. By far the best woodmeal is that produced from white Baltic pine; as for hardwood species, use is made of maple, oak, and birch.

Mineral fillers find application for melamine-formaldehyde resins, thus enabling compositions to be obtained with heat resistance in excess of 200 °C.

Mineral fillers include asbestos, glass fiber, stone meal.

Asbestos-filled compositions have high chemical, mechanical and thermal resistance and outstanding dielectric properties.

Glass fibre gives compositions high mechanical strength, thermal resistance, and chemical resistance. By using asbestos or glass fibre with additions of organic fillers we may change the properties of aminoplasts ever a wide range.

Cuttings of cotton fabrics are mainly used as fillers for carbamide resins in obtaining compositions endowed with a high impact strength. The composition in this case is in the form of lumps or flocs.

Introduced as additives to the compositions based on aminoresins are lubricating substances (Zn, Ca stearates), methanol, dyes [3].

1.3 Types of Aminoplasts Made in the USSR

According to composition and designation, aminoplasts in the USSR are subdivided into different types and grades. Aminoplasts are available in the form of powder, granules, fibrous mass or crumb.

The KFA type is a general-purpose aminoplast based on carbamide-formaldehyde resin and alpha-cellulose (fillers).

The aminoplast of the KFA-2 grade is the most extensiveley produced one. It is available in powder form. Produced on the basis of KFA-2 is the KFA2-PRG grade in granulated form. The casting grade KFA4 is likewise in the form of granules.

The MFB type is an aminoplast based on a melamine-formaldehyde resin and alpha-cellulose.

The MFV type is an aminoplast based on a modified melamine-formaldehyde resin with an organic (cotton cellulose) or a mineral (asbestos, talc) filler.

The MFD type is an aminoplast based on melamine-formaldehyde resin with an asbestos (filler) with a small addition of cotton cellulose.

The MFE type is an aminoplast based on melamine-formaldehyde resin and glass fibre.

Table 1 lists the technological properties of aminoplast produced in the USSR [4].

1.4 Modification of Aminoplasts

For a purposeful variation of the properties of compositions based on carbamide and melamine resins with cellulose fillers use is made of various agents which modify the resins. Particularly good results are attainable by using benzoguanamine as a modifier. The content of benzoguanamine may vary from 5 to 50% [5].

Table 1. Technological properties of USSR-made aminoplast

Nos.	Grade	
	KFA2	KFA2-PRG
1. Length of the plastic-viscous state, at 120 °C	60–150	90–180
140 °C	10–50	30–70
160 °C	–	–
2. Effective viscosity, Pa s at 120 °C	$(90–270)^5 \times 10$	$(105–210) \times 10^5$
140 °C	$(60–230) \times 10^5$	$(70–180) \times 10^5$
160 °C	–	–
3. Hardening time, s at 140 °C	80–140 up to $\sigma = 4$ MPa	110–150 up to $\sigma = 3$ MPa
160 °C	–	–
4. Flowability, mm	70–160	110–190
5. Hardening in test core,	$\leqq 80$	80–90
6. Bulk density, g/cm^3	$\geqq 0.3$	0.5–0.7
7. Density, g/cm^3	1.35–1.45	–
8. Moisture, %, not more than	3.0	3.0
9. Residue after screening, % on mesh No. 018	$\leqq 3.0$	> 90
No. 1.0	–	< 5.0
No. 2.0	–	0

Aminoplasts based on urea-benzoguanamine-formaldehyde oligomers when compared to urea-formaldehyde press-materials are distinguished by an increased water- and heat-resistance, improved physico-mechanical properties, and insignificant shrinkage.

Table 2 gives properties of compositions based on modified resin. As seen from these data, with respect to water- and heat-resistance, as well as the electric resistance after boiling in water the compositions based on the modified resin are superior to those based on the carbamide-formaldehyde oligomer and approach materials based on the melamine-formaldehyde oligomer.

Grade					Testing method
KFA4	MFB	MFV	MFD	MFE	
200–280	120–220	200–320	–	–	GOST 15882-79
20–60	–	–	–	–	
–	25–60	30–110	–	–	
up to 150×10^5	$(70-160) \times 10^5$	$(25-90) \times 10^5$	–	–	GOST 15882-79
$(40-130) \times 10^5$	–	–	–	–	
–	$(40-90) \times 10^5$	$(20-50) \times 10^5$	–	–	
120–160 up to $\sigma = 2\,MPa$					
–	90–120 up to $\sigma = 6\,MPa$	–	–	–	
150	80–180	140–195	90–150	120–190	GOST 9359-80
–	$\leqq 90$	–	–	–	GOST 9359-80
0.55–0.70	$\geqq 0.25$	–	–	–	GOST 9359-80
–	1.45–1.55	1.6–1.85	1.7–1.9	1.95–2.05	GOST 15139-89
2.0–3.5	4.0	4.5	–	–	GOST 9359-80
>90	<3.0	–	–	–	GOST 9359-80
–	–	–	–	–	
<5.0	–	–	–	–	

In Ref. [6] results of modifying aminoresins with phenol and polyester resins to give them impact resistance, moisture resistance and dimensional stability for applications in electrical engineering.

An investigation into the effect of certain organic silicon compounds on the properties of aminoplasts was made in Ref. [7]. It was shown that the modification of aminoplasts with siloxanes makes it possible to improve their strength and performance indices, such as the collapse strain in bending, thermal resistance, and to reduce water absorption. The maximum effect is obtained by the use of 1–1.5% hydroxyloxane as a modifying additive.

Table 2. Properties of aminoplasts based on modified resin

Nos.	Indice	MBF5	MBF15	MBF25	MBF35	MBF50	MF	M1F
1.	Water absorption after boiling, %	2.1	1.8	1.2	1.0	0.8	6.2	0.8
2.	Thermal resistance, °C	120	120	120	130	130	110	140
3.	Electric resistance after boiling Ohm	10^{7-8}	10^{8-9}	10^9	10^{9-10}	10^{9-10}	10^{6-7}	10^{9-10}
4.	Arc resistance, s	90–100	90–100	100–120	100–120	120	100	117
5.	Test for weather resistance	−	−	−	−	−	+	+

MBF5, MBF15, MBF25, MBF35, MBF50 — urea-benzoguanamine-formaldehyde compositions containing 5, 15, 25, 35, 50 wgt.% benzoguanamines;
MF   − urea-formaldehyde composition;
M1F − melamine-formaldehyde composition;
− not weather resistant;
+ resistant

An overall trend towards an improvement of the properties of polymeric materials is also observed in the development of modified melamine-formaldehyde materials.

The shortcomings which are characteristics for melamine-formaldehyde compositions are due to the same factors as for urea-formaldehyde polymers. The formation of a rigid three-dimensional structure takes place similarly to urea-formaldehyde polymers due to the interaction of reactive methylol and amino-groups. The residual reactive groups contribute to the progress in solidified polymers of further reactions with the release of water and formaldehyde, thus impairing the properties of the polymers. Chemical modification of melamine-formaldehyde polymers is one of the most promising ways for improving the properties of these materials.

Particularly effective is the use of guanamines for the modification of melamine-formaldehyde compositions. Melamine-guanamine-formaldehyde aminoplasts are produced chiefly in the USA and Japan. The main purpose of these compositions is the manufacture of crockery. In this case, alongside thermal and water resistance, the hygienic and toxicological properties are highly important: resistance to the appearance of stains from foodstuffs, surface lustre, elimination of the release of formaldehyde in use.

Reference [8] contains a description of the development of commercial grades of the compositions BG-M50 and BG-M80 based on melamine-benzoguanamine-formaldehyde oligomers. The process of producing these materials does not radically differ from the technological scheme adopted in the production of aminoplasts.

The materials BG-M50 and BG-M80 have technological properties enabling their processing to be achieved by the method of pressure casting, as well as improved service properties. What has been said can be illustrated by the data describing the properties of these materials, as follows:

Density, g/cm^3	1.45–1.46
Strength in static bending, MPa	0.94–1.27
Charpy impact ductility, kJm^{-2}	2.3–2.8
Rockwell hardness, scale M	111–120
Shrinkage, cm/cm	7.2×10^{-3}
Water absorption, %	0.060
Insulation resistance, Ohm:	
under ordinary conditions	$7.4 \times 10^{10} - 7.0 \times 10^{11}$
after boiling	$3.2 \times 10^{9} - 6.0 \times 10^{10}$
Electric resistance, kV/mm	13.7–19.9
Arc resistance, S	123–124
Thermal resistance, °C	181.5

The work [9] discusses publications analysing physico-chemical indices and the specifics of the chemical structure of cast aminoplasts modified with 20–50% (with respect to the mass of melamine) of caprolactam and other lactams. Materials of this type, havin an increased impact strength, thermal stability, strength and elasticity modulus in bending, are manufactured in Czechoslovakia under the trade mark UmaLur MK-1, MK-2, MK-3.

Apart from investigations devoted to variation of the properties of the compositions due to resin modification, works have been published describing the use of various combinations of fillers (by way of partial substitution of the main filler). Thus, in Ref. [10] for the purpose of increasing the resistance of aminoplasts to impact loads either a partial or complete substitution of the cellulose filler with polyamide fiber was untertake. Use was made of discards of capron fibre, staple, tangle, caprowool, combings of man-made fur. Impact ductility in this case was found to be 2–3 times higher than for the aminoplasts with a cellulose filler. Table 3 gives the magnitude of the impact viscosity of compositions according to fiber length.

Table 3. Impact ductility versus type and length of fiber

Type of fiber and its length	Impact ductility, kJ/m^2
Staple: 25–27 mm	28
Tangle: 10–15 mm	20–26
5–7 mm	12–15
Tangle (80 parts/wt) — 5–7 mm	21–22
Cellulose (10 parts/wt)	
Combings — 10–15 mm	24–30
Powder-like polycaproamide	9–12
Combings of man-made fur and talc	25

In recent years, high-strength articles made from plastics reinforced with directed organic fibers or with fabrics produced with these fibers habe become very popular. The impact strength of the specimens increases in this case up to 90–240 kJ/m², and the deformational thermal resistance up to 200 °C. Such an aminoplast can be used in articles which are expected ot meet increased requirements as regards impact loads.

Compositions have been prepared on the basis of carbamide resins with 30–50% addition of powder clay. The resulting polymer is an elastic material undergoing deformation by compression without brittle collapse [11].

Fillers have been studied for their effect on the rheological properties of aminoplasts. Investigations of the rheological properties of aminoplasts filled with cellulose diacetate (CDA) wastes and into the effect of minor quantities of plasticizers on the viscosity of compositions have been carried out [12]. With a CDA content increase from 40 to 60% the viscosity of the composition declines. With the introduction into the aminoplasts of a lubricant (stearic acid) containing 50% CDA the viscosity of the compositions also declines. The same is observed when introducing 1–2% of plasticizer (unsaturated PEF).

An optical composition, as regards technological properties, is the containing 45% of melamine-formaldehyde oligomer, 55% of CDA, 1% of lubricant and 2% of PEF, P-509.

The process of preparing compositions based on high-porosity aminoresins that can be used as heat-insulating materials for buildings is reported in Ref. [13].

2 Application of Aminoplasts

For the manufacture of a diverse range of articles, the following properties of aminoplasts are very important: possibility of being coloured, absence of small, superior hardness, good dielectric properties, a satisfactory resistance to the impact of water in the case of urea-formaldehyde aminoplasts and an excellent one for melamine-formaldehyde aminoplasts.

Articles made of aminoplasts are practically unaffected by fats, oils, alcohol, acetones, and weak acids; they retain dimensional stability, are not easily combustible and are endowed with good resistance to solvents and utility chemicals. The service temperature for aminoplasts of the KFA type is up to 90 °C, for type MFB aminoplasts — up to 120 °C.

Aminoplasts based on urea- and melamine-formaldehyde resins with a cellulose filler are used for the manufacture of article in electical engineering, and for technical and utility applications. The cheap KFA type aminoplast is used to manufacture the following articles for the electrical engineering industry: plug-and-socket connections, plugs, sockets, switches, electric shaver components, keys, handles, as well as diverse articles for domestic application, such as toy parts, table games, knobs, watch frames, glasses, dishes, souvenirs. Furthermore, the list of manufactured products comprises minor fancy goods and stationery articles, covers for small receptacles, receptacles for cosmetics, brooches, pens, pencils, shoe polish receptacles, flower stands, ash-trays [14], as well as larger articles: tables, children's chairs, cases for radio sets, and wall clock cases.

Melamine-formaldehyde materials cost 30–50% more than urea-formaldehyde ones. Today the world output of melamine-formaldehyde materials is about 25% of the total output of aminoplasts with a tendency to increase further because of the lowering of melamine prices.

Aminoplasts based on melamine-formaldehyde oligomers have properties superior to urea-formaldehyde compositions and are widely used for the manufacture of domestic articles, electrotechnical products, machine components, etc. These articles have a hard, scratch-resistant surface and are stronger than those made of the KFA type aminoplasts.

Buttons on underwear that are expected to possess high resistance to boiling solutions of washing agents and crockery (plates, cups) which must be resistant to the effects of hot water, soap and other washing agents are produced from aminoplasts based on melamine-formaldehyde resin with a cellulose filler [4].

Melamine-formaldehyde aminoplasts filled with alpha-cellulose are suitable for the manufacture of electrotechnical articles operating under rigorous temperature conditions and in a atmosphere of high humidity under load.

Melamine-formaldehyde materials with textile fabrics are notable for high mechanical strength and are therefore used in the manufacture of gearboxes and other machine components. Melamine-formaldehyde materials filled with short-length asbestos fibers contain, as a rule, a smaller amount of resin (25–40%) than the cellulose-filled press-materials do. A high content of the filler imparts to them a greater density (about 2.0 g/cm^3), and insignificant subsequent shrinkage and good dielectric properties, though their mechanical strength is impaired.

Articles made from type KFA and MFB aminoplasts are noted for their high resistance to the effects of eddy currents. A far as dielectric properties are concerned, these press-materials are superior to phenoformaldehyde ones and surpass all other materials with respect to arc resistance and resistance to eddy currents.

MFB type aminoplasts are being applied in the manufacture of electrotechnical articles, including mine equipment.

Aminoplasts of the MFE type are characterized by high mechanical strength and dimensional stability and find application in electrical equipment units (blocks, plates, etc.), in components endowed with electrical strength, thermal and arc resistance. They have proved to be an excellent material for making electric devices in rocket engineering and can be incorporated into equipment operating under tropical conditions.

Aminoplasts of the MFD type are characterized by a high thermal and arc resistance. They are mainly being used for the manufacture of electrotechnical articles.

The main purpose of compositions based on melamine-guanamine-formaldehyde oligomers is the manufacture of crockery. Crockery made from the modified material is notable for its beautiful appearance, long service life, resistance to the appearance of stains from foodstuffs, and surface lustre. In use, no formaldehyde is released because the sanitary-hygienic properties of the material are high.

Structural glass-reinforced plastic based on aminoplasts is being used in aircraft construction, machine- and ship-building, in electrical and radio engineering (switchgear, switchboards, panels).

Glass-reinforced plastics, on the basis of aminoplasts, are being used in the manufacture of radio and electrotechnical items; light partitions in ships, ship's furniture and ship's equipment components; items with a high thermal resistance and mechanical strength [15].

Improved physico-mechanical indices of fibrous press-materials (above all, a high impact viscosity) permit them to the successfully used as structural elements in machine building (components for textile looms, flywheels, panel, etc.). Laminated plastics based on aminoresins with paper fillers are used as a trimming facing material in vehicle and ship building (railway coaches and ship berths' lining), in construction, in the woodworking and furniture industries.

3 Methods for Appraising the Technological Properties of Aminoplasts

The knowledge of the technological properties of aminoplasts is necessary both for the right choice of the method of material processing and for maintaining optimal conditions at all processing stages. Each processing method is characterized by its physico-chemical processes proceeding at different temperatures and times and by some specific methods for undertaking mechanical treatment of the material.

In this case during the treatment process the initial aminoplasts (powder, granules, tablets, etc.) are subject to far-reaching changes associated with the phase state and the very structure of the substance.

In view of the continuity and simultaneity of all the processing stages the indicates to be determined will characterize not some specific physical process, but a combination of the processes.

Appraisal of the technological properties of aminoplasts involves a large-scale application of diverse rheological and physico-chemical measurement techniques.

3.1 Rheological Methods

Particularly interesting are the various rheological methods enabling us to determine both the qualitative and quantitative regularities in the processes of shear flow and solidification of thermoactive materials. Essentialy these methods are associated with the realization in the measuring apparatus of such types of flow that make it possible to register in the simplest possible way the parameters characterizing the relationship between the magnitude of shear velocity and tangential stress. The given relationship of the plastic flow of polymer materials is nonlinear and can be characterized by a viscosity coefficient. This coefficient depends not only on the nature of the polymer, but also on the parameters of the flow itself, especially on shear velocity.

3.1.1 Rotational Methods

The simplest type of flow of a medium that yields itself to an analytical description within the framework of the precise hydrodynamical equations of viscous liquids (Navier-Stocks equations) is the Couette flow. This flow occurs under the impact of tangential stresses generated in a viscous liquid by a solid surface moving in it. The magnitude of the force that has to be applied to this surface to securse its movement in the viscous medium characterizes the tangential stresses and the velocity of its movement — the shear velocity.

At the basis of rotational methods lies the setting up of a shear stressed and deformed state of the material being tested in the gap between two surfaces of the instrument's working element, one of which is stationary and the other is rotating at a velocity varying according to a given law. The form of the measuring surfaces in the working element determines the distribution of stresses and shear velocities in the working gap. In practice rotary-type instruments with coaxially-cylindrical measuring surfaces have become very popular. Given small gaps between the latter, we may achieve a high homogeneity of shear velocities, as well as secure favourable conditions for heat exchange between the material tested and the heat-exchange surface of the measuring element.

The specific character of rotary-type instruments lies in the fact that the viscosity measurements in it can be combined with a great number other rheological measurements and measurements of specific characteristics of the material. Moreover, the properties of the material are determined in absolute physical quantities. In practice the relationship between stresses and deformations of a polymer in rotary-type instruments is established after the method of a constant deformation rate ($\dot{\gamma} = $ const). The latter condition is equivalent to a constant rotation frequency ($\Omega = $ const) of one measuring surface relative to the other one. Directly from actual practice we obtain torque versus time — M(t), from whence we find tangential stress versus time — τ(t). Given an invariable velocity of movement of one measuring surface relative to the other one, it is easy to realize a transition from tangential stress versus time to tangential stress versus the amount of deformation ($\dot{\gamma} = \Omega t$) [16]. The former characterizes the kinetics of the deformation process and the latter enables the behavior of different compositions under similar deformations to be compared, and thus we can determine a number of important parameters characterizing the deformational properties of materials.

In some countries recognition has been accorded to the method of estimating the viscous-plastic properties and solidification kinetics of reactoplasts, among them aminoplasts, on the Kanavets-type plastometer [17] with the working unit of the "cylinder — cylinder" type.

The Kanavets plastometer represents in principle two coaxial cylinders, into the gap between them is placed the mass under pressure. The inner cylinder (in the form of a bar) is set in rotation. On the inner surface of the matrix and at the outside surface of the bar grooves are made preventing the melt from slipping along the wall.

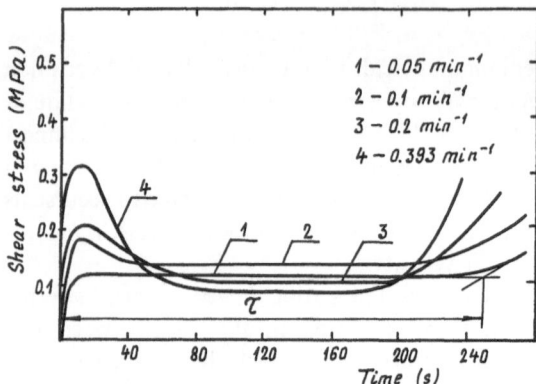

Fig. 1. Shear stress versus deformation time when the rotor rpm is 0,05–0,393 min^{-1}*

The plastometer secures deformation of the material tested at shear velocity from 4×10^{-3} s^{-1} to 3×10 s^{-1} (in 12 working ranges) at temperatures up to 300 °C.

Figures 1, 2 show typical curves of shear stress versus deformation time of aminoplastic at 120 °C for different r.p.m. of the plastometer rotor, from $\Omega = 0.05$ to 50 min^{-1} (i.e. for shear rates from 1.4×10^{-2} to 1.5×10^1 s^{-1}).

Curve 1 in Fig. 1 under low rotor r.p.m. corresponds to the case when the rate of the stress buildup under the effect of deformation is commensurate with the rate of their relaxation. The development of time-delayed elastic deformations determines the final rate of attainment of the steady-state regime of viscous flow, under which the stationary value of the shear stress is recorded. Starting with the moment of time τ, the stress will increase and the period of the viscous-plastic state comes to an end.

With an increase in the deformation rate the curves are seen to show stress maxima at the initial stretch, this being associated with the fact that following the

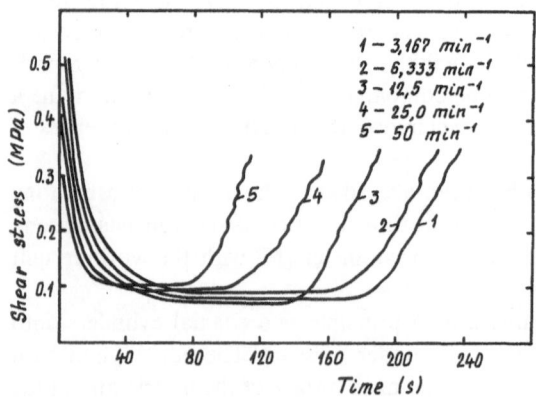

Fig. 2. Shear stress versus deformation time when the rotor rpm is 3,167–50,0 min^{-1}

---

* In all figures decimal commas are used instead of decimal points.

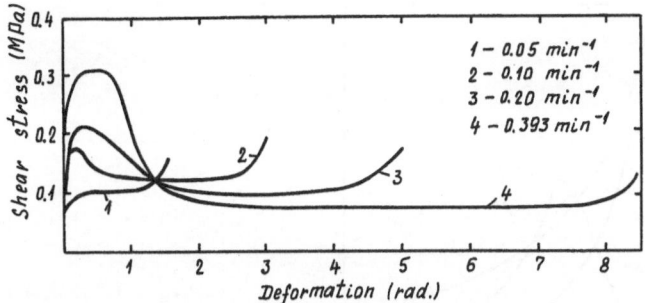

Fig. 3. Shear stress versus deformation the rotor rpm is 0,05–0,393 min⁻¹

beginning of deformation the growth rate of shear stresses supercedes the rate of their relaxation. Thereafter the tangential stress is observed to decline gradually because elastic deformations are subject to the effect of flow, and the curves begin to correspond with the steady-state regime of flow (tangential stress being constant). The value of the stationary shear stress under the viscous-flow regime of the polymer will likewise increase intially, and thereafter will decline down to values even lower than those under low deformation rates (curve 4, Fig. 1).

Figure 2 shows the experimental curves obtained with rotor r.p.m. from 3 to 50 min⁻¹. As seen from the curves, as the deformation rate increases the time of the viscous-plastic state perceptibly declines. This is mainly due to an acceleration of the hardening process through the dissipation of mechanical energy. However, in spite of this the absolute value of deformation with the growth of deformation rates increases. This is illustrated in Figs. 3, 4 where the same shear stresses versus the magnitude of deformation are given.

The curves shown in Figs. 1–4 describe the particularly characteristic deformation processes of aminoplastics that are of importance for a correct choice of both the method and regimes of processing the material. Under other prescribed

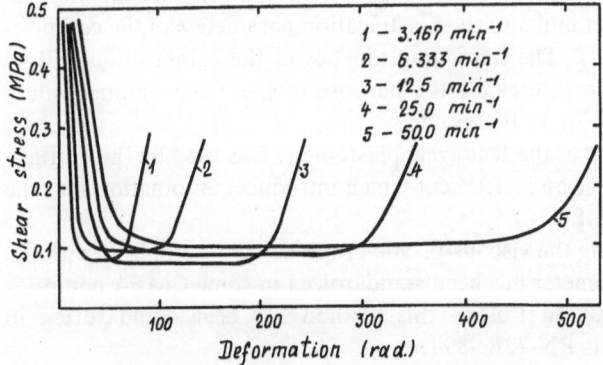

Fig. 4. Shear stress versus deformation the rotor rpm is 3,167–50,0 min⁻¹

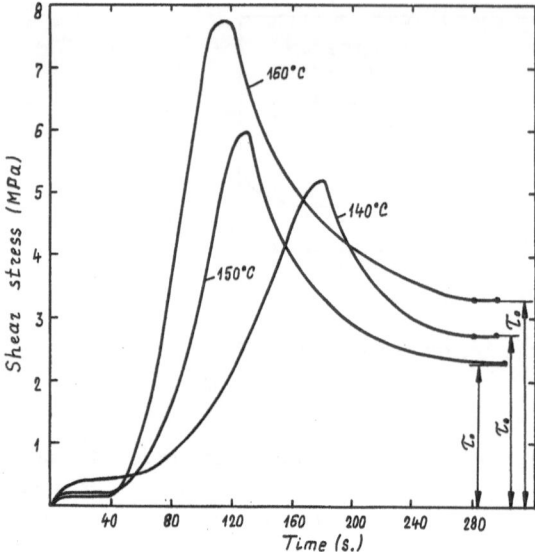

Fig. 5. Hardening of type KFA 2 aminoplast and relaxation of stresses

temperatures (up to 150 °C) the same processes take place, and qualitatively the character of the curves remains the same.

Upon hardening of the composition the plastometer will record the process of relaxation of the stresses with the working bar being idle. Stresses arising in the material with the progress of time fall off to an equilibrium value τ_0. Figure 5 shows the hardening process of the aminoplasts at temperatures 140 and 150 and 160 °C under a continuous deformation and the relaxation of stresses which attains the maximum value τ_0.

The procedure for estimating the viscous-plastic properties on the plastometer has been standardized in the USSR by GOST (State Standard) 15883-79.

It is recommended that the assessment of the technological properties of aminoplasts should be started with determining the temperature of softening T_s. Thereupon at a temperature 20–30° above T_s one should estimate the viscous-plastic properties which set limits to the plastification parameters of the composition under pressure casting. The hardening kinetics of the composition will be determined at higher temperatures for the purpose of specifying optimal values of temperature and seasoning in the mould.

In the FRG the principle of the Kanavets plastometer was used by the Gottfert Messtechnik Co. to produce an instrument which introduces automation into the processing of test results [18].

The method of estimating the viscous-plastic properties of reactoplasts with the aid of the Kanavets plastometer has been standardized in some CMEA countries (according to PC 2491-70). In Poland this method has been standardized in accordance with the norms PN-72/C-8903.

In [19] use was made of a modified testing mould on the Kanavets plastometer for the study of the effect of wall thickness and temperature of the mould on the hardening kinetics of aminoplasts. In the course of testing the authors recorded the closing-up time of the mould, initial shear stress, the length of the viscous-flow state of the material and the minimal time of its hardening in the mould until the attainment of the shear stress 2.0 MPa. It was found that the wall thickness of 2–4 mm did not directly affect the reactive capacity of the aminoplast specimens studied, but it did affect the hardening rate because of the changes in the heat-exchange conditions. The hardening dynamics are mainly determined by the temperature of the mould and may be represented by the exponential function.

The main sources of error in estimating the viscous-plastic properties with plastometers are fluctuations of the mould temperature and of the specimen mass of the material tested. Reducing the specimen mass will account for an inadequate density of the specimen, and, consequently, for a decline in the resistance to shear. An insignificant increase of the specimen mass above the optimal one will not practically affect the results of the tests.

The plastograph of the Brabender Co. enables similar experiments to be carried out at higher speeds: the r.p.m. of the working units varies from 20 to 200 min^{-1}. The working assembly of the instrument is a heated chamber with two shaped rotor-rollers. A theoretical analysis of the flow is in this case rather difficult.

The mass specimen is loaded into the chamber and by means of a special plate, upon which are set weights, the material is put under pressure.

Modern designs of plastographs manufactured by the Brabender Co. (for instance, PlastiCorder Informatique PL 2000) are equipped with IBM personal computers which enable the processes of measurement and analysis of test results to be completely automated [20]. In testing the materials on the plastograph obtained plastograms are, i.e. graphs of the torque on the rotors versus time. These dependences are obtained in the process of an elastic-deformational impact on the material at all the stages of the latter's transformation in the plastograph's working chamber: melting, viscous-flow state, hardening and subsequent destruction. The plastograms can be used to achieve a qualitative assessment of the technological properties of the compositions, associated with the characteristics of effective density in the processes of melting, shear flow and hardening. The plastograms are, moreover, employed to determine the duration of the above processes.

Figure 6 shows a typical plastogram in the form of curves showing torque variations versus time. Also shown here is the curve of the variation of the material temperature versus time in the instrument's working chamber. We can see the following characteristic regions. At point A, the lumping process and the formation

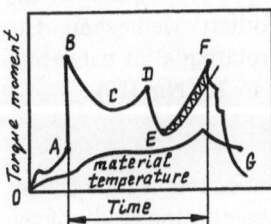

Fig. 6. Variation of torque with time

of the melt begins, the torque sharply increasing (AB). On the BC part of the curve the viscosity of the melt is declining, this is due to the dissipation of mechanical energy and to the heat transfer from the chamber walls. On the CD part of the curve the melt viscosity is rising in the process of hardening. An abrupt variation of the torque at DE is accounted for by a transition from the shear flow to sliding. On the EF part of the curve, torque fluctuations associated with the overpressing of the high-viscosity melt are observed. At point F the collaps of the hardened material occurs. A collapse of this kind takes place under those conditions when the medium is subject to pressure with shear with simultaneous cooling. This involves the realization of the elastic-deformational mechanism of collapse associated with the accumulation in the material of elastic energy [21]. It has been found [22] that the progress of the crushing process in this case is not a gradual process with the material being progressively crushed to smaller and smaller particles. Crushing proceeds abruptly, within a limited spatial and time interval, thus pointing to a spontaneous progress of the destruction processes and enabling us to assume the presence of a branched mechanism for the development of destruction foci.

Elastic-deformational crushing at the terminal stages of tests in the plastograph are characterized by extremely low energy consumption which are 2–3 orders below those of the crushing carried out on impact-type machines. Further detailed investigations of the given mechanism of elastic-deformational crushing are likely to foster a new attitude to certain aspects of the theory of strenth of the compositions concerned. Furthermore, the plastograms at the stage of hardening of the reactoplast and its subsequent elastic-deformational crushing can obviously be employed for a qualitative assessment of the strength properties of the compositions concerned in the solid state. Such an estimate could be useful from the viewpoint of predicting the quality of the articles obtainable as a result of processing the polymers.

Assessment of the viscous-plastic properties with the aid of the Brabender plastograph suffers from some significant shortcomings. The rheological interpretation of experimental data obtained with the instrument is made difficult because of the complexities involved in determining the deformation rates. The conditions of the hardening of the material in the instrument's testing chamber are different from those of the real technological processes of pressing and pressure casting. The material is given additional energy that is spent on overcoming the external and internal friction. As a result, hardening proceeds at a more rapid rate than the ordinary conditions of pressing and pressure casting.

A similar application, the measurement principle and design are incorporated in the "Record" instruments produced by the Haacke Co. (FRG), and in the extrusiometer "Torsiograph" manufactured by the Gottfert Meßtechnik Co. (FRG). The latter has worm shafts of 20 mm diameter, rotating at at between 1 and 140 r.p.m. The region of torque variations is from 5 to 200 Nm [18].

3.1.2 Capillary Method

The capillary method based on setting up shear deformations in the movement of viscous media in channels is being used on a large-scale in the study of the rheological characteristics of low-molecular and high-molecular weight liquids. This is associated with the fact that the steady-state flow in channels yields itself to a sufficiently accurate analytical description not only for the Newton, but equally for nonlinearly-viscous and viscous-plastic media. However, using the results of capillary viscosimetry in aminoplast technology, particularly in the case of highly-filled compositions, is fairly complicated because of the departure of the real conditions of flow from the theoretical ones. In particular, because of the wall slipping the true velocity of shear cannot be determined without a great deal of error, and therefore the flow curves and using them to determine the constants of rheological equations fail to provide the required accuracy.

Nontheless, the capillary methods can be extremely useful both for the qualitative and the comparative evaluation of the flow of aminoplasts. Such an evaluation is necessary for a technological characterization of the flow of the material through the casting gates in moulding articles from aminoplasts by the pressure casting method. By using capillaries of different dimensions we may observe the effect of the instrument's geometrical parameters on the resistance to the flow. This is simultaneously accompanied by the manifestation of the effect of the slipping of the material over the capillary walls, as is observed in real production processes. The capillary method enables us, moreover, to study the entrance effects in the flow of the material through different channels [23].

The method of squeezing through different channels is amply illustrated by using the instrument made by the Zwick Co. (FRG).

The Zwick-made instrument has become very popular in the FRG. The material is squeezed in it under constant pressure through channels of an specified length (l = 150 mm) and standard cross-section (4 mm × 10 mm) under constant temperature. The length of inflow is recorded in the diagram "path of flow − time of flow" (Fig. 7). The flowability characteristic comprises the total path and time of flow, the movement velocity of the weight at the softening and flow stretches. By making use of the expression for determining the volume flow (Q) of nonlinearly-viscous liquid through a chanel of rectangular cross-section with the height 2 h and width b we may calculate the viscosity coefficient [24]:

$$Q = \frac{bh^{\frac{2n+1}{n}} \cdot n}{2n+1} \left(\frac{\Delta p}{h\mu}\right)^{1/n} \times F\left(\frac{2h}{b}\right) \tag{1}$$

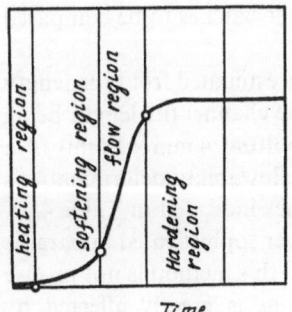

Fig. 7. The "flow path − flow time" relationship

where $F\left(\dfrac{2h}{b}\right)$ is a correction coefficient depending on the chanel width; Δp — is

the pressure differential over the length; n — is the exponent of the power law of the flow; μ — is the viscosity coefficient.

However, on account of the cork-like character of the flow in the channel the error of the design scheme is great, whilst the reproducibility of the results may unsatisfactory because of friction losses in the punch-and-die mating in the chamber. Other shortcomings of this method are due to difficulties of extracting the specimen, a long preparatory period prior to testing.

According to the Rossi-Picesa technique standardized in the USA by ASTMD 569-82, ("The Method of Measuring Flow Properties of Thermoplastic Molding Materials), the flowability of thermoreactive materials is determined by measuring the effort and time of flow in a vertical cylindrical channel of 3.18 $\pm$ 0.013 mm diameter and 38 mm in length. Pressure on the material in the 25.4 mm diameter die amounts to 0.62 — 1.03 MPa.

The use of the capillary method for the study of aminoplastics suffers from certain shortcomings: laboriousness of the experiments because of the possibility of the material solidifying in the capillary; the presence of entrance effects influencing the collapse of the material structure; the impossibility of characterizing the material in the steady-state regime of flow; the impossibility of attaining a pure shear because of the slipping of the material over the capillary walls and the appearance of slipping effect on the walls (in the limiting case there arises the "core-like" flow with the velocities of all the particles being similar).

3.2 The Primary Assessment of Rheological Properties

Technological methods permit us to estimate the capacity of thermoreactive compositions for the moulding of articles and to obtain comparative indices for the behaviour of the material. These methods are widespread in practical work of industrial enterprises and by no means have they lost their significance. In the USSR among such methods are the flowability tests on reactoplasts according to the Raschig technique and determination of the hardening rate from the appearance of the article in the form of a test core. These two methods have been standardized by GOST (State Standard) 9359-80 for aminoplasts. In this case the flowability tests are made not in absolute but rather in conventional units, thus permitting the rheological behaviour of different materials and their batches to be compared under comparable conditions.

According to the Raschig technique the flowability is estimated from the length of a thin core of material, that has been obtained in the channel (its length being 200 mm and the cross-section both at the top and bottom 4 mm $\times$ 1 mm) of a special press-mould under standard conditions. In the flowability determinations the aminoplast specimen is pressed by using the regimes indicated in Table 4.
The Raschig method is simple and does not require either sophisticated apparatus or long-term processing of experimental data. However, this method is not precise because the accuracy of the flowability determinations is greatly affected by

Table 4. Flowability determination regimes after Raschig

Type	Pressing regime		
	Specific pressure MPa	Temperature °C	Seasoning time min
KFA	30 ± 2.5	143 ± 2	3
MFB	30 ± 2.5	150 ± 2	3
MFV	30 ± 2.5	165 ± 5	3
MFD	30 ± 2.5	150 ± 1	3
MFE	50 ± 2.5	150 ± 5	5

the fluctuations of the moulding pressure and temperature, by the condition of the channel surface in the mould, etc.

The method of flowability determination has also been standardized in Romania with STAS 5991-68. According to the Polish standard P-61/C-8936 the reactoplastics in accordance with this method are subdivided into 4 classes: with a core length of up to 50 mm — poor press-material; 50–80 mm — press-material with a low flowability , 80–120 mm — press-material with an average flowability; 120–180 mm — press-material with a high plasticity.

No functional relationship between the flowability index according to Rashig and the other indices of the viscous-plastic properties has yet been established. The flowability index determined by this method serves as a preliminary assessment of the suitability of the material for processing.

The hardening rate of an aminoplast is determined by the amount of time necessary for a complete hardening of the specimen in the form of a conical test core. The specimen is pressed under conditions specifically indicated in the standards on the aminoplast:

Type of aminoplast	Specific pressure MPa	Temperature °C
KFA	25.0 ± 2.5	145 ± 3
MFB	40.0 ± 2.5	160 ± 5

The time of hardening is assumed to be the time of seasoning of a pressed test core which, judging by its appearance, seems to be of proper quality.

The test-core method is used in Britain according to BS 2782; in the USA according to ASTM D731-57; in the FRG according to DIN-53465.

The duration of flow, in the course of which the mould filled with press-material is being closed up, is measured.

The press-mould has to be filled with enough material to exceed by 2 g the amount necessary for pressing the test core, and the time elapsing from the moment pressure is applied to the moment when burrs cease to be formed at the upper edge of the test core is noted.

The Czechoslovak Standard CSN 640214 recommends that the test core method should be employed for determining the flowability of aminoplasts. The test core is stepped in shape. The flowability criterion is the time required to fill up the mould. The five-point scale for assessing the flowability is used. The average pressure on the material amounts to 54 ± 1 MPa, the free-stroke velocity of the crosspiece of the press is 20 ± 5 mm/s.

The test core technique has failed to find large-scale acceptance because of the unsatisfactory reproducibility of the results. The reasons for this may be due to differences in the force-velocity characteristics of the presses, as a result of which the recorded efforts and movements versus time largely reflect the properties of the equipment used rather than the properties of the material being tested. To achieve a better reproducibility of the test results the test must be carried out on equipment with calibrated metrological parameters.

Flowability determinations on reactoplasts according to the method set out in ASTMD 31 23-72 (USA) involve the forming of a spiral of a specified cross-section. Use is made of a pressing machine for transfer pressing. The 20 ± 0.1 g of the material is placed in a chamber 25.4 mm in diameter. Tests are carried out at 150 ± 3 °C, at a pressure of 6.9 ± 0.17 MPa, and a piston velocity of $25-100$ mm/s. The cross-section of the spiral formed is a half-circle with a radius of 1.6 ± 0.5 mm. The maximum length of the spiral is 1270 mm.

All the above methods are based on obtaining conventional and relative characteristics of the material, and this does not allow us to judge with sufficient adequacy the flowability of aminoplasts, and so they can be employed solely for manufacturing articles under similar conditions. They are suitable mainly for estimating the materials for their adherence to standards. As regards control over the treatment process, they are not very informative. These methods suffer from the following disadvantages: a) the accuracy of the methods is insufficient; b) no homogeneous temperature field is secured during testing; c) the tests fall to determine the duration of the viscous-plastic state; d) insignificant variations of the channel's cross-sectional dimensions, in the surface condition, in pressure and temperature account for a significant variation of the length of inflow (for instance, of the length of test core in the Rashig press-mould); e) an insufficient reproducibility of experimental results.

3.3 Chemico-Physical Methods

More accurate, though labour-consuming and lengthy chemico-physical methods can be used for determining the hardening kinetics of reactoplasts.

Chemical analysis permits us to control the loss of any kind of groups participating in the hardening of binders either directly or upon the release of the byproducts of the reaction (by observing the loss of the methylol groups or the release of such low-molecular products as water and formaldehyde).

The main difficulty in using chemical analysis is due to the insolubility of the material at the hardening stages after the gel-forming point. The analysis is conducted during the swelling of a highly-reticular polymer in solvents, thus prolonging the time and reducing the assessment accuracy.

There are known chemical methods for estimating the hardening rate, for instance, according to the bromine number. The intricacy of performing tests by using these methods does not allow them to be recommended for the quality analysis of the raw material under factory conditions. These methods prove helpful in carrying out research work and in working out new recipes of raw materials.

Today in the USSR there is no universal procedure for determining the quantity of extracted substances for reactoplasts, including aminoplasts, because this factor is omitted in the standards on materials. The industrial establishments conduct determinations by different methods and this hinders with obtaining of comparable results.

Methods for determining the amount of hardening after the substances are extracted have been standardized in the FRG by DIN 53700, in the USA by ASTMD-494-46, in Britain by BS: 771-1938. The extent of the progress of the hardening reaction at the terminal stage is estimated, as a rule, after the quantity of the functional groups having entered in the reaction by using different chemical or physical methods of analysis. However, the results of the analysis are affected by the presence of modifiers, fillers and other ingredients.

The amount of hardening of the binder can be determined from the rate of occurrence of chemical nodes in the polymer network (according to network density). Chemical nodes of polymer networks are the points of chain branchings or the points at which the chains are bonded together by chemical (covalent) bonds resistant to destruction [25].

The physical methods of analysis are based on the measurement of electrical and thermal characteristzics.

The infrared spectroscopy method (IR) is being successfully employed for checking the hardening of phenol-urea-melamine-formaldehyde resins and materials derived from them.

IR-spectroscopy enables polymer materials to be qualitatively estimated according to the interdependence between functional groups and the observed absorption bands [26].

IR-spectroscopy was used in [27] to study the process of hardening of the urea-formaldehyde oligomers. It was found that the content of methylol groups versus time was of a nonmonotonous character. A possible mechanism of the appearance of fluctuations in the manifestation of this effect has been proposed, associated with the suggestion about simultaneous progress of the reaction of condensation and hydrolysis.

Acoustic methods of testing enable the degree of hardening of the materials studied to be determined from the velocity of souind in the hardened and nonhardened material. Determination of the velocity of sound in thermoreactive materials is achieved by using the DUK-20 instrument [28].

The rate of hardening can also be assessed from a variation of the dielectric properties of the composition, as well as by using calorimetric methods [29]. However, all the studies devoted to these methods have, as a rule, a descriptive character and do not solve the problem of a quantitative description of the hardening process of the composition concerned.

4 Processing Methods of Aminoplasts

The principal processing methods of aminoplasts are: pressing, pressure casting and extrusion.

4.1 Pressing

Pressing is one of the widespread methods for the manufacture of articles from aminoplasts. There exist varieties of pressing: direct (compressional) and casting (transfer).

In direct pressing the pressure directly affects the material confined within the space of a heated mold. In the press-mold the effect of pressure and heating accounts for the fact that the material is compacted, softened, plasticized and thereafter hardened to the required degree.

In compaction the particles of the material come so much close to one another that there arise between them the forces of intermolecular interaction, and thus a compact body is formed.

The interrelationship between the density of the material (ϱ) and pressure in pressing (P) is fairly complex and is expressed by an exponential function. Investigations [30] have revealed that under the actually used pressures such a function can be substituted with a linear one of the type $\varrho = P/RT (A + BP)$, where R is a universal gas constant; T is the temperature of pressing, °C; A and B are constant coefficients.

Aminoplasts being distinguised by a low thermal conductivity, their heating in the technological process equipment is accompanied by the appearance of great temperatures gradients across the thickness of the article. In studying the temperature fields in the melt, use is made of the approximate analytical solutions of the equation of thermal conductivity for constant physical properties of the material and the medium with the use of boundary conditions of the 1st kind. The calculation formula for the time of heating across thickness of an article in the form of a plate can be represented in the form [31]:

$$\tau_{heat} = \left(\frac{2h}{\pi}\right)^2 \cdot \frac{1}{a} \ln\left[\frac{4}{\pi} \cos\left(\frac{\pi x}{2h}\right) \cdot \frac{1}{\theta}\right] \tag{2}$$

where 2h — is the thickness of the plate;
    a — is the coefficient of thermal conductivity.

$$\theta = \frac{t_m - t_0}{t_m - t_f}$$

t_m — is the temperature of the mold;
t_0 — is the initial temperature of the material;
t_f — is the final temperature of the material, it is assumed that ($t_m - 4$ °C).

Figure 8 shows the character of the curves obtained in heating the material in the mold with a constant temperature of the wall. As the material is being heated,

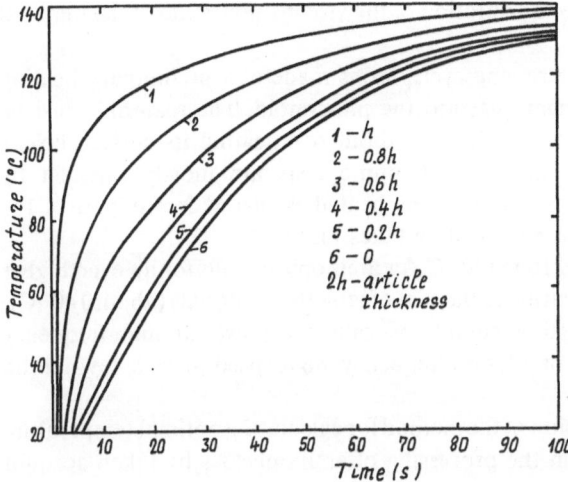

Fig. 8. Variation of temperature versus time for different points across the thickness of an article from a KFA2 aminoplast

the temperature in every cross-section approaches that of the mold asymptotically. Subject to a particularly rapid variation is the temperature of the points lying close to the surface of the mold. As the time of heating increases, the difference between the mold temperature and the cross-section of the aminoplastic melt gradually declines. With further seasoning of the article in the mold, a progressive equalization of of the temperature field takes place and, givern a significant seasoning time, there comes the moment of a quasi-stationary temperature field. In practice the removal of the article takes place earlier than when the material attains a uniform temperature field. Accordingly, the degree of polymer hardening within the material of the article is dissimilar. Furthermore, temperature gradients are at to arise also in the process of cooling of the article to normal temperature because the surface layers of the article cool at a higher rate than do the interior ones. As a result, the different areas of the article have different thermo-elastic characteristics. The fields of stresses arising in the material are nonsymmetrical and may cause a distortion of the shape of the article.

In the course of pressing, as the material is subject to hardening, volatile products are released, i.e. low-molecular weight reaction products. The removal of gaseous reaction products is achieved by means of subpressing, the latter involving a short-term lifting of the punch and the subsequent closing-up of the mould. This partly relieves inner stresses and adds to the homogeneity of the aminoplast structure. Subpressing proves to be particularly effective in manufacturing thick-walled articles. It is practicable to carry out several subpressings of short duration for 2–5 s in place of a single long one. Subpressings are undesirable for pressing of articles with reinforcement, in the presence of removable inserts in the mould, and in pressing thin-walled articles of great height (depth).

An optimum temperature interval of the mold in the processing of aminoplasts is 135–160 °C. The main condition for the choice of the press-mold temperature within the above limits is a termination of the process of material compaction during the time not exceeding the duration of the viscous-flow state of the material at this temperature.

To reduce the time of the pressing cycle, use is made of a preliminary heating of the material prior to its charging into the press-mold. The material must be heated uniformly throughout its volume without overheating its surface layers. The above requirements in the case of aminoplasts are mostly satisfied by high-frequency heating. The material to be heated is placed in the gap of the high-frequency capacitor where its heating takes place.

High-temperature pressing, 160–170 °C, for aminoplasts calls for an even higher temperature of preliminary heating in the case of tabletted materials (about 140 °C).

Introduction of such pressing conditions calls for powerful high-frequency generators with an increased oscillation frequency, high-speed presses, low-inertia thermoregulators [32].

In Ref. [33] there is a discussion of the method for calculating optimal temperature regimes and seasoning time in the processing of aminoplastics by taken account of the heating and hardening rates of the materials for different temperatures of preliminary heating (60 and 80 °C).

Since the processing of aminoplastics is realized under nonstationary heat-transfer conditions, an accurate calculation of the seasoning time in the mold ($\tau_{seas.}$) proves to be a complex problem.

For the sake of simplification $\tau_{seas.}$ is regarded as a sum of the consequent processes of heating ($\tau_{heat.}$) and hardening ($\tau_{hard.}$) of the aminoplastic. The time of heating is determined from the equation of heat transfer i.e. Eq. (2).

To determine the hardening time of aminoplastics, use is made of the Kanavets plastometer. An equation is deduced for determining the seasoning time in the mould:

$$\tau_{seas.} = \frac{h^2 F_0}{4a} + 0.9 \, (\tau'_{hard.} - \tau_{heat.})^{v(t'_0 - t_m - 4)} \tag{3}$$

where $\tau'_{hard.}$ — is the hardening time found from the plastogram under the temperature of the standard aminoplastic test (t_0) 140 °C;

t_m — is the mold temperature, °C;

F_0 — is the Fourier criterion;

v — is the temperature coefficient for aminoplastic, $v = 0.28$.

The authors of Ref. [34] optimized the pressing technology for electrotechnical articles with the use of powder based on melamine-formaldehyde resin (MLFR) and cellulose. To this end, a complex investigation was undertaken into the physico-mechanical properties of the article versus temperature (140–180 °C) and pressing duration (20–160 s) in the automatic pressing device Bucher Ka-100. The choice of optimal pressing parameters was made (160 °C and 90 s instead of 150 °C and 120 s) making it possible to save 2600 h of machine time and to increase labour productivity.

At a Hungarian works for the manufacture of articles from plastics [35] automatic production lines have been set up — Bucher KA-100 and Tavannes 50/10 and

72/8 to carry out the pressing of articles from phenoplasts and aminoplasts. Once the automatic production equipment had been installed, labour productivity rose three-fold.

Compressional pressing is distinguished by its technologicl simplicity. A shortcoming of the processing method by direct pressing is long-time seasoning to secure the hardening of the material in the mold. In this case the composition is subject to irregular thermal impact, as a result of which there arise inner stresses that are likely to cause deformations of the article in the course of service.

4.2 Cast Pressing

The method of cast pressing can be employed for the manufacture of articles of complex profile, with greatly different thickness, with thin fixtures, and with threaded signs. The article is shaped in a closed mold, the requisite pressure being set up in the charging chamber and transmitted through the material. The pressure in the charging chamber amounts to 100–150 MPa and is required for filling up the molding space in the press-mold and for compacting the material. Pressure in the press-mold is set up at the very termination of the molding stage. The process of compaction of the composition under the condition of the incompressibility of the mass is described for cast pressing by an exponential function of the type: $Q = K(P - P_m)^n$ (4).

Taken as the compaction parameters is the voluminal flow Q of the plasticized mass through the gasting gate. This flow is proportional to the difference between the pressing pressure P and the pressure in the molding space P_m raised to a power the exponent of which n is dependent on the rheological parameters of the material. The proportionality coefficient K characterizes the geometrical dimensions of the gating system.

As it flows through a narrow channel, the material is intensively intermixed, is additionally heated up due to friction by 20–30 °C on average. In this case the hardening of the material takes place to a greater depth and is more uniform than in the case of direct pressing. The articles obtained are notable for a greater homogeneity of their properties.

In the case of cast pressing of differently-thick articles of great cross-section the length of the cycle can be significantly reduced, as much as up to 50%, as compared to direct pressing. Aminoplasts notable for high flowability and rapid hardening are particularly suitable for cast pressing. For cast pressing the preheating of tablets is indispensible.

The method of transfer pressing with trough-type plastification is described in Ref. [36]. The material is heated in the plastification cylinder to 60–120 °C according to the type of the material. The trough-type plasticiser can supply a number of presses in succession. The heated mass goes from the plastification cylinder into the casting cylinder and is then pushed by a piston into the mold. In the mould the hardening of the article takes place. Once the article is removed, the piston moves downward and a new portion from the plastification cylinder can be introduced into the casting cylinder and thereafter into the closed mold.

A review is given in Ref. [37] on the principles of the pressing and transfer molding technology of profile articles from compositions based on amino-, pheno- and epoxy resins. The designs of the press-molds in use and the technological equipment of commercial processes are described.

4.3 Pressure Casting

A significant increase in labour productivity as against pressing can be attained in the processing of aminoplasts in the casting machines.

The principal technological schemes of pressure casting as applied to articles from reactoplasts, as well as the analysis of the processes taking place in this method of treatment have been described in much detail in Ref. [38]. We would like to mention only certain specifics in the treatment of cast aminoplasts. To fill up the molding space with the aminoplast a pressure of 100–200 MPa is necessary; at the terminal stage of the filling-up of the molding space, the pressure should be reduced to 70 MPa to avoid overcompaction of the material [39].

The use of pressure as high as this is due to the high viscosity of the composition melt. In the state of the melt the aminoplasts are high-filled suspensions.

The minimum pressure in the mold is determined principally by the viscosity of the composition and the plateau time, as well as by their dependence on temperature.

The maximum pressure developed in the molding space determines the minimally required effort for closing the mold.

The movement of the material in the region of the plastification cylinder, lying directly behind the charging zone, is of a block-type character without significant relative displacements of the granules. The working of this zone is determined by the laws of dry friction. Although the material in the through channel converts into the viscous-flow state, its movement is accompanied slipping on the wall in the practical absence of shear deformations. A change-over from a regime of dry friction to that of liquid slipping takes place at that section of the trough length equal to 0.8–1.3 of its diameter. These sections of the trough are known as the plastification and transportation zones.

A significant heating-through of the material begins after it comes from the transportation zone into the plastification zone, because of contact with the hot surface of the gate channel, as a result of compaction and the appearance of dissipative heat release [38]. The process of conversion into the melt (plastification stage) of the composition must be carried out in such a manner that chemical transformation at this stage are practically absent.

To secure maximum productivity, the temperature of the melt must be as close to that of the mould as possible, while the progress rate of the hardening reaction must be such that the beginning of a perceptible increase in the melt viscosity should coincide with the termination of the mold-filling stage. However, in this case the time that the material is in the viscous-flow state declines. Therefore, the choice of the processing parameters and conditions of the casting method is essentially associated with the optimization of the temperature-time requirements for plastification and molding.

For cast aminoplasts, according to the viscous-plastic properties of the material, specific temperature is to be maintained in the various cylinder zones, as follows:

	Materials based on urea-formaldehyde resin	Materials based on melamine-formaldehyde resin
1st zone, °C	85–100	95–110 (T_1)
2nd zone, °C	65– 75	70– 80 (T_2)
3rd zone, °C	35– 45	45– 55 (T_3)
T of the mould, °C	140–150	150–160

The charging zone is water-cooled and its temperature is 18–25 °C.

The temperature of the casting mold should be chosen by keeping in mind that the filling time of the mold must not exceed the time during which the material remains in the viscous-plastic state at this temperature. With regard to cast aminoplasts, the recommended temperature range for the mold is 135–160 °C. This distribution of temperatures has a stepped character. Indeed, the temperature distribution in the material being plasticized has a more complex character both over the length of the portion and as regards the time the material is present in the injection cylinder.

For practical estimates and deductions of approximate relationships for determining the temperature-time conditions for reactoplast treatment, we may use a simplified model of the thermal field, namely by assuming that that the temperature of the material in each of the zones, the mold included, is constant for the entire period of the presence of a portion of the material in a specific zone.

The distribution of temperature of the material between the cylinder and mold zoneswill have a stepped character. Such a model of the temperature profile enables us to obtain the requisite design relationships for a coordinated choice of both the temperature and the time that the material being treated in the various zones of the injection cylinder and the mould is present.

For this purpose we may use an assessment criterion of the viscous-plastic state in any temperature-time regime of the polymer hardening reaction, as given by the following relationship [23]:

$$A = \int_0^{\tau_{vps}} \exp\left[-\frac{\Delta U}{RT(t)} \right] dt, \tag{5}$$

where τ_{vps} — is the time of the viscous-plastic state;

A — is a parameter characterizing the quantity of products of the polymer composition, capable of entering in the reaction;

ΔU — is the activation energy of the polycondensation reaction;

R — is a universal gas constant;

$T(t)$ — is the polymer temperature versus time.

The values of A and ΔU are the physical characteristics of the material and can be easily computed from two viscosimetric experiments in the isothermal regime under different temperatures T according to the expression:

$$\tau_{vps} = A \exp\left[\frac{\Delta U}{RT}\right] \tag{6}$$

For the chosen model of the stepped distribution of the material temperature, Eq. (5) is written in the form:

$$A = \sum_{i=1}^{3} \tau_i \exp\left[-\frac{\Delta U}{RT_i}\right] + \tau_m \cdot \exp\left[-\frac{U}{RT_m}\right] \tag{7}$$

where i — is the number of cylinder zones; τ_i, τ_m — is the presence time of the material in the various zones of the injection cylinder and in the mold at the temperatures T_i, T_m.

Equation (7) for the accepted assumptions concerning the character of distribution of the material temperature is accurately fulfilled whenever the total time that the material is present in the injection cylinder (plastification time) and in the mold corresponds to the time of the viscous-plastic state of the material in any given temperature-time regime, i.e. when

$$\tau_{vps} = \tau_1 + \tau_2 + \tau_3 + \tau_m. \tag{8}$$

Equation (8) is, to a certain degree, an optimization for estimating the minimum time of treatment of the material to produce the finished article. For a given temperature-time regime of plastification, chosen according to various technological considerations, we may use Eq. (7) to obtain an expression for determining the duration of the viscous-plastic state, according to the temperature of the material in the mould:

$$\tau_m = \left(A - \sum_{i=1}^{3} \tau_i \exp\left[-\frac{\Delta U}{RT_i}\right]\right) \cdot \exp\left[\frac{\Delta U}{RT_m}\right]. \tag{9}$$

The relationships obtained (7–9) linking up the temperatures and the time that the material is present in the various zones of the plastification and molding systems constitute a basis for a further optimization of the temperature-time regimes of pressure casting of reactoplasts, including granulated cast aminoplasts.

Heating of the casting machine cylinder is achieved by various systems. The greatest advantages are inherent in liquid heating of the cylinder, where a gentle heating of the material is set up and a rapid withdrawal of heat once the material is overheated is made possible. The temperature inhomogeneity of the melt is nearly twice as low as with electric heating. The possibility of a noncontrolled hardening of the melt in the cylinder is thereby reduced.

In the casting mold exhaust vents are arranged opposite the inlet casting vents. They measure 0.03 mm in depth and 3–12 mm in width and are intended to enable the withdrawal of volatiles.

While the reactoplasts are hardening in the mold, structural transformations associated with the physico-mechanical properties of the material are taking place. In contrast to thermoplasts, the reactoplasts are not characterized by a widely varying morphological structure. Fundamental structural elements of polymers with a compact reticular structure are globules [40].

In Ref. [41] are the processing parameters for aminoplasts by using different methods according to the thickness and overall dimensions of the articles moulded are given. In this case the seasoning of aminoplast articles in the mold in pressing needs 20–40 s per 1-mm thickness of the article, for cast pressing − 30–190 s, and in pressure casting − 15–80 s according to the thickness of the article. Given therein are also the service properties of different grades of aminoplasts.

Reference [42] describes the main parameters characterizating pressure casting of aminoresin-based compositions and includes temperatures, injection pressure, degree of compression, viscosity variations of the material, length of the molding cycle in casting. Cited therein is the formula for calculating the injection dose, types of defects in molding and their causes, design specifics of casting molds.

At present the progress in the pressure casting method as applied to aminoplasts involves the automation of control for the casting process, as well as the elimination of the hardening of the composition melt in the distribution gates.

Reference [43] discusses the methods of wasteless pressure casting enabling losses of thermoreactive material to be eliminated and the cost of the product to be reduced due to technological improvements. Losses for central and distributive gates amount to 30% on average. Processing of reactoplast waste is difficult and they are usually not utilized. The author proposes a molding method which is similar to pressure casting of thermoplasts with a system of heated gates. In this case the loss of material for gate formation declines to 9%.

For the manufacture of large articles proposed a variant of gateless molding has been proposed. A new method of "casting-pressing" (casting into a semiclosed mold with the subsequent molding) is discussed in Ref. [44].

4.4 Extrusion

The aminoplast intended for extrusion must contain an insignificant amount of moisture and a large quantity of lubricant (to be added during crushing in the ball mill) so as to have good plasticity.

Westlane Plastics (USA) [45] uses an extrusion method to produce rods of up to 31.75 mm diameter, strips, pipes with 9.52–12.7 mm wall thickness and other profile articles from compositions based on phenol, melamine, carbamide, and alkyd resins. Fillers used incluse wood crush, cellulose, chalk, talc, clay, mica, silica, coke, graphite, and carbon fibers.

Lubricants used are polytetrafluorethylene, molybdenum disulfide (MoS_2), graphite, silicone oil, and stearates. Extruded reactoplasts are endowed with some specific properties: excellent dimensional stability, very high thermal resistance and high mechanical strength. Owing to a high pressure and the homogeneity of the melt, a uniform moulding of the reactoplast takes place, as a result of which

Table 5. Indices of mechanical properties of USSR-made aminoplasts

No.	Name of index	Measurement units	Types of aminoplasts	
			KFA2	KFA2-PRG
1.	Density	g/cm^3	1.4–1.5	1.4–1.5
2.	Breaking stress in bending	MPa	70–95	80-95
3.	Elasticity modulus in bending	MPa	9×10^4	9×10^4
4.	Impact ductility without undercutting	kJ/m^2	6–8	7.5–11.0
5.	Breaking stress in compression	MPa	170	–
6.	Relative compressive strain in failure	%	9	–
7.	Ball indentation test	MPa	385	290–370
8.	Softening temperature in bending under 1.8 MPa stress	°C	160	160
9.	Thermal stability after Martens	°C	100–125	95–125

the material has a high density, its chemical and dielectric properties improve, hygroscopicity and shrinkage decline. Bars made from melamine resins can be processed into balls for bearings on a special machine combined with an extruder. The possibility for continuous molding of complex-profile articles with a high dimensional accuracy from aminoplasts based on urea- and melamine-formaldehyde resins is discussed in Ref. [46]. The manufacturing program includes transistor bodies, insulators, air inlets for aircraft, and wire coatings.

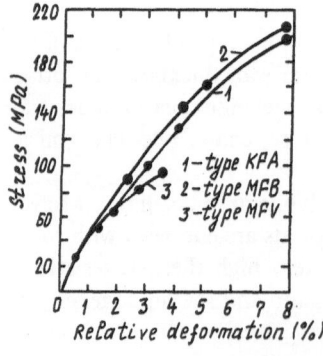

Fig. 9. "Stress-strain" curves in compression of aminoplasts at 20 °C

Types of aminoplasts					Testing methods
MFB	MFV	MFD	KFA4	MFE	
1.5	1.6–1.9	1.7–1.9	–	1.95–2.05	GOST 15139-69
80–100	not below 60	not below 40	85–115	80	GOST 4648-71
1×10^5	1.3×10^5		–		GOST 9550-81
6–10	4–5	no below 8.0	6–7.5	30.0	GOST 4647-80
135	110	–	–	–	GOST 4651-82
8	4	–	–	–	GOST 4651-82
375	360	–	–	–	GOST 4670-77
200	180	–	–	–	GOST 12021-75
120–150	140–185	200	90–110	180	GOST 21341-75

5 Physico-Mechanical and Dielectric Indices of Aminoplasts which are of Importance for the Performance of Various Articles

Physico-mechanical properties of aminoplasts are mainly determined by the type of resin and filler and by the amount of the filler. Deriving formulas expressing the relationship between mechanical properties and the content of the ingredients in the composition involves great difficulties. Furthermore, varying the majority

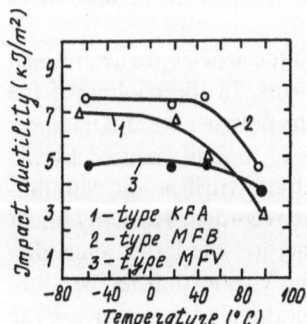

Fig. 10. Impact ductility versus temperature

Table 6. Electric properties of aminoplasts

Nos.	Name of index	Measure-ment unit	Types of aminoplasts	
			KFA2	KFA2-PRG
1.	Specific surface electric resistance	Ohm	$10^{12}-10^{15}$	$10^{12}-10^{15}$
	after 24-h in water		$10^{13}-10^{14}$	$10^{13}-10^{14}$
2.	Specific volume resistance	Ohm · cm	$10^{11}-10^{15}$	$3.5 \times 10^{13}-2 \times 10^{14}$
	after 24-h in water		$10^{10}-10^{13}$	$10^{10}-10^{13}$
3.	Internal resistance	Ohm	$10^{12}-10^{13}$	$10^{12}-10^{13}$
	after 24-h in water		2×10^{12}	2×10^{12}
4.	Electric strength	kV/mm	17–19	16–18.5
	after 24-h in water		17–19	16–18
5.	Loss tangent of a dielectric under the frequency			
	50 Hz		0.02–0.03	0.02–0.07
	10^3 Hz		0.02–0.05	0.01–0.02
	10^6 Hz		0.02	0.02
6.	Dielectric constant under the frequency			
	50 Hz		7–8	7–8
	10^3 Hz		6–8	6–8
	10^6 Hz		6–7	6–7
7.	Arc resistance	s	9	9

of mechanical characteristics is possible only within narrow limits. Thus, the impact viscosity of aminoplasts is rarely more than $5-8$ kJ · m^{-2}, whilst the collapse stress in bending is seldom more than 60–80 MPa.

The physico-mechanical properties of aminoplasts in the articles are determined by the degree of hardening and macrostructural defects. In the cooling of the articles down to room temperature reactive groups in the polymer are still retained, but their interaction is made difficult due to the loss of mobility caused by the molecules of the reticular polymer because of the latter's vitrification. Simultaneously a nonequilibrium supramolecular structure is recorded. Heat treatment of the articles does not alter the supramolecular structure, the latter remaining invariable. Heat treatment at a temperature below the vitrification temperature may only cause either a certain additional hardening of the binder or increase the

| Types of aminoplasts | | | | | Testing method |
KFA4	MFB	MFV	MFD	MFE	
$4.5 \times 10^{13} - 6.6 \times 10^{15}$	$10^{12} - 10^{14}$	$10^{12} - 10^{14}$	10^{12}	10^{12}	GOST 6433.2-71
–	$10^{13} - 10^{14}$	$10^{13} - 10^{14}$	–	10^{10}	
$6.3 \times 10^{12} - 7.0 \times 10^{14}$	$10^{11} - 10^{14}$	$10^{13} - 10^{14}$	10^{12}	10^{12}	GOST 6433.2-71
–	$10^{10} - 10^{13}$	$10^{10} - 10^{13}$	–	–	
–	$10^{11} - 10^{13}$	$10^{11} - 10^{12}$	10^{9}	10^{9}	GOST
–	$(1-4) \times 10^{11}$	$10^{9} - 3 \times 10^{10}$	10^{7}	10^{7}	6433-2-71
16.5–19.4	12–20	12–16	5	10	GOST
–	18–20	13–15	4	–	6433.2-71
0.02–0.086	0.07–0.5	0.2–0.5	0.8	0.065	GOST
–	0.02–0.05	0.2–0.3	–	–	6433.2-71
0.02–0.086	0.03	0.15	–	–	
6.3–8.0	10–11	20–25	–	6.7	GOST 6433.4–71
–	8–9	14–16	–	–	GOST 22372-77
6.0–7.3	5–8	6–7	–	–	GOST 22372-77
	10	10–12	180	120	GOST 10345-78

structural defectiveness of the material. Therefore, heat treatment of the articles made of aminoplasts is not always practicable [40].

To draw a comparison between various grades of aminoplasts the physico-mechanical properties of the material are determined on standard specimens. Table 5 presents some highly important measurements of the properties of the USSR-made aminoplasts, obtained in tests on standard specimens [4]. Martence thermal resistance of the various aminoplast grades amounts to 100–200 °C. Aminoplasts are strong materials with a high elasticity modulus in bending $(9-13) \times 10^4$ MPa and a high surface hardness (360–390) MPa.

The stress-strain relationships and the stress and strain values in collapse in the KFA and MFB type aminoplasts are practically similar, whilst in the MFV type the stress and strain values in collapse are nearly twice as low (Fig. 9).

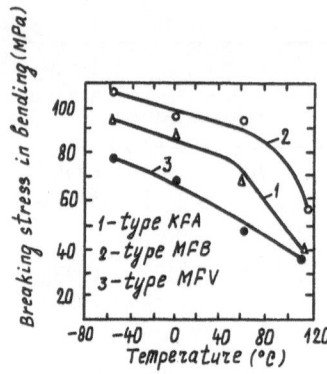

Fig. 11. Breaking stress in bending versus temperature

The collapse of aminoplasts is brittle both at low and high temperatures. In the 20 to 60 °C temperature range the impact viscosity of aminoplasts is subject to significant variation. At 100 °C the impact viscosity values of all types of aminoplasts are similar to one another and lie within 3–4.5 kJ m^{-2} (Fig. 10). On cooling the impact viscosity of the type KFA and MFB aminoplasts does not vary, whilst for the MFV type aminoplasts it is observed to become less. Figure 11 shows strength in bending versus temperature. The bending strength in the KFA type aminoplast is reduced by nearly 50% with a temperature increase from 20 to 120 °C, whereras for the MFB grade it varies insignificantly within the temperature range from −60 to 80 °C, the strength index for this grade at 120 °C is 40% less. In the MFV aminoplast the bending strength is observed to monotonously decrease to 50% with a temperature variation from −60 to 100 °C.

Table 6 lists values of the electrical properties of aminoplasts. Aminoplast articles are noted for their superior electrical properties and arc resistance and, therefore, they are widely used for electrotechnical articles [4].

When subject to moisture the insulating properties of aminoplasts vary insignificantly.

Table 7. Results of combustibility tests on aminoplasts

Name of index	Type of aminoplast			Testing method
	KFA	MFB	MFV	
Oxygen index, %	31.5	39.7	40.9	GOST 21793-76
Mean burning time, s	25	0	0	GOST 21207-81
Mean length of burnt portion, mm	50	40	22	GOST 21207-81
Category of combustibility	VI	VO	VO	Procedure UL-94

Arc resistance of the KFA-, MFB- and MFV-type aminoplasts declines only slightly with long-term (up to 120 days) increased temperatures (up to 100 °C). At 120 °C arc resistance is observed to be reduced by a half after 120 days in KFA- and MFB-type aminoplasts and remains practically invariable in the MFV type aminoplasts.

Combustibility of aminoplasts is calculated by means of three procedures: GOST 21793-76, GOST 21207-81 and UL-94. All of them yield a comparative estimate of the capacity of plastics to catch fire under specific conditions of the tests.

Table 7 gives the results of testing aminoplasts for combustibility.

Physico-mechanical values of different grades of aminoplasts the specimens of which were obtained in accordance with DIN 7708 are to be found in Ref. [14]. Also shown graphically are bending strengths of certain grades of aminoplasts (TYPI 31.5; 130; 150; 154) versus temperature (from 0 to 200 °C). With temperature increase the bending strength declines, at 100 °C half that at 20 °C.

6 References

1. Garabar MI et al. (eds) (1967) Manual on plastics, Khimiya Publishers, Moscow, vol 1 p 462
2. Chem and Eng News (1984) 10: 25
3. Encyclopedia of polymers (1972) Moscow, 1: 110
4. Aminoplasts (1985), Prospect NIITEKHIN, Cherkassy, p 17
5. Jap pat 18448, 1968
6. Nichira Masao (1976) Plast Age 2:. 56
7. Rendar BM et al (1982) Dep 2688-82, All-Union Institute for Scientific and Technical Information (VINITI)
8. Tsuneo Tsubakimoto, Kenji Minami (1973) Japan Plast Age, 7: 45
9. Marek P, Marek O (1987) Plasty a Kauc 12: 353
10. Gorbunov VN et al. (1979) Shock-resistant aminoplasts reinforced with organic fibre, (NIITEKHIM), Moscow, 1: 33
11. Pashkov DN (1975) Plastmassy, 2: 21
12. Kononenko SK et al. (1988) Plastmassy, 1: 21
13. Kutoritdinov RN et al. (1986) Plastmassy, 11: 39
14. Plastverarbeiter (1983) 1. 53
15. USSR catalogue P-2705 (1982) NIITEKHIM, Cherkassy, p 12
16. Belkin IM et al. (1968) Rotary-type instruments, Mashinostroenie Publishers, Moscow, p 68
17. Danilkin NI, Kanavets IF (1969) Plastmassy, 5–71
18. Gottfert's Prospect (FRG) (1987)
19. Plaste und Kautschuk (1987) 11: 413
20. Informations Chemie's Prospect (1988) 294: 225
21. Kazane A, Porter R (1983) Reactions of polymers under the impact of stresses, Khimiya Publishers, Leningrad, p 440
22. Yenikolopyan NS, Fridman ML (1986) Fizicheskaya khimiya, Dan, 2: 379
23. Leonov AI et al (1977) Principles of processing reactoplasts and rubers by means of pressure casting, Khimiya Publishers, Moscow
24. Torner RV (1972) Main processes in the treatment of polymers, Khimiya Publishers, Moscow, p 456
25. Plastics for structural uses (1974), ed Trostyanskaya EB, Khimiya Publishers, Moscow, p 93
26. Schmolke R et al (1987) Acta polym, 10: 574

27. Pshenitsina VP et al (1986) High-molecular compounds, 6: 403
28. Gagina LV et al (1987) Zavodskaya laboratoriya, 2: 58
29. Telezhkin VV (1987) In. Proc Moscow Forestry Inst, 192: 83
30. Braginskii VA (1979) Pressing, Khimiya Publishers, Leningrad, p 19
31. Lykov AV (1967) Theory of thermal conductivity, Vysshaya shkola Publishers, p 989
32. Encyclopedia of polymers (1977), Moscow, 3: 170
33. Telesheva MS et al (1976) Plastmassy, 1: 34
34. Baldogz Lasio et al (1986) Muanyag es gumi, 11: 327
35. Muanyag es gumi (1982), 5: 147
36. Sokolev AD, Shvets MM (1975) Casting of reactoplasts, Khimiya Publishers, Leningrad, p 11
37. Martin Vicente L (1986) Rev Plast Mod, 365: 611
NI, Kazankov YuV (1984) Cast moulding of polymers, Khimiya Publishers, Moscow, p 62
39. Terent'eva ZhP, Tunkel VI, Fridman ML (1988) Production and processing of plastics and synthetic resins, NIITEKHIM, Moscow, 4: 38
40. Sokolov AD (1985) Plastmassy 2: 48
41. Plastverarbeiter (1983) 34: 53
42. Laguma Castellanas O (1984) Rev plast mod 334: 397
43. Kagosima K (1976) "Koge dzaire" 5: 57
44. Parker FI (1977) Norsk plast 6: 4
45. Plastics Design Forum July–August (1986) p 82
46. Purasuticcusu, Jap Plast (1974) 3: 47

Editor N. S. Enikolopyan
Received August 15, 1990

Functionalized Oligomers and Polymers as Stabilizers for Conventional Polymers

Jan Pospíšil

Institute of Macromolecular Chemistry, Czechoslovak Academy of Sciences, 16206 Prague 6, Czechoslovakia

The low physical persistence limits the exploitation of the inherent chemical efficiency of stabilizers in polymers used under demanding conditions in domestic, engineering or medical applications. The problem can be solved by means of stabilizers having an increased molecular weight and a proper molecular architecture. Methods of synthesis, examples of structures and data dealing with properties and activity mechanisms of polymeric stabilizers are summarized with the aim of exploiting in an optimum way the knowledge of the chemistry of polymeric stabilizers for practical stabilization, to attract to a greater extent physical and physicochemical attention to solve involved open problems, and to increase by this approach the complex understanding for a more proper exploitation of physically persistent additive systems.

Advances in Polymer Science 101
© Springer-Verlag Berlin Heidelberg 1991

List of Abbreviations and Symbols

ABS   Acrylonitrile-butadiene-styrene copolymer
AF   Antifatigue (anti-flex crack) agent
AIBN   2,2'-Azobis(isobutyronitrile)
AO   Antioxidant
AOZ   Antiozonant

AR	Antiradiant
ATR-IR	Attenuated total reflection infrared spectroscopy
BIO-S	Biostabilizer
BR	Polybutadiene, butadiene rubber
CB AO	Chain-breaking antioxidant
CR	Chloroprene rubber, poly(2-chlorobutadiene)
CTC	Charge transfer complex
EPDM	Ethylene-propylene terpolymer
EPM	Ethylene-propylene copolymer
ESCA	Electron spectroscopy for chemical analysis
FR	Flame retardant
FT-IR	Fourier transform infrared spectroscopy
GTP	Group transfer polymerization
HALS	Hindered amine light stabilizer
HD AO	Hydroperoxide decomposing antioxidant
HD PE	High density polyethylene
HMW	High molecular weight
IIR	Isoprene-isobutylene rubber, butyl rubber
I-PS	Impact polystyrene
IR	Polyisoprene, isoprene rubber
LD PE	Low-density polyethylene
LS	Light stabilizer
MD	Metal deactivator
MMA	Methyl methacrylate
$\bar{M}_n$	Numerical average molecular weight
NBR	Acrylonitrile butadiene rubber, nitile rubber
NR	Natural rubber
PA	Polyamide
PC	Polycarbonate
PE	Polyethylene
PET	Poly(ethylene terephthalate)
PMMA	Poly(methyl methacrylate)
PP	Polypropylene
PPO	Poly(phenylene oxide)
PS	Polystyrene
PUR	Polyurethane, urethane elastomer
PVC	Poly(vinyl chloride)
SAN	Styrene-acrylonitrile copolymer
SBR	Styrene-butadiene rubber
SEM	Scanning electron microscopy
SRF black	Semi-reinforcing furnace black
tert-butyl	1,1-dimethylethyl
T_g	Glass transition temperature
UV	Ultraviolet

Preface

The rapid accumulation of data dealing with polymer stabilization in recent years has created a need to treat separately the chemistry and physical aspects of using physically persistent stabilizers for polymers. Use of polymers in demanding environments has been the main driving force in the development of this group of stabilizers. It became apparent that the extent of data dealing with the chemistry exceeds that of mechanistic and physical studies. There is a call for extended elucidation especially in the area of physical relations between stabilizers and host polymers with the aim of exploiting more efficiently the described stabilizing systems. Methods of synthesis and examples of the structures of stabilizers are given. A selection of references dealing with data published on polymeric stabilizers up to the end of March 1990 was made. Abbreviations for conventional polymers were used according to [1, 2].

1 Stabilization of Polymers Against Environmental Attacks

Plastics, coatings and rubbers are attacked during processing, storage or in use to varying extents with physical (heat, radiation, mechanical stress), chemical (atmospheric oxidants like oxygen, ozone, peroxyacetyl nitrate, NO_x) and microbial deteriogens. Commercial polymers vary in their inherent sensitivity to individual degradation processes and are generally exposed during their lifetime to combinations of deteriogens varying in intensity. New chemical structures or reactive intermediates are formed in trace concentrations in stressed polymers. New systems formed may — together with defect structures present in commercial polymers and with metallic impurities introduced during fabrication, processing, or application of polymers — sensitize, catalyze or initiate consecutive degradation processes in polymers.

Most degradation processes of hydrocarbon polymers are of oxidative character [3]. Autoxidation, metal catalyzed oxidation, high energy and sunlight radiation initiated or photosensitized oxidation, mechanochemically induced oxidation, ozonation, oxidative pyrolysis and/or burning are involved. Depending on the polymer structure and on the presence of other compounds, like softeners or fillers, thermal processes and/or biodeterioration can contribute to the overall process. To maintain the long-term polymer properties under expected application conditions, the use of proper stabilizers is mandatory in most commercial polymers.

1.1 Additive Systems as Stabilizers

Polymers are generally doped with low amounts of preservatives during fabrication, processing or confectioning. The individual classes of stabilizers involve chain-breaking (CB-AO) and hydroperoxide decomposing antioxidants (HD-AO) [3–5], antifatigue agents (AF) and antiozonants (AOZ) [6, 7], metal deactivators (MD) [8], light stabilizers (LS) with properties of quenchers of excited states, UV absorbers, hydroperoxide decomposers, and radical scavengers [9, 10], antiradiants (AR) [11], flame retardants (FR) [12], thermal stabilizers for PVC [13] and biostabilizers (BIO-S) [14].

1.2 Inherent Chemical Efficiency of Stabilizers

The chemical structure of stabilizers, above all the presence of proper functional groups or elements, determines primarily the activity in chemical, physical and microbial stabilizing processes [6]. During the last three decades, the chemistry of stabilizers, involving both syntheses and activity mechanisms, have been studied very intensively and most mechanistic phenomena have been explained.

Chain-breaking antioxidants consist of hindered phenols and secondary aromatic amines [5, 6]. Trivalent phosphorus compounds and organic compounds of sulfur are HD-AO. As AF and AOZ, secondary aromatic diamines and heterocyclic amines are used almost exclusively [6, 7]. Activity mechanisms of various AO, AF and AOZ have been described in detail [15–21]. Most of commercialized LS are derivatives of 2-hydroxyacetophenone or 2-hydroxybenzophenone, 2-(2-hydroxyphenyl)-2H-benzotriazole, nickel containing salts or complexes, metal complexes of thiocompounds and derivatives of hindered piperidine or piperazine. The activity mechanism of all classes of LS was reviewed in Refs. [9, 10]. Antioxidant type stabilizers are mostly used as AR too [11]. Organic FR, important for various electrical, electronic, construction and transportation applications consist mostly of compounds having a high halogens and phosphorus content. These systems should be involved in the formation of noncombustible gases, inhibition of free radical processes in the gas phase and in the char formation [12]. Salts of carboxylic acids, sulfurless or sulfur containing organotin compounds are the most important commercialized thermostabilizers of PVC [13]. Due to the high toxicity for human beings and animals, BIO-S must be applied very carefully. Organic compounds containing arsenic, antimony, copper or tin are mostly applied [14].

1.2.1 Cooperation Between Different Stabilizing Functions

Deterioration of commercial polymers proceeds in practice by several distinguishable parallel or consecutive mechanisms. Effective protection against a complex of deteriogens requires a proper selection of two or more stabilizers matching exactly the degradation conditions. It is customary to use combinations of primary antioxidants (i.e. CB-AO) with secondary antioxidants (i.e. HD-AO) in processing stabilization of polyolefins [4–6], or of CB-AO with various LS for long-term protection of polyolefins against weathering [9]. Most mechanistic data deal with these two combinations of stabilizers [22]. It cannot be expected that various stabilizers act simultaneously. The chronology of exploitation may be very different. However, even in systems like this, e.g. in combinations of AO with FR [12], the cooperation between various stabilizers should have a supporting character, resulting in additivity or − in more favourable cases − in synergism.

Antagonistic relations should be fully eliminated (unfortunately, this phenomenon was observed by some authors in combinations of efficient hindered phenolic AO and hindered piperidines [23]).

The optimum exploitation of physical mixtues of effective monofunctional stabilizers results in the ultimate complex protective effect having a favourable cost/performance relation. By means of a proper alteration of the molecular architecture of the stabilizer molecule, bifunctional stabilizers were synthesized e.g. intramolecularly cooperating systems having in one molecule CB-AO and HD-AO functions or AO and LS functions [22]. However, one of these functionalities is mostly dominant in these systems, the second functionality supports inherently the dominant function in a more or less concerted mechanism.

Physical mixtures of stabilizers or bifunctional stabilizers represent a very successful exploitation of the knowledge of the inherent chemical efficiency of stabilizers in both intermolecularly and intramolecularly cooperating systems. Moreover, there are data available dealing with the exploitation of transformation products of stabilizers in supporting mechanisms [24].

1.3 Physical Factors Limiting the Inherent Chemical Efficiency of Stabilizers

Plastics, rubbers and coatings have been exposed to chemical and physical deteriogens, many times under very aggressive conditions. Stabilizers have been stepwise chemically depleted as a result of the active participation in the stabilization mechanism [15–17, 21]. The chemical consumption is proportional to the concentration and reactivity of chemical deteriogens. This is however not the single mode of the loss of stabilizers from polymers used in the both domestic and engineering applications in an aggressive physical environment. The latter involves a high environmental temperature, air flow, pressures lower than that of the atmosphere, intensive irradiation and/or contact with extracting media.

Stabilizers are physically lost from the polymer by evaporation and leaching and their concentration in the polymer drops. As a consequence, polymer articles are no more chemically protected. Physical factors limiting the lifetime of stabilizer doped polymers due to the physical impermanence of stabilizers are explained in Sect. 1.3.1. Factors influencing the efficacy of the two permanent and impermanent stabilizers due to the physical interaction with the polymer matrix are given in Sect. 1.3.2.

Examples of conditions which increase the physical losses of stabilizers include processing of fibers, including high-temperature post-spinning treatment, dry cleaning of textiles, cyclic attacks on textiles by aqueous solutions of detergents and subsequently by hot air during drying, contact of rubber articles with hot oils, gasolines containing new polar additives (e.g. methyl *tert*-butyl ether) and aromatic hydrocarbons, hydraulic fluids or aqueous acids, contact of polyolefins with streaming hot water, engineering applications of polymers in hot air atmospheres or under space conditions. The physical loss of stabilizers accelerates the ageing of polymers more than thermal oxidation [25] or photooxidation [26].

A premature failure of polymer articles is a consequence. This reveals that the observed stabilizing effect in the aggressive environment has more to do with the ability of stabilizers to survive physically in the host polymer than with the true intrinsic chemical efficiency under application conditions.

It must be therefore taken into consideration that the physical loss of stabilizers may influence the results of the accelerated ageing test. This is true mainly in tests where the polymer degradation is accelerated by physical factors having very different intensities to those applied under practical conditions.

Generally, the physical loss of stabilizers is dependent on the geometry of polymer samples. This was proved by comparison of the results of accelerated weathering tests [27]. The form of tested samples must be therefore always specified. The most serious physical losses were observed from articles having high surface to volume ratios (monofilaments, thin films, coatings).

It is understandable that due to the increased use of polymers under demanding conditions, an increase in the permanent resistance of polymers to degradation has been deemed necessary. Articles are now expected to withstand more severe attacks of chemical and physical deteriogens. Moreover, a safety performance for longer periods of time than ever before has been called for in materials for the construction and automotive industries.

Another problem arising from the low physical persistence of stabilizers must be taken into consideration. This involves a potential contamination of the surrounding environment with stabilizers or their transformation products considered to be dermatiogenic and/or acting as primary irritants or equivocal tumorigenic agents [28]. Problems like this arise in articles coming into contact with human beings. This involves various areas of technical application and the use of polymers as packaging materials for food, pharmaceuticals and cosmetics, toys and various medicinal uses.

The problem of the physical loss of stabilizers has been considered as a very serious one since the early days of polymer application [29]. It has always been a subject of great continuous interest and subjected to theoretical treatment and interpretation [27, 30].

1.3.1 Physical Losses of Stabilizers due to Volatility and Leaching

The physical loss of stabilizers from polymers involves evaporation from the polymer surface, leaching into liquids and exudation to the polymer surface as a precipitate [27, 30]. The knowledge of the relationships between the chemical structure and the physical permanency of stabilizers in polymers has been of growing practical importance.

The volatility of pure stabilizer is dependent on the vapor pressure of the latter. However, the volatilization from a polymer is a more complex problem. It can be controlled either by the rate of the stabilizer evaporation or the rate of the stabilizer diffusion to the polymer surface [27]. The latter problem has been expected to dominate if leaching is also involved in the physical loss.

The surface area of a polymer, the surface concentration, thermal properties and solubility of stabilizers in a polymer, environmental pressure and temperature and the air flow intensity are factors affecting the evaporation rate. The latter changes in the area of the glass transition of a polymer. This must be taken into consideration if measured values are to be extrapolated into another temperature interval.

The volatility of a particular stabilizer is certainly related to its molecular weight. It is however difficult to define unambiguously the limit value of the molecular weight of a stabilizer assuring a low volatility under specific conditions of application. Substituent effects in the stabilizer molecule and physical relations in the particular doped polymer matrix affect the volatility very seriously. The borderline between "volatile" and "nonvolatile" stabilizers is therefore rather diffuse. Values of molecular weights from 400 to 650 were reported to be satisfactory for solving volatility problems [31].

It is easier to dissolve a stabilizer than to evaporate it [27]. The physical loss of stabilizers due to the leaching from polymer surface layers into liquids which come into contact is therefore more serious than volatilization. Problems arise mainly in systems where the degradation process has been concentrated at the surface layer and therefore an efficient surface protection of a polymer is mandatory. This phenomenon takes place mainly in photostabilization of plastics or antiozonant protection of rubbers. The surface loss of stabilizers is extremely serious in very thin profiles or products having a very high surface/mass ratio.

Polarity of organic solvents and their ability to swell the polymer matrix increase the losses of stabilizers due to the leaching. Water is generally considered to be less serious than organic solvents. In spite of this fact, serious damage may be caused during long-term contact with water. Extraction of aromatic amines from tires [32] or of phenolic antioxidants from PE [25, 33] were reported. The ease with which water leaches various derivatives of phenylenediamine from NR and SBR stocks changed inversely with the molecular weight of the respective phenylenediamine and with the leachant acidity [32]. An intensive extraction of stabilizers can take place during laundering of textiles with aqueous solutions of detergents.

Due to the extended application of hindered amine light stabilizers (HALS) in stabilization of foils and fibers, a great deal of attention has been paid to HALS extractability by water and aqueous solutions of detergents or solvents used in dry cleaning [34].

Determination of extractability of stabilizers should be a substantial part of the rating of stabilizer efficiency where leaching is a common deterioration effect or where legislation requires quantitative data dealing with extractability of stabilizers from packaging materials by food simulants.

The stabilizer solubility is influenced by the polarity and the molecular weight [31]. The limit of the molecular weight of a stabilizer assuring its persistency against leaching is much higher than that mentioned for volatility and is considered to be approximatively 3000. The physical relations in the system stabilizer/polymer matrix, however, influence the leaching very substantially. The decreased surface concentration of stabilizers due to volatilization and leaching is replenished by diffusion from the polymer bulk. A concentration gradient of the stabilizer is created by means of migration between the polymer surface and the polymer bulk [27]. The diffusion of a stabilizer is temperature dependent. As a consequence, the most serious physical losses take place in systems cyclically stressed by extraction and hot drying: after an efficient surface extraction, stabilizer can be replenished during drying and then repeatedly extracted from the surface. The ultimate failure of the stabilized polymer is thus hastened [29].

Numerous methods can be used for quantitative determination of stabilizers in extracts from polymers [35]. Great difficulties may arise in analyses of agedpolymers. In practice, the amount of an extractable stabilizer was determined only exeptionally. Most experimental data discussing consequences of the physical loss of a stabilizer due to the leaching from polymers have been based on comparison of the residual stability of polymers after extraction with data obtained before extraction.

1.3.2 Physical Relations Between Stabilizers and the Polymer Matrix

Physico-chemical aspects of the molecular structure of all components of the system polymer/stabilizer are reflected in solubility of stabilizers, their compatibility with the host polymer in the solid state and migration in the polymer matrix [27].

1.3.2.1 Solubility and Compatibility of Stabilizers

High solubility of stabilizers is an essential requirement for a good physical retention of a stabilizer in a polymer [27]. Molecules of most stabilizers have relatively high polarity and their solubility in unpolar hydrocarbon polymers is, therefore, only low. Microscopic domains consisting of aggregated polar stabilizers and surface exudates can be formed and are one of reasons for the uneven distribution of a stabilizer in the host polymer as well as for the physical loss of a stabilizer.

General rules valid for low molecular weight compounds govern both the solubility and distribution of stabilizers in the polymer matrix [27]. Stabilizers usually have a reasonable solubility in melt semicrystalline polymers at processing temperatures. During cooling, polar stabilizers are expelled from spherulites into their border areas and to the amorphous part. The reverse diffusion of stabilizers into spherulites is negligible. The distribution of stabilizers in solid semicrystalline polymers is, therefore, always nonuniform and their actual local concentration is dependent on the content and the distribution of the amorphous phase. A nonuniform distribution and solubility of stabilizers can limit the long-term performance of the otherwise chemically efficient stabilizers and result in a quick start to oxidation in insufficiently stabilized amorphous areas [36]. Oxidation spreads then to the residual bulk of the polymer.

In multiphase systems, like in rubber modified plastics, stabilizers can partition between different phases of the system [37]. The actual stabilizer concentration in either elastomer or thermoplastic phases may differ very significantly from the average stabilizer concentration declared for the whole multiphase system. In polymers like ABS, the partitioning of stabilizers may exert a controlling influence on the final stabilization effect. A proper chemical modification of the stabilizer molecule can enhance its affinity to a phase of the multiphase system which is more sensitive to degradation.

At stabilizer concentrations lower than that corresponding to a saturated state in a solid polymer, an equilibrium state is created stepwise. Volatility and leaching represent the main modes of the physical losses of stabilizers in these systems. Due to a good solubility of stabilizers in melt polymers, an uniform distribution is reached. The solubility of stabilizers decreases dramatically during cooling, however, and an oversaturated polymer characteristic of a metastable state is created. A stabilizer redistribution takes place as a consequence. A diffusion controlled equilibrium state is slowly formed and is accompanied by stabilizer blooming to the surface. The bloomed stabilizer can be very easily lost by volatilization, wipeout or extraction. Oversaturated systems are sometimes manufactured in practical stabilization of polymers (e.g. polyolefins doped with high concentrations of FR).

Quantitative predictions of the stabilizer solubility in polymers by means of data obtained with model solvents are generally not possible, due to the "nonideality" phenomenon [27]. Stabilizers having low heat of fusion as well as low melting point or forming intramolecular H-bonds with the host polymer posses a better solubility. The morphology of the host polymer is an important factor. Stabilizers dissolve only in amorphous phases of semicrystalline polymers. Within concentration limits used in practice, the dissolved low molecular weight stabilizers do not affect the crystallinity degree of the host polymer. An increased permanency of stabilizers observed in polar polymers is due to the more intensive creation of H-bonds and dipole-dipole interactions, having an increased solubility and lower diffusion rate as a consequence.

In amorphous systems, e.g. in rubbers, a better solubility of stabilizers and a more uniform distribution than in polyolefins can be reached. The equilibrium stabilizer concentration should not exceed the saturation state to prevent stabilizer blooming after 1 year storage at ambient temperature. Complications may arise with insoluble stabilizers, as with N,N'-diaryl-1,4-phenylenediamine in rubbers [16]. The limited solubility does not allow a sufficient antiozonant protection to be achieved.

Cohesion and adhesion forces and complex surface interactions among all components of the system influence the compatibility of stabilizers with the polymer matrix. The compatibility may be related to differences between the halftimes of crystallization of the pure and stabilizer doped polymer, to the solubility of stabilizers or volatility differences between pure and in polymer dissolved stabilizers. Experimental data confirm that the compatibility of AO and LS is an important factor for the finally observed stabilization effect [30].

1.3.2.2 Diffusion of Stabilizers

The diffusion rate determines how easily can be a stabilizer physically lost by extraction or volatilization. Changes in the mobility of stabilizers or of species with which a stabilizer interacts (e.g. $RO_2^{\cdot}$) are reflected in the finally observed stabilization effect [27].

The structure of both the stabilizer and polymer are factors determining diffusion characteristics. Substituent effects influencing the polarity and consequently the solubility of stabilizers, the morphology and the processing history of polymers, polymer chain branching and crosslinking, structural factors influencing the flexibility of a polymer chain and the orientation degree of a polymer (e.g. the influence of stretching of fibers) are involved [27, 30, 31].

The Arrhenius plot is valid for the temperature dependence of the diffusion coefficient D in a particular combination polymer/stabilizer. The value of D is independent of stabilizer concentration and was mostly determined by quantification of data dealing with the transfer of a stabilizer from a doped into a virgin polymer. The values of D of antioxidants in PP decrease approximately with increasing molecular weight of AO, with branching of substituents, increasing difference between the polarity of the polymer and that of stabilizer. A generalization is, however, very difficult [27, 30].

Diffusion requires cooperative motions of both the polymer and the diffusant [27] and is therefore only low below T_m and severely restricted below T_g. The diffusion of various stabilizers was elucidated in amorphous and semicrystalline polymers and in multiphase systems. In semicrystalline polymers, diffusion takes place almost exclusively in the amorphous phase and the value of D is sensitive to the total crystallinity and the morphology. It is difficult to predict D within a homologuous series of stabilizers. For a given molecular weight, long and flexible molecules diffuse more rapidly than more rigid and compact structures. For a given stabilizer, the value of D is usually lower in PP and HDPE than in LDPE. It was demonstrated that typical AO molecules have a very restricted mobility in polymers. They are, however, insufficient experimental data to correlate the AO mobility and the AO efficiency.

The polymer geometry also influences the importance of the diffusion in the control of the physical loss of a stabilizer. It was reported that the diffusion-controlled loss dominates in thick sections. From the point of view of the stabilising efficiency, a low diffusion rate for thin samples is desirable. A high diffusion rate may be of advantage for thick samples if chemical or physical depletion of deteriogens near the sample surface is required.

2 Enhancement of the Physical Persistency of Stabilizers

Under moderate operating conditions, chemically efficient commercial stabilizers having molecular weights in the range 350 to 500 daltons ensure sufficient service stability of polymers. Physical factors mentioned in the Sect. 1.3.1 limit the efficiency in aggressive conditions, however. The serious problem of the physical loss of stabilizers can be solved by a chemical modification of the structure of physically non-persistent stabilizers. This involves the application of salts or metallic complexes of stabilizers or an increase of the molecular weight of stabilizers.

2.1 Salts and Metallic Complexes

Numerous AO and LS containing ions of metals have been described in the patent literature. The presence of some ions (e.g. of Ca(II) or Ba(II)) only improves the physical persistency. Other metals, like nickel and tin, participate in the stabilizing mechanism too. The improvement of the physical persistency of stabilizers by means of this approach is not within the scope of this paper. Compounds 1 (M = Ni, Irgastab 2002, Ciba-Geigy, and M = Ca, Irganox 1425, Ciba-Geigy) and **2** [A = ⬡– N(CH$_2$CH$_2$OH)$_2$, Ferro AM 101, Ferro Corp., and A = C$_4$H$_9$NH$_2$, Chimassorb N-705, Ciba-Geigy, Cyasorb UV 1084, American Cyanamid] are examples of commercialized stabilizers of this class.

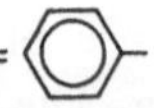

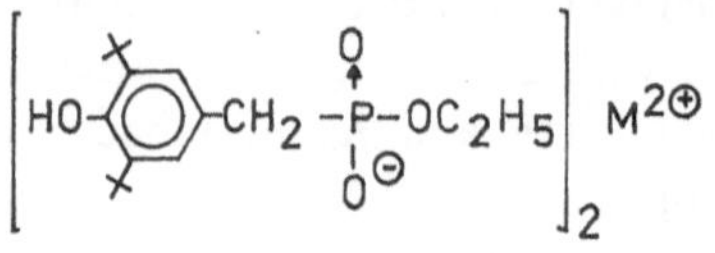

$$\left[HO{-}\hspace{-2pt}\bigcirc\hspace{-2pt}{-}CH_2-\overset{\overset{O}{\uparrow}}{\underset{\underset{O^{\ominus}}{|}}{P}}-OC_2H_5 \right]_2 M^{2\oplus}$$

1

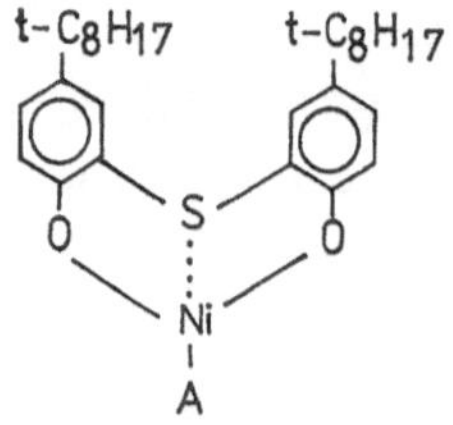

2

2.2 Stabilizers with Increased Molecular Weights

Synthesis of compounds having an increased molecular weight represents the most natural way to produce physically persistent stabilizers. Long-chain substituents can be attached to the efficient but volatile low molecular weight stabilizers (the molecular weight is increased by this modification, but the relative content of the active moiety in such a molecule is diminished) or two or more functionalized moieties may be combined with various alkylidene, aromatic or heteroatoms containing bridges. Many various efficient mono or bifunctional stabilizers, synthesised in this way and listed as nonvolatile high molecular weight (HMW) stabilizers have been commercialized. Most of them fulfill requirements expected from the physically persistent additives. The concept of the exploitation of compounds having an increased molecular weight has been therefore correct. With an analysis of application of HMW stabilizers, Minagawa [38] reports the following data dealing with the range of molecular weights of the contemporary commercialized stabilizers:

Stabilizer type	Molecular weight
Phenolic AO	200–1178
Organic sulfides	515–1160
Organic compounds of phosphorus	216–1829
UV absorbers	182– 849
Nickel chelates	501– 713
HALS	261– 800

Compounds **3** (Irganox 1010, Ciba-Geigy, Mark AO-60, Asahi Denka), **4** (Hostanox 03, Hoechst), **5** (Naugard 445, Uniroyal), **6** (Wingstay SN-1, Goddyear Chemicals), **7** (Mark PEP-24, Asahi Denka, Ultranox 626, Borg Warner and General Electric), **8** (Ferro AM 320, Ferro Chemical), **9** (Mark LA-34, Asahi Denka; Sumisorb 300, Sumitomo; Tinuvin 327, Ciba-Geigy), **10** (Sanol LS-770, Sankyo; Tinuvin 770, Ciba-Geigy), **11** (Mark CDA-6, Asahi Denka), **12** (Sandoflam, Sandoz), **13** (Estabex, AKZO Chemie; Irgastab, Ciba-Geigy); Hostastab, Hoechst; Stan, Sankyo; Cyastab, American Cyanamid) and **14** (Vinyzene, Ventrone Europe) are selected examples of commercialized HMW stabilizers having properties of AO (**3–7**), LS (**8–10**), MD (**11**), FR (**12**), thermostabilizers for PVC (**13**) and BIO-S (**14**) respectively. Many other mono and bifunctional stabilizers are available [6, 39].

The tendency to synthetize stabilizers with higher molecular weights is evident. The performance of stabilizers in respect to their physical persistency can also be improved by physical adsorption on surfaces of reinforcing fillers, e.g. of carbon black or amorphous microground silica [40] or by formation of associates with ionexes [41]. However, a more efficient approach to persistent stabilizers having

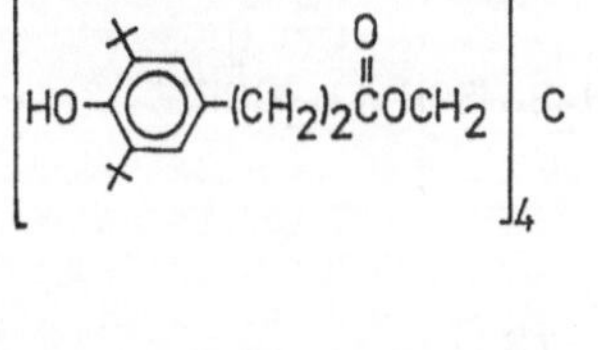

3

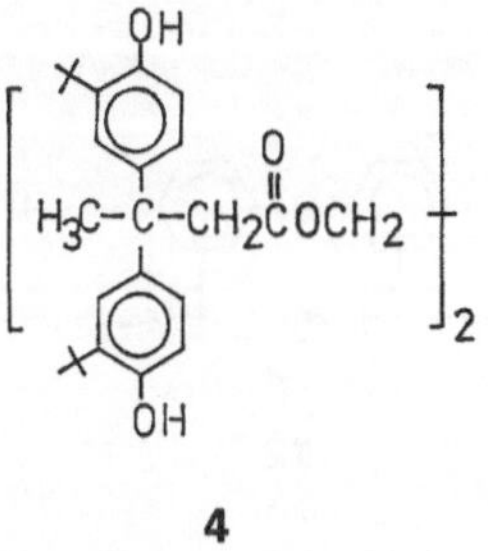

4

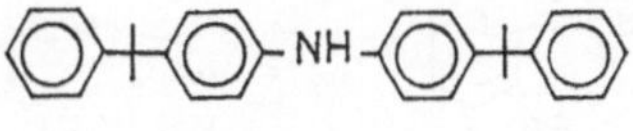

5

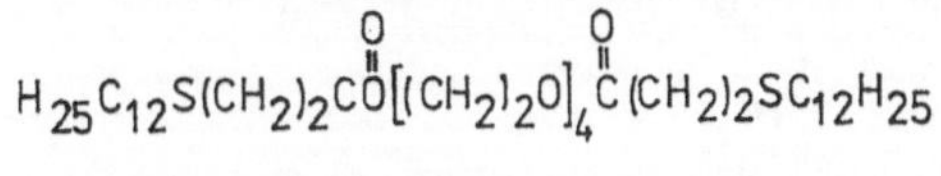

6

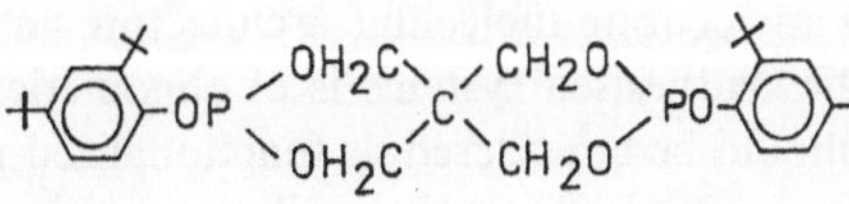

7

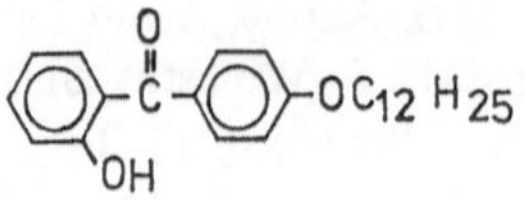

8

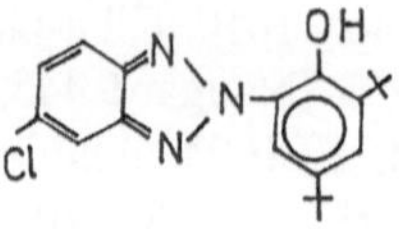

9

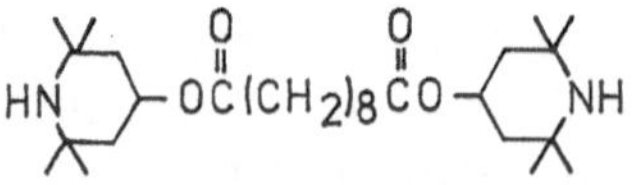

10

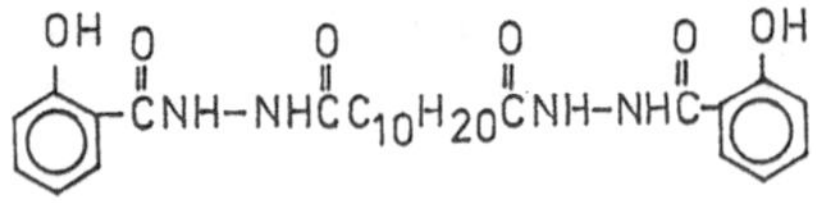

11

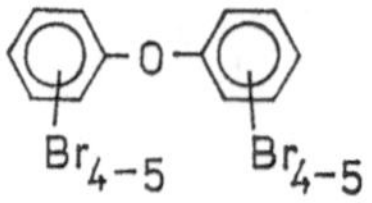

12

$(n\text{-}H_9C_4)_2Sn(SCH_2CH_2\overset{O}{\overset{\|}{C}}OR)_2$

13

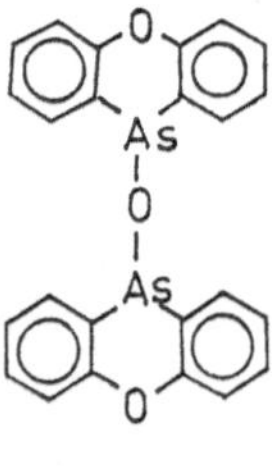

14

a relatively high content of active moieties involves the synthesis of organic macromolecular stabilizers. Conventional synthetic methods for polymers as well as versatile methods of organic chemistry have been exploited with the aim of preparing sytems with an accurate molecular architecture and functional properties. A great part of the synthesised systems is of oligomeric character. Polymer supported stabilizers should be considered as functionalized polymers for special applications. Both the concentration and distribution of functional groups in a macromolecule are of great importance. The stabilizing moiety Ⓢ can be a part of a repeating unit and built into the polymer backbone (Type A, Scheme 1), attached directly (Type B) or by means of a spacer ⊗ (Type C) to the polymer backbone. The functionalized moiety ⑤ can be attached to every unit (homopolymers) or can alternate with conventional monomer units (copolymers, polycondensates). Types D and E respectively represent attachement of ⑤ to side chains in grafted copolymers and in crosslinks respectively. Types F and G are models of polymers endcapped with stabilizing moieties.

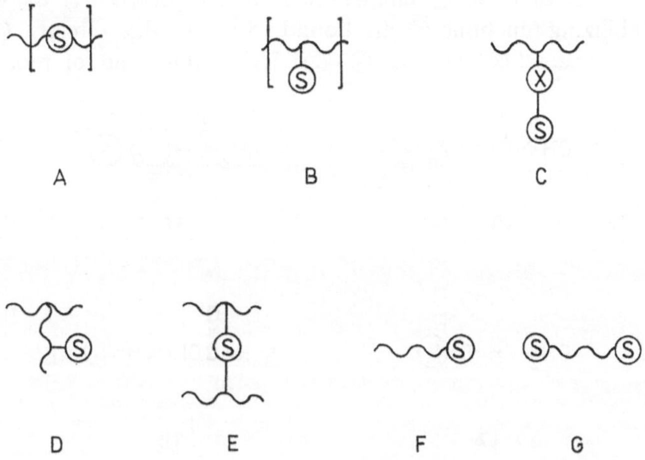

SCHEME 1

3 Synthesis of Physically Persistent Stabilizers Based on Functionalized Oligomers and Polymers

Physically persistent stabilizers classified as functionalized oligomers or polymers can be synthesized by polyreactions of functionalized monomers and/or by polymer analogous reactions exploiting the reactivity of functionalized reactive low molecular weight compounds with polymeric substrates. For a detailed classification of synthetical approaches, the classification principle used in Houben-Weyl [42] was adopted. The nomenclature of functionalized polymers is based on the monomer's unit principle. Abbreviations of conventional polymers are used as recommended by [1, 2].

3.1 Polyreactions

Various kinds of polymerization, polyaddition, substitution and polycondensation reactions of monomers bearing stabilizing moieties have been involved. Almost every kind of synthetical approach has been tested. However, only some of them are of practical interest. Examples of monomers and/or of the respective macromolecular systems proposed as stabilizers were selected from many literature data to show the diversity of structures. Only representative references are reported.

3.1.1 Polymerization

Functionalized compounds containing multiple bonds C=C and some heterocycles have been used as monomers.

3.1.1.1 Functionalized Monomers with Polymerizable Multiple Bonds C=C

All typical groups of stabilizers were prepared in the form of monomers containing
a polymerisable function Ⓕ characteristic of the presence of the double bond
C=C. Stabilizing functions Ⓢ are bound with Ⓕ either directly, Ⓢ–Ⓕ, or by
means of a spacer Ⓧ, i.e. Ⓢ–Ⓧ–Ⓕ. Using this kind of monomer, linear

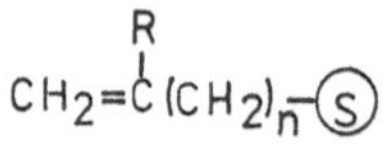

$$CH_2=\overset{R}{\underset{}{C}}(CH_2)_n\!-\!Ⓢ$$

15

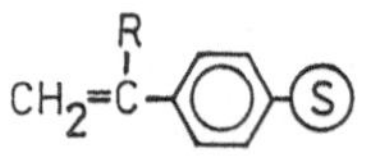

$$CH_2=\overset{R}{\underset{}{C}}\!\!-\!\!\langle\bigcirc\rangle\!-\!Ⓢ$$

16

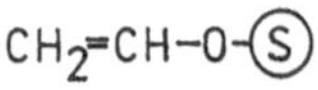

$$CH_2=CH-O-Ⓢ$$

17

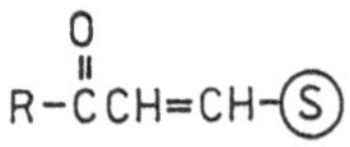

$$R-\overset{\overset{\textstyle O}{\|}}{C}CH=CH-Ⓢ$$

18

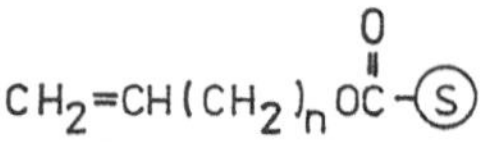

$$CH_2=CH(CH_2)_n\overset{\overset{\textstyle O}{\|}}{OC}-Ⓢ$$

19

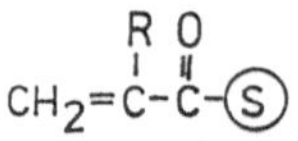

$$CH_2=\overset{R}{\underset{}{C}}-\overset{\overset{\textstyle O}{\|}}{C}-Ⓢ$$

20

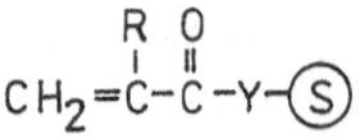

$$CH_2=\overset{R}{\underset{}{C}}-\overset{\overset{\textstyle O}{\|}}{C}-Y-Ⓢ$$

21

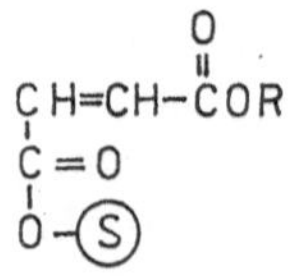

$$\begin{array}{l} CH=CH-\overset{\overset{\textstyle O}{\|}}{C}OR \\ \underset{}{\overset{|}{C}}=O \\ \overset{|}{O}-Ⓢ \end{array}$$

22

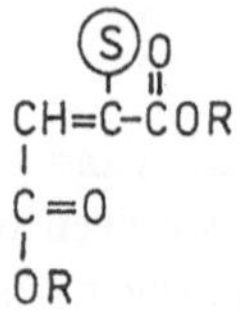

$$\begin{array}{l} \overset{Ⓢ}{\overset{|}{CH}}=\overset{\overset{\textstyle O}{\|}}{C}-COR \\ \overset{|}{C}=O \\ \overset{|}{O}R \end{array}$$

23

$$\begin{array}{c} CH=CH \\ O=C\diagup\;\;\diagdown C=O \\ \diagdown\!\!\underset{N}{}\!\!\diagup \\ \overset{|}{Ⓢ} \end{array}$$

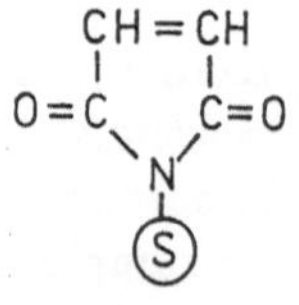

24

homopolymers and statistical copolymers of the type B and C (Scheme 1) or graft copolymers (type D) were prepared. Two different functions Ⓢ can be included in the same monomer. Monomers containing two or more polymerisable functions Ⓕ can be exploited either for the synthesis of functionalized polymers of the type B or C, or for stabilizers of the type E containing crosslinks carrying a functional moiety Ⓢ. The most common types of polymerizable functions attached to the moiety Ⓢ and used successfully for the synthesis of polymeric stabilizers are exemplified (**15–24**, spacers Ⓧ are not given). Monomers bearing vinyl (**15**, R = H, n = 0), isopropenyl (**15**, R = Me, n = 0) or allyl (**15**, R = H, n = 1) groups, styrene type functions (**16**, R = H, Me), vinyloxy groups (**17**), α, β-unsaturated ketone moiety (**18**, R = alkyl), vinyl or allyl esters of carboxylic acids (**19**, n = 0, 1), acryloyl or methacryloyl (**20**, R = H, Me), acryloyloxy or methacryloyloxy (**21**, R = H, Me, Y = S) and acryloylamino or methacryloyl-amino groups (**21**, R = H, Me, Y = NH), moieties of cinnamic (**18**, R = OR^1, NHR^1) or maleic (**22, 23**) acids (similarly, moieties of itaconic or mesaconic acids may be included) or maleic imide (**24**) are characteristic compounds. Other various polymerizable groups can be introduced by alkenylation of functionalized moieties with linear dienes, cyclic dienes or isoprenoid alcohols [43]. Aliphatic or aromatic hydrocarbon links or various functionalized links like esters or amides can form a spacer Ⓧ between Ⓕ and Ⓢ (Scheme 2)

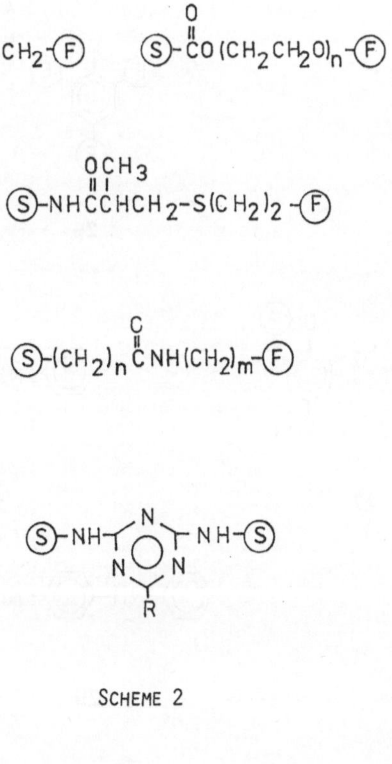

$$\text{Ⓢ-CH}_2\overset{\overset{\displaystyle OH}{|}}{C}\text{HCH}_2\text{-Ⓕ} \qquad \text{Ⓢ-}\overset{\overset{\displaystyle O}{\|}}{C}\text{O(CH}_2\text{CH}_2\text{O)}_n\text{-Ⓕ}$$

$$\text{Ⓢ-NHC}\overset{\overset{\displaystyle O}{\|}\,\overset{\displaystyle CH_3}{|}}{C}\text{HCH}_2\text{-S(CH}_2)_2\text{-Ⓕ}$$

$$\text{Ⓢ-(CH}_2)_n\overset{\overset{\displaystyle O}{\|}}{C}\text{NH(CH}_2)_m\text{-Ⓕ}$$

$$\text{Ⓢ-NH}\underset{N}{\overset{N}{\bigcirc}}\text{NH-Ⓢ}$$
R

SCHEME 2

Moieties Ⓢ involve active groups characteristic of a specific stabilizing property. Consequently, chemical structures of functionalized monomers are very variegated. Some selected examples of the latter are given (without any specification of Ⓧ and Ⓕ). These are CB-AO having the structure of substituted phenols, e.g. **25–27**, aromatic amines **28** and diamines **29**, HD-AO containing sulfide groups **30** or sulfur containing heterocycles (benzoxozolethione **31**, benzothiazolone **32** or benzothiazole **33**), metal deactivators, e.g. **34**, **35**, light stabilizers containing 2-hydroxyacetophenone **36**, 2-hydroxybenzophenone **37–39**, 2-(2-hydroxyphenyl-2H-benzotriazole **40–44**, alkyl or arylsalicylate **45**, α-cyano-β-phenylcinnamate **46**, **47** or dibenzoylmethane moieties **48**; photoantioxidants having the structure of hindered piperidine **49–54**, piperazinone **55** or 7,15-diazadispiro[5.1.5.3]hexadecane **56** (or of its 14,16-dioxoderivative); thermostabilizers of PVC containing the benzothiazolonyl moiety **32**, flame retardants containing brominated aromatics **57** or phosphate moieties **58**, biocides, exclusively compounds containing organo tin moieties **59**, **60**, and antistatica **61**. An attempt has been made to synthesize monomers bearing two antioxidant stabilizing moieties acting by different mechanisms. Monomers **62** and **63** containing hindered phenolic (aminophenolic) and activated sulfidic functions are given as examples. An aspect of bifunctionality is considered to be involved in anilino-substituted phenylthiazole [45] **64**. Ⓢ-Functionalized monomers containing C=C double bonds can be used for the synthesis of polymeric stabilizers by means of polymerization, copolymerization, polysubstitution and grafting processes, respectively, exemplified in Sects. 3.1.1.1 and 3.2.1.

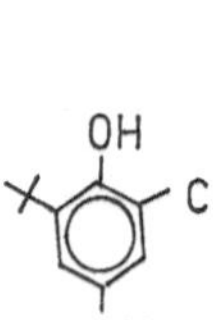

25

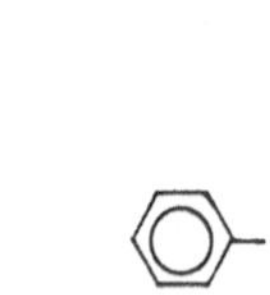

26

27

28

29

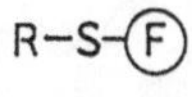

30

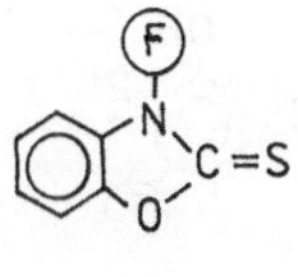

31

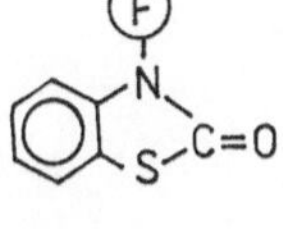

32

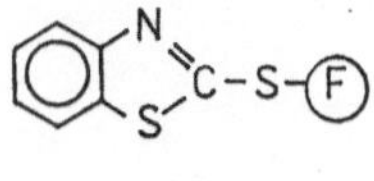

33

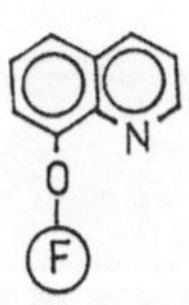

34

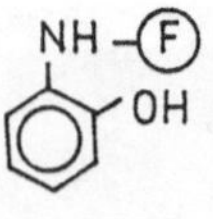

35

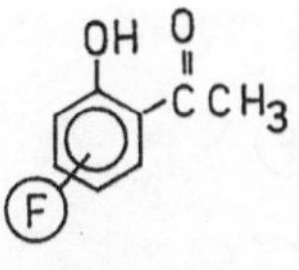

36

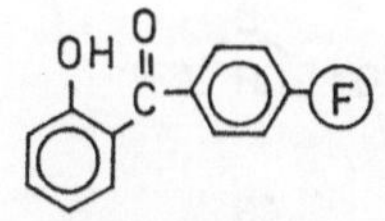

37

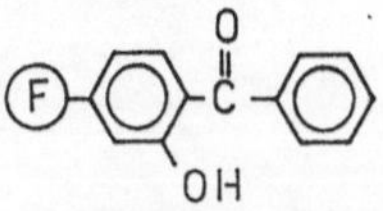

38

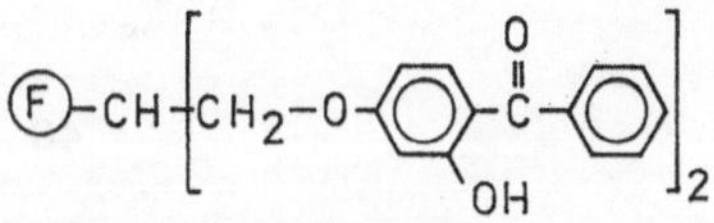

39

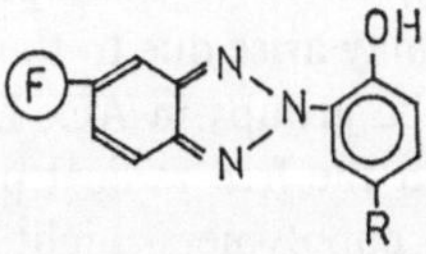

40

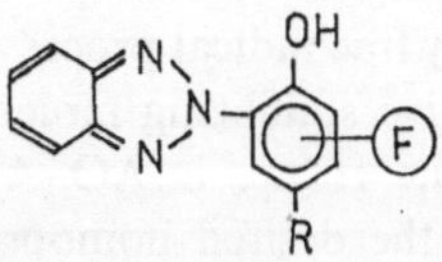

41

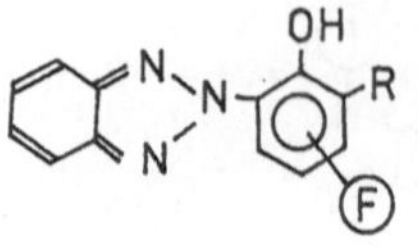

42

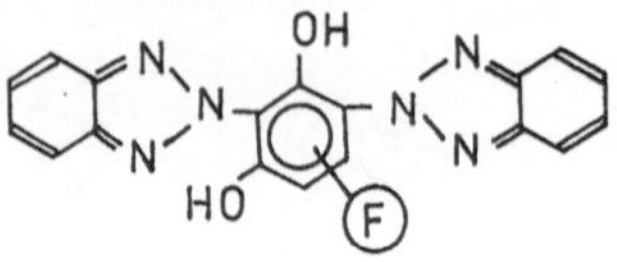

43

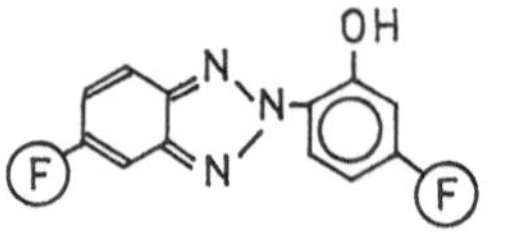

44

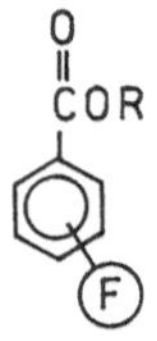

45

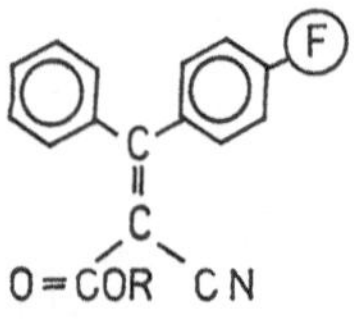

46

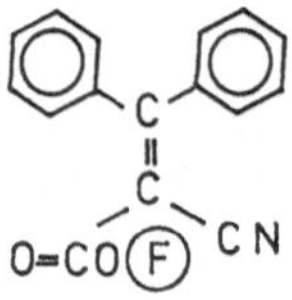

47

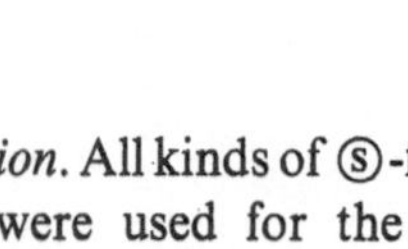

48

Free-Radical Polymerization. All kinds of ⑤-functionalized monomers described in the preceding Section were used for the synthesis of homopolymers and copolymers by free radical processes. Difficulties may arise due to the interference of some reactive stabilizing moieties, like phenolic groups in AO, LS or FR or aromatic amine groups in AO, AF or AOZ with the free radical process. As a consequence, the desired homopolymers and/or copolymers might not be synthesized, the desirable molecular weight or its distribution is often not attained, the concentration level of bond-in stabilizing active groups is low or blocks of

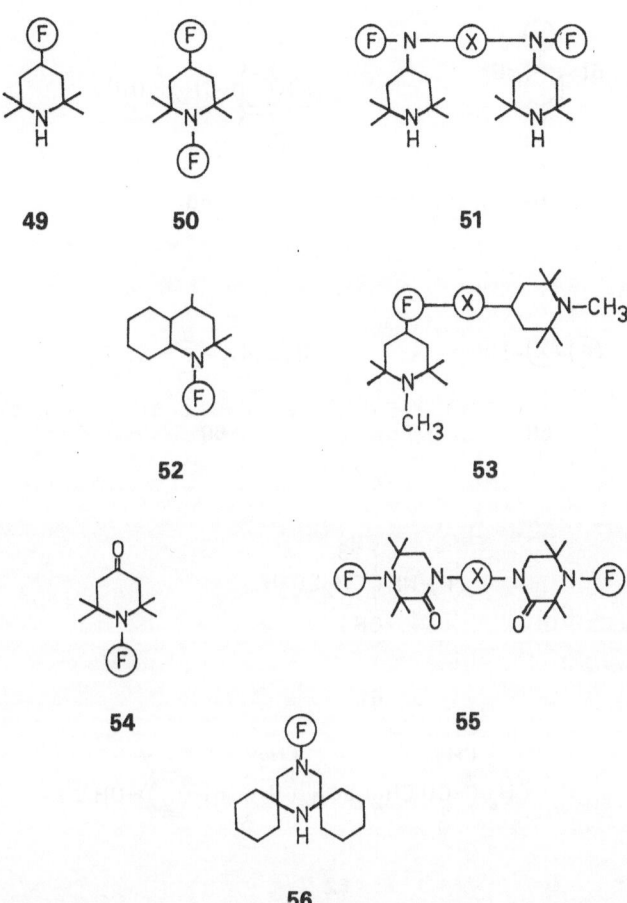

49 50 51

52 53

54 55

56

functionalized units having an unfavorable block length in copolymers are created [46].

Features of the free-radical initiation processes are similar for both the homopolymerization of functionalized monomers and copolymerization of the latter with conventional monomers. Common chemical initiators were applied. Azo-bis(isobutyro nitrile) was mostly used in bulk polymerization. No interference with phenolic hydroxy groups was observed in polymerization of 2-hydroxybenzo-phenones, acetophenones, salicylates and of their derivatives [47]. The most rigorous elimination of oxygen from the reaction mixture was necessary to achieve polymerization of monomeric hindered phenolic antioxidants or derivatives of 2-(2-hydroxyphenyl)benzotriazole [48]. An oxygen-free atmosphere is also an advantage for aromatic amines. A higher initiator level and/or increased temperature appear to be necessary to achieve normal polymerization rates with Ⓢ-functionalized monomers [46].

To avoid difficulties due to the interference of phenolic hydroxyls with the free radical process, the phenolic group may be blocked before radical polymerization by acetylation or silylation. The phenolic OH must be restored in the formed

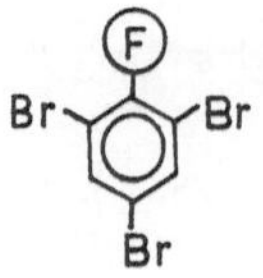

57

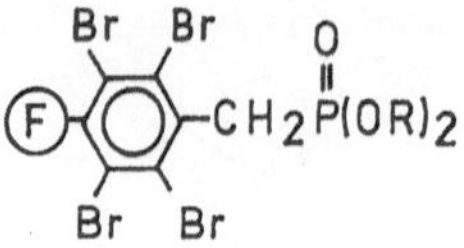

58

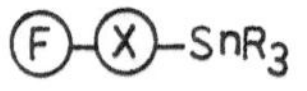

59

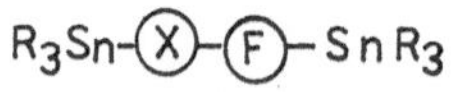

60

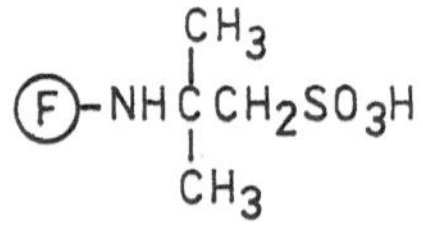

$$\text{(F)}-NH\underset{\underset{CH_3}{|}}{\overset{\overset{CH_3}{|}}{C}}CH_2SO_3H$$

61

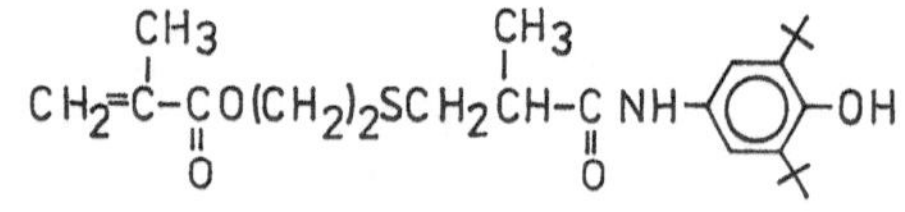

$$CH_2=\underset{\underset{O}{\|}}{\overset{\overset{CH_3}{|}}{C}}-CO(CH_2)_2SCH_2\underset{\underset{O}{\|}}{\overset{\overset{CH_3}{|}}{CH-C}}\,NH-\text{(O)}-OH$$

62

$$CH_2=\underset{\underset{CH_3}{|}}{C}-\overset{\overset{O}{\|}}{C}O(CH_2)_2NH\overset{\overset{O}{\|}}{C}O(CH_2)_2O\overset{\overset{O}{\|}}{C}(CH_2)_2S-\text{(O)}-OH$$

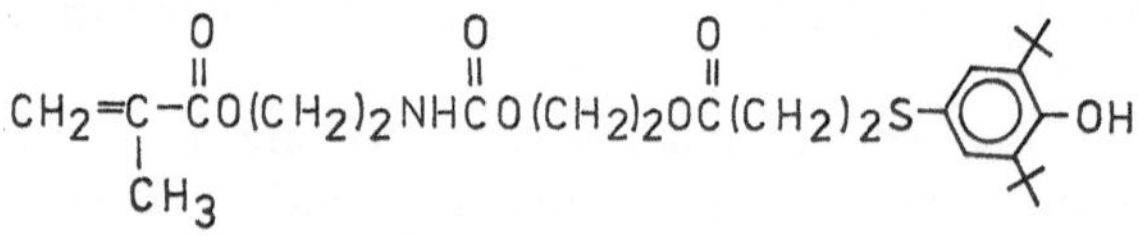

63

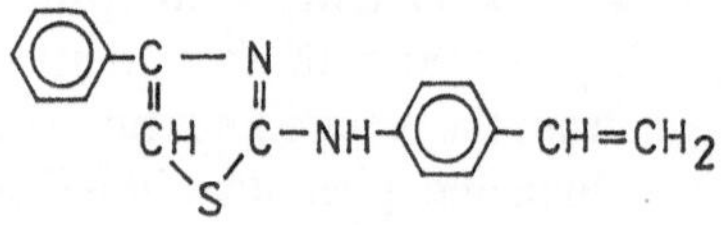

64

polymer. Some monomers, e.g. 2-vinylhydroquinone, cannot be polymerized without this pretreatment at all [49]. Processes like this are however too expensive for commercialization. Fortunately, polymerization in an oxygen-free atmosphere is mostly successful.

Peroxidic initiators, like dibenzoyl peroxide, are too reactive with phenolic and aminic moieties and can be applied only exceptionally [50]. Inorganic peroxides, like potassium persulfate, were used for emulsion polymerization of functionalized monomers [51]. Phase transfer catalysis may be also applied using persulfate initiation. To circumvent problems with the peroxide initiation, 4,4'-azobis(4-cyanovaleric acid), a water soluble initiator, was successfully used [48].

The earliest data on solution of stabilizer permanency problems by means of polymerization of functionalized monomers having double bonds C=C were published in the late 1960s. They deal with results concerning stabilization of rubber [50–53] (experience from the chemical modification of rubber were successfully exploited in this case) and with light stabilization of polyolefins [54].

Examples of homopolymers are given. Poly(4-vinylphenol) was prepared as a prepolymer for the subsequent alkylation [55]. Poly[2-(4-vinylbenzyl)hydroquinone] 65 is an example of the unhindered phenolic antioxidant for rubbers. Many homopolymers bear a hindered phenolic moiety. Homopolymer 66 was proposed for blending with BR and IR [56]. Other examples are poly[vinyl-3-(3,5-di-*tert*-butyl-4-hydroxyphenyl)propionate] [57] (67), poly(3,5-di-*tert*-butyl-4-hydroxybenzyl methacrylate) [58] (68) or poly[N-3,5-di-*tert*-butyl-4-hydroxybenzyl) maleimide] [59] (69). Numerous polymeric antioxidants are functionalized with aromatic amine groups. Poly(4-anilinophenyl methacrylate) [53] (70) serves as an example.

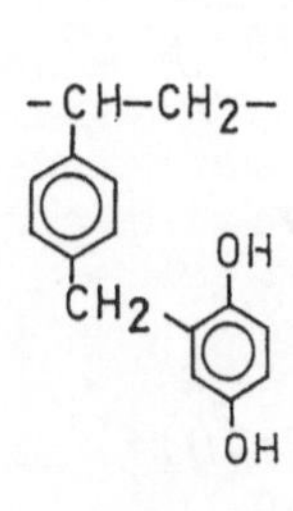

65 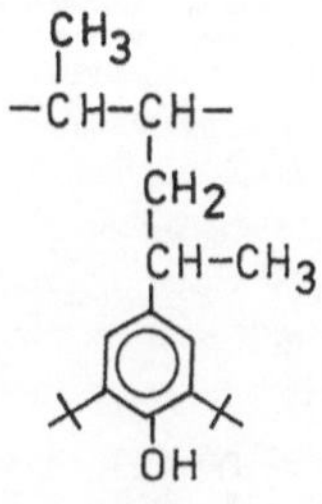   66 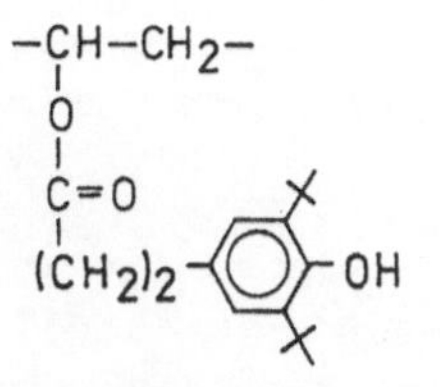   67

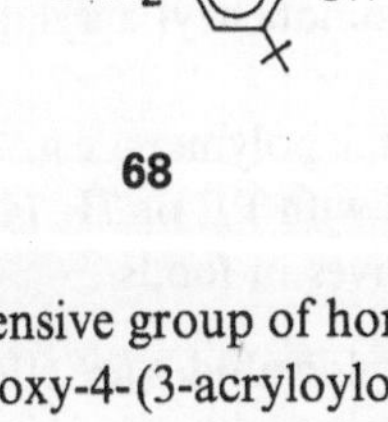

68 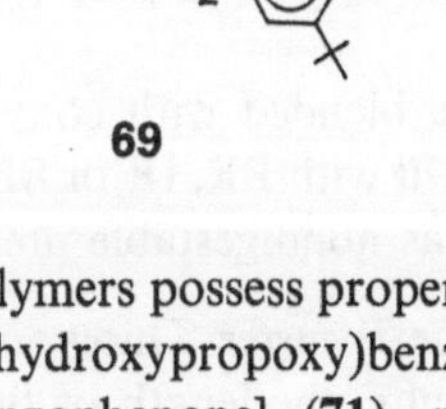   69 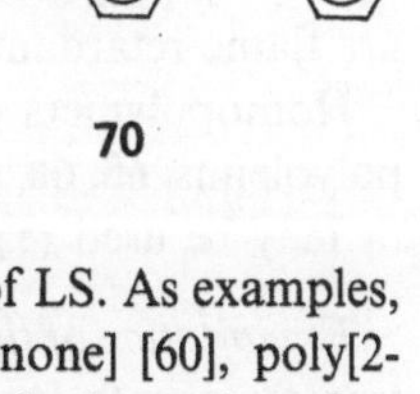   70

A comprehensive group of homopolymers possess properties of LS. As examples, poly[2-hydroxy-4-(3-acryloyloxy-2-hydroxypropoxy)benzophenone] [60], poly[2-hydroxy-4-(2-acryloyloxyethoxy)benzophenone] (71), poly(2-hydroxy-4-methacryloyloxybenzophenone) [61] (the latter was prepared using an initiation system containing dioxane hydroperoxide/cobalt(II)acetylacetonate), poly[2-(2-hydroxy-5-vinylphenyl)-2*H*-benzotriazole] [62] (**72**), polymers of salicylates functionalized in positions 3, 4 and/or 5 with vinyl or acryloyl groups, e.g. poly(methyl 5-vinylsalicylate) [63] (**73**). Various homopolymers contain hindered piperidine groups, e.g. poly[2-(diallylamino)-4-methoxy-6-(2,2,6,6-tetramethyl-4-piperidyl-amino)-1,3,5-triazine] [64] and its analogues, poly[bis(1,2,2,6,6-pentamethy-4-piperidyl) itaconate] [65] (**74**) or poly[1,1'-ethylenebis(4-acryloyl-3,3,5,5-tetra-methyl-2-piperazinone)] [66].

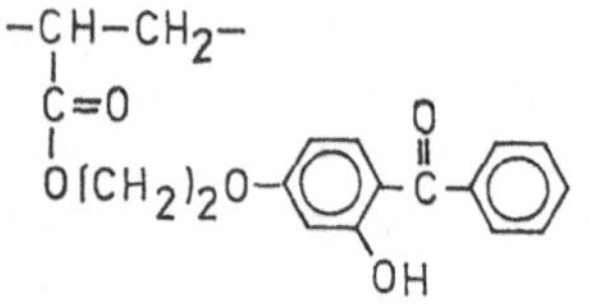

71

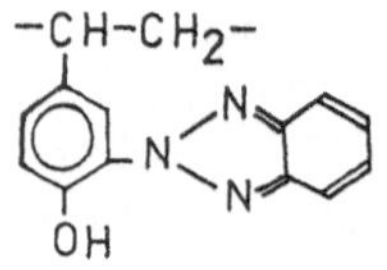

72

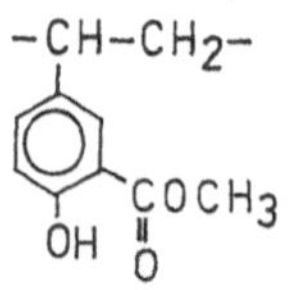

73

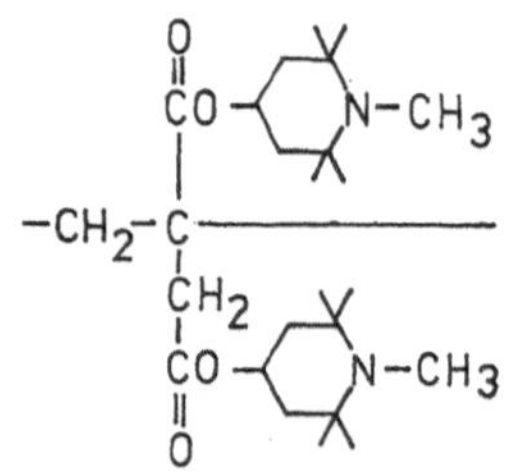

74

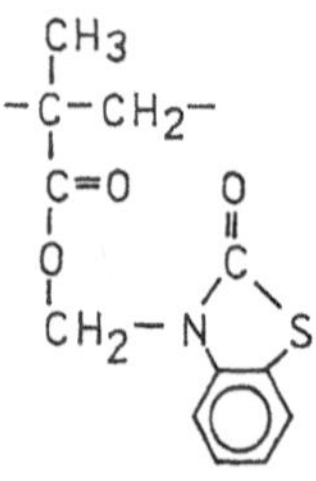

75

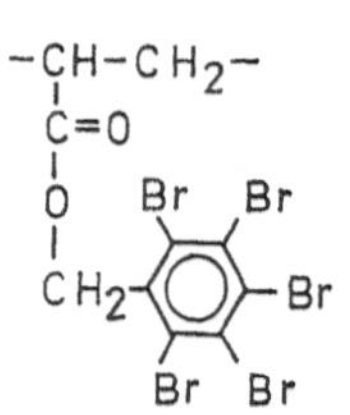

76

Poly[*N*-(α-benzothiazolonylmethyl) methacrylate] (**75**) is a thermostabilizer for PVC [67], poly(tribromostyrene) and poly(pentabromobenzyl acrylate) [68] (**76**) are flame retardants.

Homopolymers can be blended with conventional polymers, e.g. **72**, **73** with polyolefins, **65, 66, 68** or **70** with BR, IR or SBR, **76** with PE or **71, 74** with PVC, or may be used (e.g. **69**) as nondigestable preservatives in foods.

Exploitation of the Chain Transfer. Thiols and diaryldisulfides are effective chain transfer agents. As a results, the length of the polymer chain is shortened, and formed thiol radicals AS· are bound in the polymer chain (A = moiety carrying a stabilizing group, Scheme 3).

$$\text{ASH, ASSA} \xrightarrow{P·} \text{AS·}$$

$$\text{AS· } + n\text{M} \rightarrow \text{ASM}_{\overline{n}} \xrightarrow[\text{M}]{\overset{\text{ASH}}{\text{ASSA}}} \text{AS· } + \text{AS-polymer}$$

Scheme 3

This approach was exploited for the synthesis of functionalized telechelic polymers using functionalized iniferters (initiators/transfer/termination/agents) [69]. End groups may be introduced in a step reaction process involving a monofunctional reactive species. For example, an α-ω-bifunctionalized oligo-styrene 77 or analogous oligo acrylate can be synthesised using tetramethylthiuram disulfide. The oligomer 77 possess properties of a HD-AO. Moreover, 77 may be hydrolyzed to a dithiol 78, useful for functionalization of unsaturated hydrocarbon polymers via reactivity of thiyls (see Sect. 3.2.2).

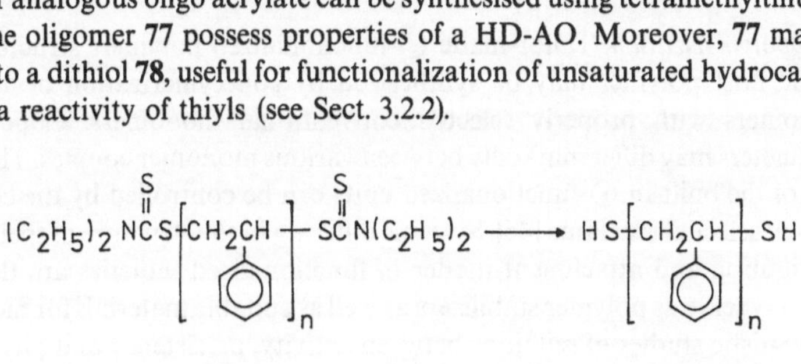

77   78

ABS polymers terminated with HD-AO moiety RS- were prepared by means of polymerization of butadiene, styrene and acrylonitrile in the presence of *tert*-dodecylmercaptan [70]. Synergism with phenolic AO was observed. A radical copolymer active as a thermostabilizer for fluoroalkyl methacrylates [71], was prepared from methyl methacrylate (MMA), 2,2,2-trifluoroethyl methacrylate and pentaerythritol tetrakis(thioglycolate).

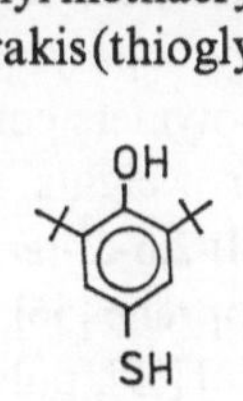

79 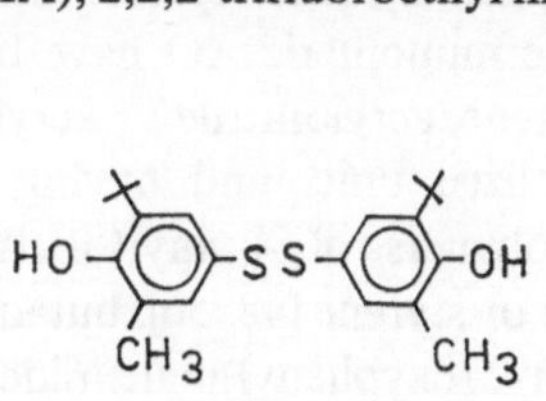   80

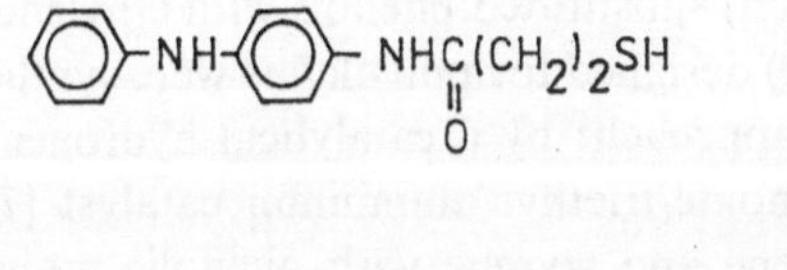

81 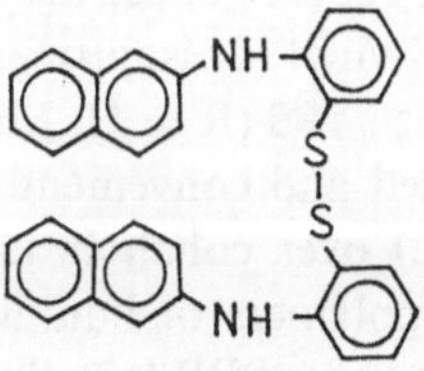   82

Using 1,2,2,6,6-pentamethyl-4-piperidyl-*tert*-amylperoxy carbonate as an initiator in the copolymerisation of alkylmethacrylate, 2-hydroxyethyl acrylate, methacrylic acid and styrene, copolymers having numerical average molecular weight <4000, a narrow molecular weight distribution and a terminal built-in initiator derived HALS moiety were obtained and used as thermoset coatings with high durability to weathering [72].

The mechanistic principle of the chain transfer exploiting functionalized transfer agents was used for the synthesis of polymer bound CB AO, attached to the polymer chain via the sulfur atom. Weinstein [73, 74] used phenolic and aminic thiols **79, 81** and disulfides **80, 82** as generators of thiyls during free-radical bulk or emulsion copolymerization of butadiene or isoprene with styrene. Systems formed can be considered as bifunctional physically persistent stabilizers combining CB and HD functions.

Copolymerization. Tailor-made Ⓢ-functionalized polymers structurally related to the host polymer may be synthesized by copolymerization of functionalized monomers with properly selected conventional monomers. Copolymerization parameters may differ markedly between various monomer couples. The concentration of the built-in Ⓢ-functionalized units can be controlled by the concentration ratio of selected reactants [46]. Systems differing in the structure of their backbones, distribution and attachment modes of functionalized moieties are thus available and may serve as polymer stabilizers as well as suitable materials for more profound mechanistic studies of relations between activity, persistency and physical properties of the system additive/polymer matrix. Improvement of the compatibility with the host polymer, formation of polymers from functionalized monomers that do not homopolymerize, and polymeric stabilizers containing a proper combination of two functional groups forming cooperative systems in one molecule may be considered as the most valuable properties of copolymeric stabilizers.

Statistical functionalized copolymers are formed both in the homogeneous or heterogeneous phases. Some representative examples were selected. Phenolic and aromatic aminic monomeric AO have been mostly copolymerized with styrene, butadiene, isoprene, acrylonitrile or acrylates. Copolymers containing up to 10% of Ⓢ-functionalized units and having molecular weights up to 50,000 were obtained. Copolymers of 4-vinyl(or isopropenyl)-2,6-di-*tert*-butylphenol with *n*-butyl acrylate or styrene [48, 56], butadiene or isoprene [56], e.g. **83**, poly[*N*-3,5-di-*tert*-butyl-4-hydroxyphenyl)maleimide-co-styrene] [59], poly{*N*-(3,5-di-*tert*-butyl-4-hydroxy-phenyl)3-[(2-methacryloyloxy)ethyl)-thiol]-2-methylpropionamide-*co*-butadiene} [75], an emulsion terpolymer [44] **84** are considered as AO for conventional rubbers or for acrylates. To circumvent complications with copolymerization of vinyl (or isopropenyl) substituted phenols with ethylene, polyolefin bound stabilizers **85** (R = H, Me) designed for polyolefins were synthesised using a sophisticated and convenient approach: by a catalytical hydrogenation of **83** (R = H, Me) over cobalt hexanoate/triethyl aluminium catalyst [76]. Various emulsion terpolymers of butadiene and styrene with phenolic monomers were prepared as AO for SBR [52]. Similarly, terpolymers containing acrylonitrile have been considered as stabilizers for NBR. Copolymers of aromatic amines with

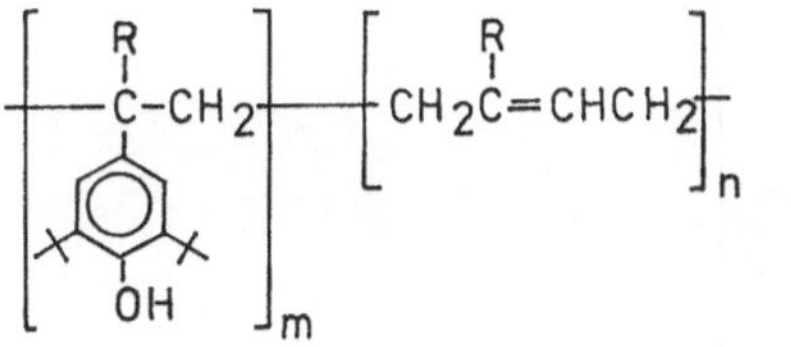

$$\left[\begin{array}{c} R \\ | \\ -C-CH_2- \\ | \\ \bigcirc \\ | \\ OH \end{array}\right]_m \left[\begin{array}{c} R \\ | \\ CH_2C=CHCH_2 \end{array}\right]_n$$

83

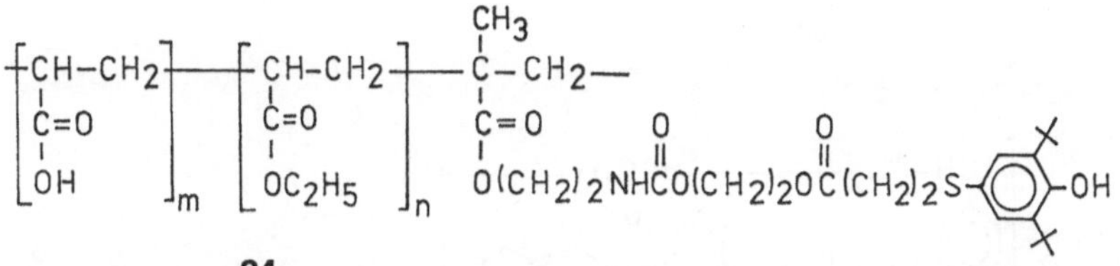

$$\left[\begin{array}{c} CH-CH_2 \\ | \\ C=O \\ | \\ OH \end{array}\right]_m \left[\begin{array}{c} CH-CH_2 \\ | \\ C=O \\ | \\ OC_2H_5 \end{array}\right]_n \begin{array}{c} CH_3 \\ | \\ C-CH_2- \\ | \\ C=O \\ | \\ O(CH_2)_2NHCO(CH_2)_2O\overset{O}{\overset{||}{C}}(CH_2)_2S-\bigcirc-OH \end{array}$$

84

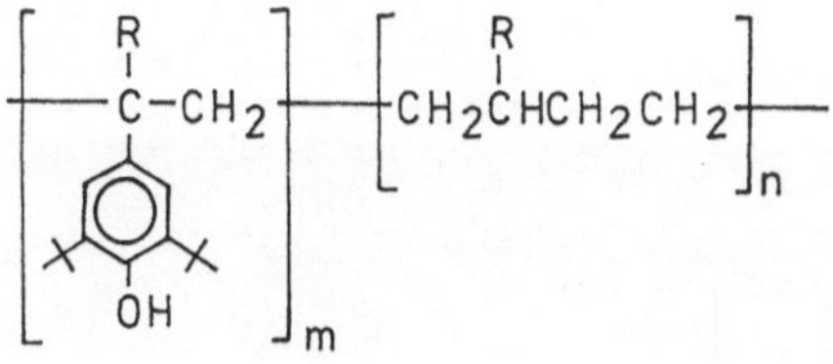

$$\left[\begin{array}{c} R \\ | \\ -C-CH_2- \\ | \\ \bigcirc \\ | \\ OH \end{array}\right]_m \left[\begin{array}{c} R \\ | \\ CH_2CHCH_2CH_2 \end{array}\right]_n$$

85

$$\left[\begin{array}{c} CH-CH_2 \\ | \\ CN \end{array}\right]_m \left[CH_2CH=CHCH_2\right]_n \begin{array}{c} CH_3 \\ | \\ C-CH_2- \\ | \\ C=O \\ | \\ NH-\bigcirc-NH-\bigcirc \end{array}$$

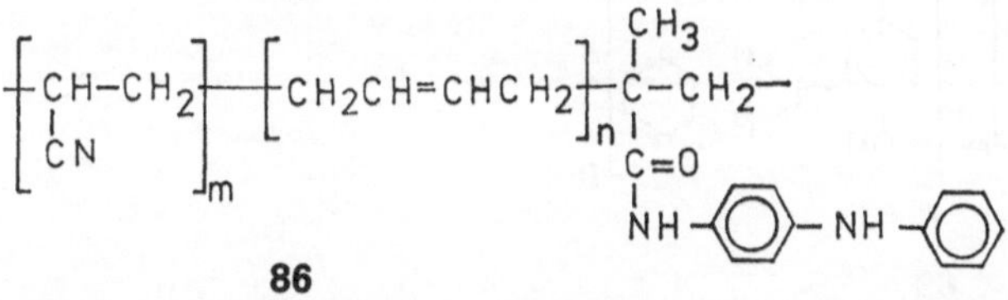

86

butadiene or isoprene or terpolymers with butadiene and styrene or butadiene and acrylonitrile are antioxidants for SBR or NBR [50, 52, 53]. Terpolymer **86** containing the *N*-(4-anilinophenyl)methacrylamide moiety [77], commercialized by Goodyear and used mainly for SRF-black filled NBR cured with a low sulfur cure system, belong among the most successful macromolecular stabilizers. Terpolymer of butadiene, styrene and 2-(4-vinylphenylamino)-4-phenylthiazole posses antioxidant properties in SBR [78]. Copolymers of styrene, MMA or alkyl itaconate with polymerizable benzothiazolones (**32**), benzoxolethiones [79] (**31**) or 2-thiobenzothiazoles [80] (**33**) are thermostabilizers for PS or PMMA. Radical copolymers of 1-vinylbutyrolactone with 4-vinylbenzylimino-*N,N*-diacetic acid posses properties of MD and are able to chelate ions of transition metals [81].

Copolymers having properties of LS form a comprehensive and very variegated group. Data have been published dealing with built-in acetophenones and benzophenones, e.g. poly(2-hydroxy-4-acryloyloxybenzophenone-*co*-styrene) [82] (**87**) designed for PS, an analogous copolymer with MMA considered for stabilization of heat curable resin coatings based on mixtures of methacrylates [83] or SAN type terpolymers with acrylonitrile and styrene, tested as stabilizers of ABS [84]. Syntheses of copolymers containing built-in (2-hydroxyphenyl)benzo-

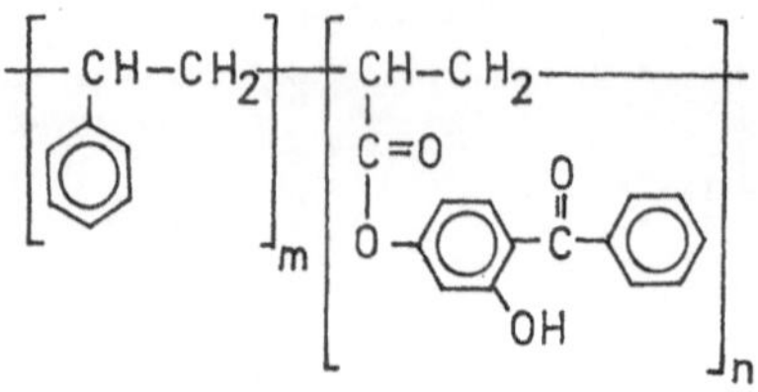

87

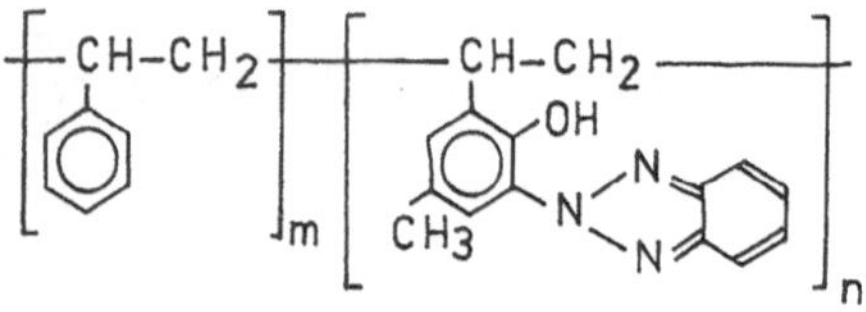

88

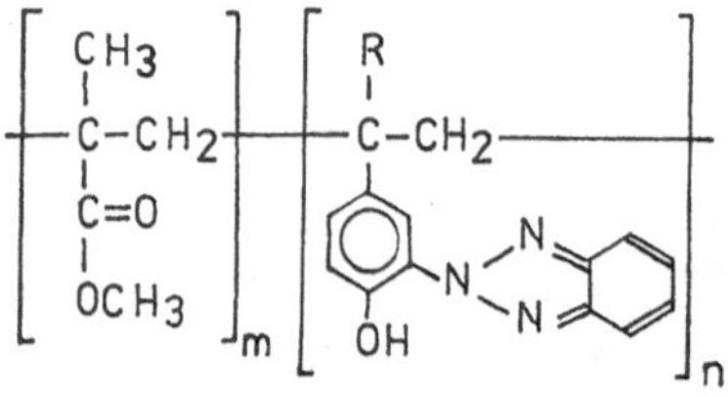

89

triazole moieties were studied intensively by Vogl [48, 56, 76]. Styrene [62, 85, 86], dibutyl maleate [87], MMA [26, 88, 89], acrylonitrile [90], butadiene [90], N-vinylpyrrolidone [91] or N,N-dimethylaminomethacrylate [92] were used as comonomers. Some of these copolymers, e.g. poly[2-(2-hydroxy-3-vinyl-5-methyl-phenyl)-2H-benzotriazole-co-styrene] [85, 86, 93], (**88**), poly[2-(2-hydroxy-5-vinyl or isopropenyl-phenyl)-2H-benzotriazole-co-methyl methacrylate] (**89**, R = H, Me), an analogue of the latter formed with styrene [26, 62, 88, 95] and poly[2-(2-hydroxy-4-methacryloyloxyphenyl)-2H-benzotriazole-co-styrene or me-thacrylate] [76, 96] were studied as stabilizers for PMMA. SAN type terpolymer prepared from styrene, acrylonitrile and 2-[2-hydroxy-4-(4-vinylbenzyloxy)]-2H-benzotriazole was designed for ABS stabilization [90]. Copolymers of ethylene, styrene or acrylonitrile or alkyl methacrylate with polymerizable alkyl or aryl salicylates were designed for polyolefin stabilization [63, 97]. Copolymers of ethyl α-cyano-β-(4-vinylphenyl)cinnamate with alkyl methacrylate were prepared as LS for application in solar energetics [98]. A copolymer of 1-vinylanthracene with dimethylsilicones containing about 0.5 mol% of vinyl groups [99] received special attention for use in medical aids.

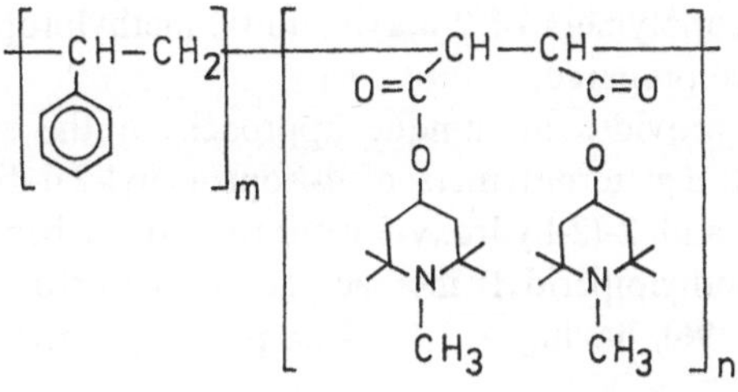

90

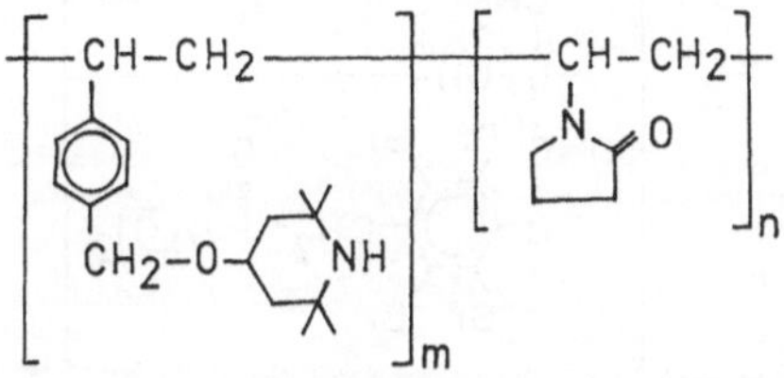

91

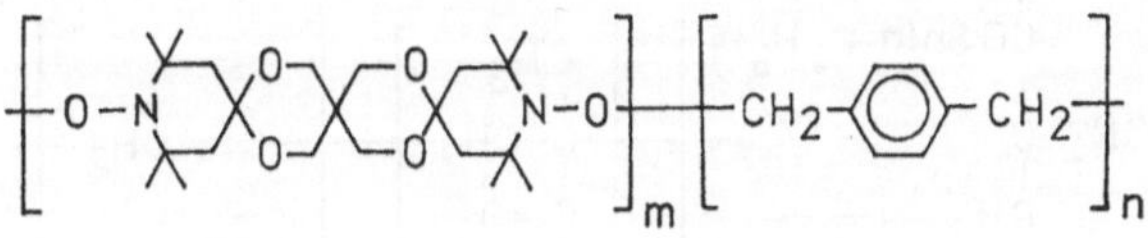

92

Numerous copolymeric LS contain a hindered amine moiety. Most of them were prepared by radical copolymerization of polymerizable 2,2,6,6-tetramethyl- or 1,2,2,6,6-pentamethylpiperidines with alky methacrylates, acrylonitrile, vinyl acetate, styrene or N-vinylpyrrolidone. Poly{bis[4-(1,2,2,6,6-pentamethylpiperidyl)]-maleate-co-styrene} (**90**) having $\overline{M}_n \sim 2,000$, a water soluble poly[$N$-vinyl-2-pyrrolidone-$co$-4-(4-vinylbenzyloxy)-2,2,6,6-tetramethylpiperidine] [100] (**91**) or poly[N-(2-methacryloyloxyethyl-2,2,4-trimethyldecahydroquinoline-co-acrylonitrile] [101] may serve as typical examples. A chemically bound HALS moiety was introduced into I-PS during copolymerization of styrene and butadiene initiated with 1,2,2,6,6-pentamethyl-4-piperidyl tert-butylmonoperoxy carbonate [102]. Copolymer **92**, having $\overline{M}_n$ up to 27,000, was formed from the respective bisnitroxide and p-xylylene generated in situ from 4-methylbenzylchloride [103].

Some other interesting copolymers having properties of PVC thermal stabilizers, like poly[N-(α-benzothiazolonylmethyl)methacrylate-co-methyl methacrylate] [45], of flame retardants like a terpolymer of styrene, acrylonitrile and a polymerizable perbrominated phenol [76] or poly[4-methacryloyloxy-2,3,5,6-tetrabromobenzyldiphenyl phosphonate-co-methyl methacrylate] [104] (**93**), biocides, mostly copolymers of monomers containing tris(n-butyl tin) or triphenyl tin moieties and alkyl acrylates, methacrylates, vinyl acetate, acrylonitrile, styrene or N-vinylpyrrolidone [105], e.g. a terpolymer of styrene, MMA and tri(n-butyl tin) itaconate [106] (**94**), or antistaticas, like copolymers of 2-acrylamido-2-methylpropane sulfonic acid and styrene [107] were prepared.

Copolymerization provides an unique approach to the synthesis of poly-functional stabilizers. E.g. terpolymers of 4-isopropenyl-2,6-di-*tert*-butylphenol, methyl methacrylate and 2-(2-hydroxy-5-isopropenyl)-2*H*-benzotriazole (**95**) or poly[4-(2,2,6,6-tetramethylpiperidyl) methacrylate-*co*-4-hydroxy-3,5-di-*tert*-butyl-benzyl methacrylate] (**96**), having $\overline{M}_n \sim 5,400$, posses properties of AO and LS [108].

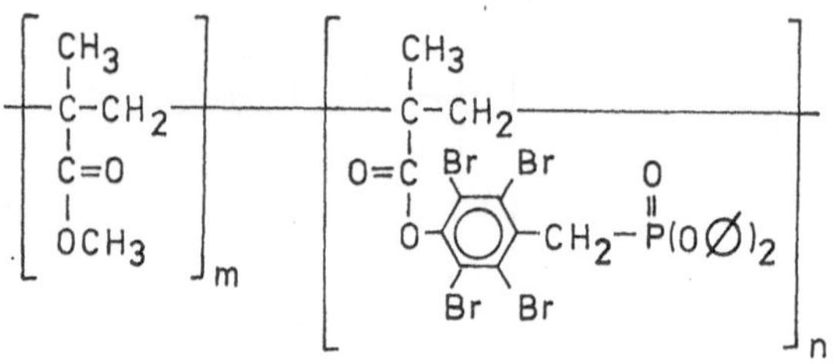

93

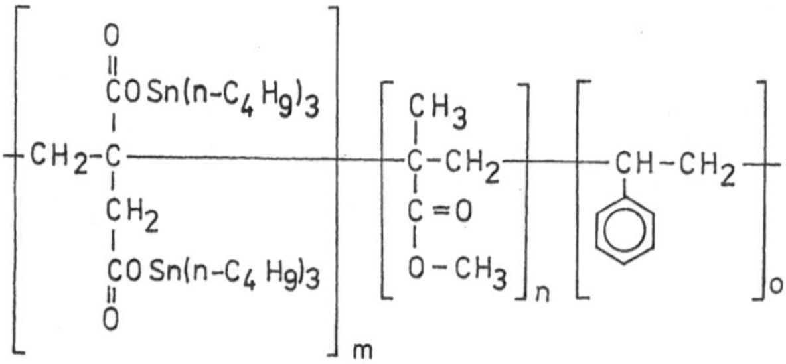

94

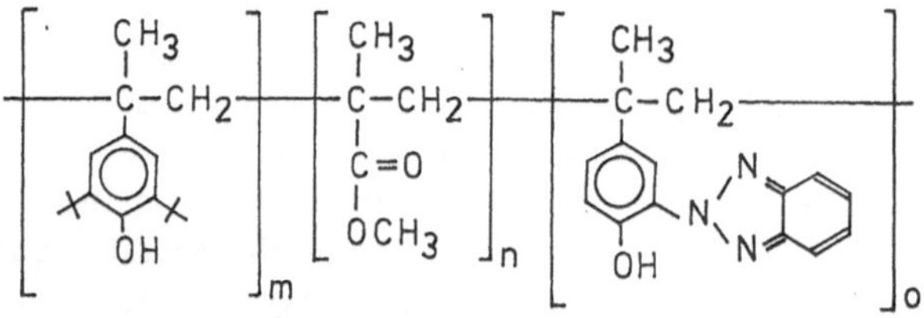

95

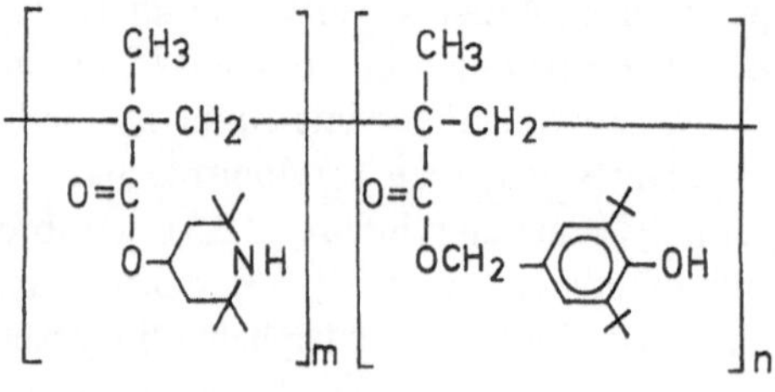

96

Polymerization Catalysed by Acids and Bases. Carbonium ions and carbanions respectively are carriers of the chain transfer in cationic and anionic polymerizations respectively. Ionic polymerization mechanism was exploited for the synthesis of polymeric stabilizers in comparison with the free-radical polymerization only exceptionally. The cationic process was used for the synthesis of copolymers of 2,6-di-*tert*-butyl-4-vinylphenol with cyclopentadiene and/or for terpolymers with cyclopentadiene and isobutylene [109]. System $SnCl_4/Et_3AlCl_3$ was used as an initiator. Poly(10-vinylphenothiazin) was prepared by means of catalysis with titanium chlorides [110]. Polymers of 4-[α-(2-hydroxy-3,5-dimethylphenyl)ethyl]-vinylbenzene [111] and 3-allyl-2-hydroxyacetophenone [112] were also prepared under conditions of cationic polymerization.

Using bifunctional agents like divinyl- or diisopropenylbenzene or 1,3-bis(1-hydroxy-1-methylethyl)benzene, polymeric antioxidants are formed from phenols or diphenylamine under alkylation conditions. Various examples have been described. Oligomeric AO for foods, having molecular weight greater than 600, were prepared from a low purity (75%) divinylbenzene and *p*-cresol in the presence of aluminium *p*-cresolate catalyst [113]. The same procedure was used for the synthesis of **97** from a mixture of phenols consisting of 4-hydroxyanisole, 2-*tert*-butylhydroquinone, 4-*tert*-butylphenol, 4-methylphenol and 4,4'-isopropylidenebisphenol [114].

A low purity (40–80%) divinylbenzene and chloromethylstyrene were used in the presence of a Friedel-Crafts catalyst ($AlCl_3$) for the synthesis of a polymeric phenolic antioxidant from *p*-cresol and 1,3-bis(1-hydroxy-1-methylethyl)benzene in the presence of *p*-toluene sulfonic acid [115]. The obtained product was either used directly as AO for thermosetting resins or was treated consecutively with α-methylstyrene [116] or *tert*-butylalcohol [117]. Other polymeric phenolic AO were obtained by reaction of phenol with *p*-menthane-1,8-diol, 1-*p*-menthen-8-ol or limonene [118] or of *p*-cresol with 3- or 4-chloromethylstyrene in the presence of BF_3-etherate or anhydrous $AlCl_3$ [119]; the product thus obtained was finally aralkylated by α-methylstyrene. Thermostabilizer and/or LS for PUR was obtained, e.g. **98**.

Dicyclopentadiene was used as another difunctional agent for the synthesis of an oligomeric AO from 4-methylphenol [120] (after *tert*-butylation, AO **99** was obtained) or diphenylamine [121]. Polymeric aromatic amines for application in rubber stabilization were obtained in the presence of acid catalysts from diphenylamine and 1,4-bis(1-hydroxy-1-methylethyl)benzene or 1,4-diisopropylbenzene [122] or from *N,N'*-diphenyl-1,4-phenylenediamine and α,ω-*p*-xylylenedichloride [123]. Polymerization of 2,2,4-trimethyl-1,2-dihydroquinoline in the presence of protic or Lewis acids [1] is very important commercially. Oligomeric product **100** is an effective AO and AF agent for rubbers.

Cationic polymerisation (via initiation with sulfuric acid or boron trifluoride) or anionic polymerization (in the presence of sodium ethoxide) were used for synthesis of polymers from 1,4-benzoquinone [124]. The number average molecular weight of products varied from 10 to 47 thousand, depending on the catalyst and its concentration. The polymerization product **101** contains hydroquinone and benzoquinone moieties and has remarkable AO activity in stabilization of SBR at 100 °C.

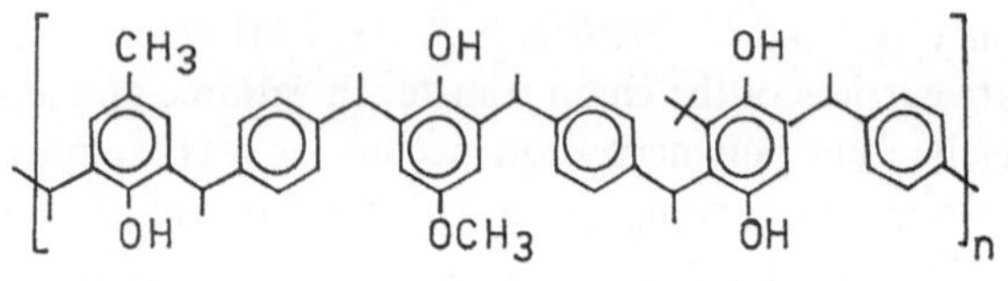

97

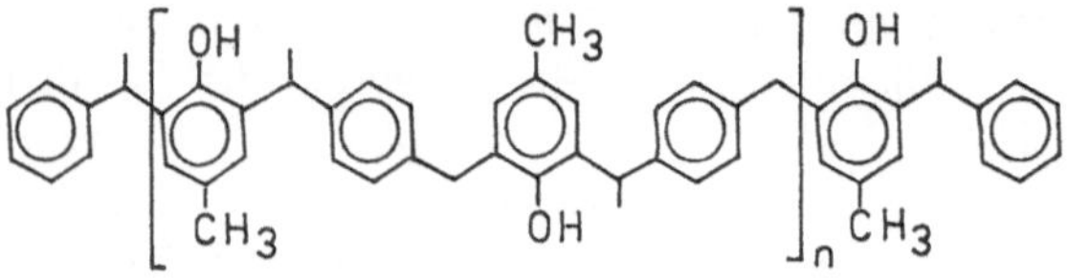

98

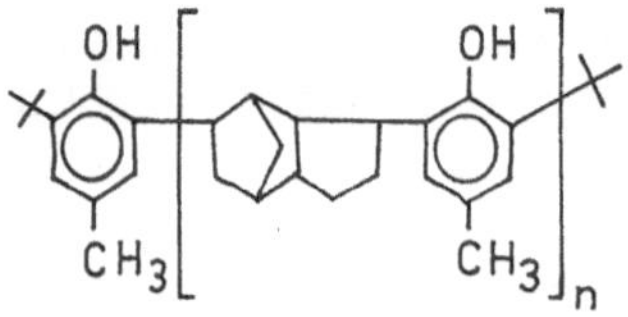

99 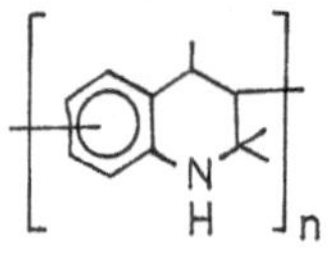   **100**

Living anionic polymerization of 4-vinylphenol was performed after transformation of the phenolic hydroxy group into trialkylsilyl ether group and removal of the protection group after polymerization [125]. *n*-Butyl lithium was used for the synthesis of poly[2-hydroxy-4-methacryloyloxybenzophenone] [61] (**102**) or HALS terminated poly(methyl methacrylate) [126]. 2-Hydroxy-4-methacryloyl-

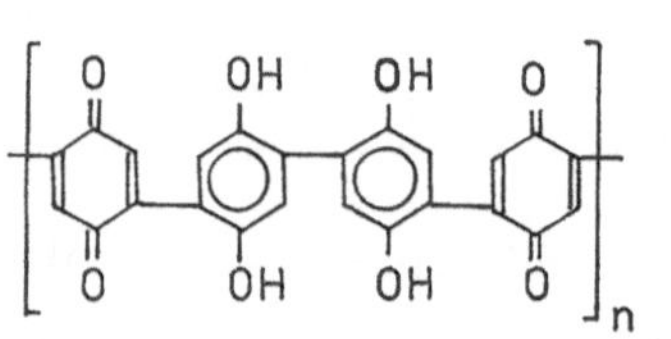

101 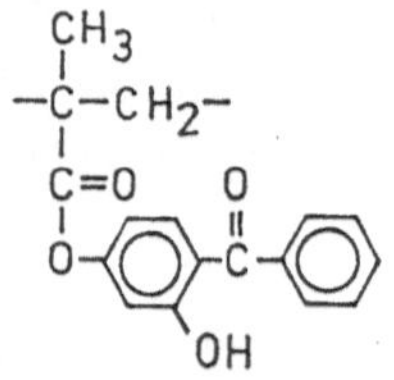   **102**

oxybenzophenone and 2-(2-hydroxy-4-methacryloyloxyphenyl)-2*H*-benzotriazole were incorporated into living PMMA in an inert atmosphere and aprotic solvents by means of the group transfer polymerization (GTP) in the presence of tris(dimethylamino)sulfonium difluoride as a catalyst and 1-methoxy-1-(trimethylsiloxy)-2-methyl-1-propene as an initiator [127]. Obtained polymeric LSs have well defined $\overline{M}_n$ and narrow molecular weight distribution. GTP method was not suitable for the synthesis of polymeric phenols or hindered piperidines due to the high activity of hydrogen atoms in the $-OH$ and $-NH$ groups.

Complex Coordination Polymerization. It is difficult to copolymerize functional monomers having a phenolic moiety, like 2-hydroxybenzophenone, 2-(2-hydroxyphenyl)benzotriazole or hindered phenolic moieties with ethylene or propylene by means of complex coordination catalysis due to the adverse effect of the free hydroxy group on the process. An economically unfavourable solution of the problem involves blocking of the phenolic function by transformation to an aluminium salt or to an acyl derivative before polymerization [128]. After polymerization, the phenolic function must be restored. Using this process, a copolymer of ethylene with 4-allyl-2,6-di-*tert*-butylphenol was prepared [128]. The process has no commercial importance.

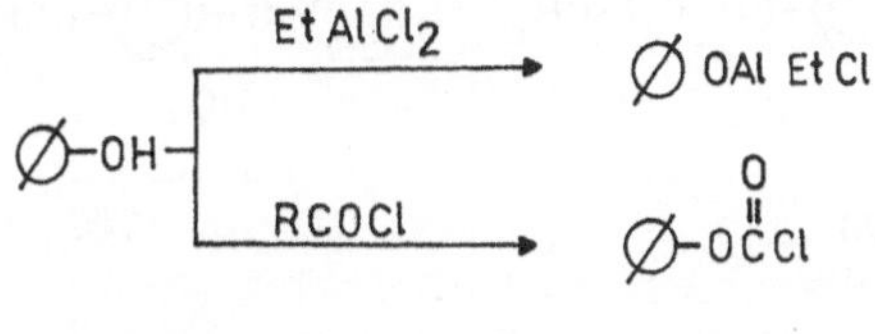

3.1.1.2 Ring Opening Polymerisation of Functionalized Heterocycles

Ⓢ-Functionalized cyclic ethers 103 were used almost exclusively from various available polymerizable heterocycles as monomers for the ring opening polymeriza-

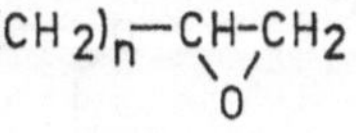

103

tion. Polyethers of type C (Scheme 1) bearing functionalized units are formed. 3,5-Di-*tert*-butyl-4-hydroxyphenyl-2,3-epoxypropylsulfide [129] (**104**), a more complicated ester type hindered phenol [123] (**105**), and various glycidylamines, e.g. 4-(2,3-epoxypropyl)diphenylamine [130] (**106**) are typical monomers for the synthesis of CB-AO. Monomers like 2-hydroxy-4-(2,3-epoxypropoxy)-5-*tert*-butylbenzophenone [131] (**107**, the free hydroxy group may participate in the reaction too, an interference like this may reduce the UV stabilizing effect of the formed polymer), 2,2,6,6-tetramethylpiperidine-4-spirooxirane [132] (**108**), 4-(2,3-epoxypropoxy)-1,2,2,6,6-pentamethylpiperidine [133], 3-(2,3-epoxypropyl)-7,7,9,9-tetramethyl-2,4-dioxo-1,3,8-triazaspiro[4.5]decane [134] (**109**) and analogous 1,5-dioxaspiro[5.5]undecane derivative [135] or 15-(2,3-epoxypropyl)-7,15-diazaspiro[5.1.5.3]hexadecane-14,16-dione [136] (**110**) were used for the synthesis of LS.

3.1.2 Polyaddition

Ⓢ-Functionalized polymers of the type B and C (Scheme 1) can be formed via polyaddition processes of bifunctional reactants, without splitting off of low molecular weight compounds. Most syntheses of stabilizers have been based on reactions of bis-isocyanates with H-acid nucleophiles. Some reactions of oxiranes may be listed here too (syntheses involving oxiranes are listed in Sect. 3.1.1.2 if polymerization aspects are more evident; syntheses of stabilizers formed via reactivity of oxirane moieties attached to an oligomeric or polymeric chain are classified as reactions on polymers, Sect. 3.2.2.2.

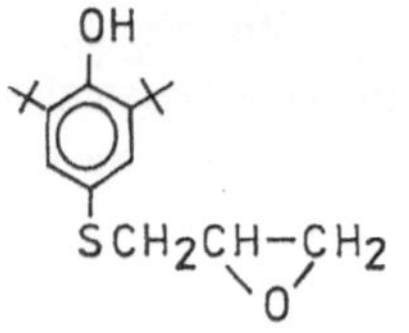

104 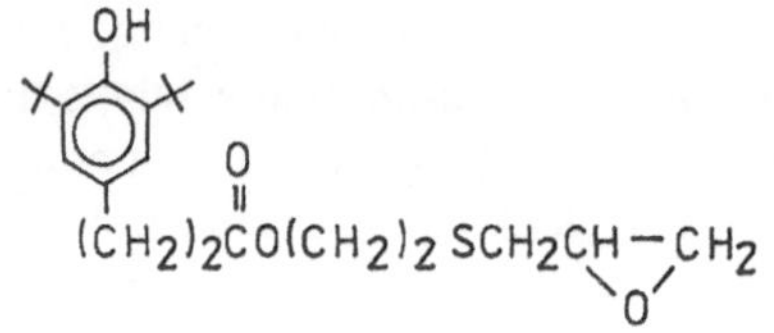   **105**

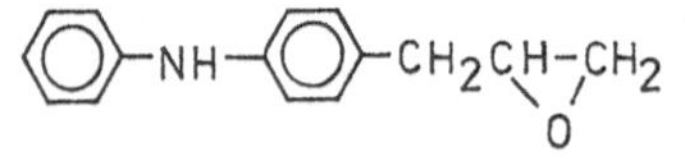

106 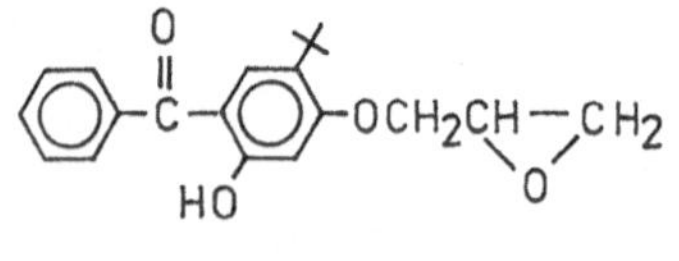   **107**

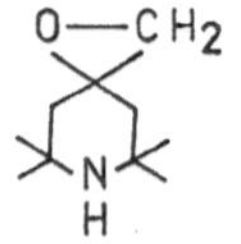

108

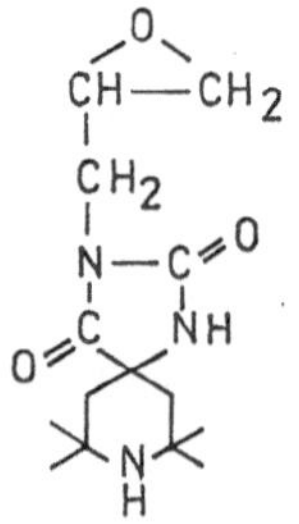

109 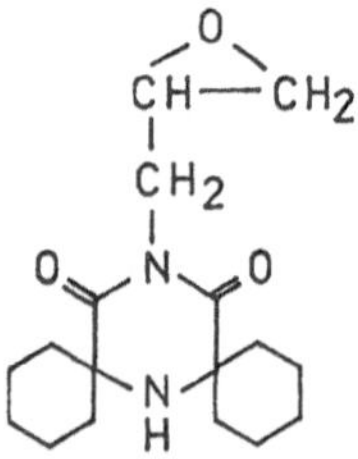   **110**

A reversible redox PUR system with antioxidant properties was formed from 2,5-bis(2-hydroxyethyl)hydroquinone and 1,6-hexamethylene diisocyanate [137]. A more complicated AO system having terminal phenolic moieties was created in a reaction mixture consisting of isophorone diisocyanate, 1,3,5-tris(3-hydroxy-2,6-dimethyl-4-*tert*-butylbenzyl)-2,4,6-trioxo-1,3,5-triazine,   1,2-ethylenediamine and methylene diisocyanate [138].

Light stable PUR **111** was created by incorporation of 4-(2,3-dihydroxyprop-oxy)-2-hydroxybenzophenone during PUR synthesis [139]. Similarly, urethane based coatings may be stabilized by immobilization of approx. 5% of 4-hydroxy-5-*tert*-butyl-3-[2-(5-chlorobenzotriazolyl)]methyl-isocyanate, added into the reaction mixture during PUR synthesis [140].

Variegated polyadducts contain HALS moieties. Functionalized polyurea **112** was formed from *N,N'*-bis(2,2,6,6-tetramethylpiperidyl)-1,6-hexanediamine and 1,6-hexamethylene diisocyanate [141]. A linear PUR **113** was synthesised from *N*-[3-(2,2,6,6-tetramethylpiperidin-4-yl)aminopropyl]diethanolamine [142]. Another functionalized PUR system was prepared from 2,4-toluene diisocyanate and 15-(2,3-dihydroxypropyl)-7,15-diazaspiro[5.1.5.3]hexadecane-14,16-dione [143]. PUR synthesised from a prepolymer **114** based on tri(*n*-butyl tin)tartrate and 2,4-toluene diisocyanate posses biocide properties [144].

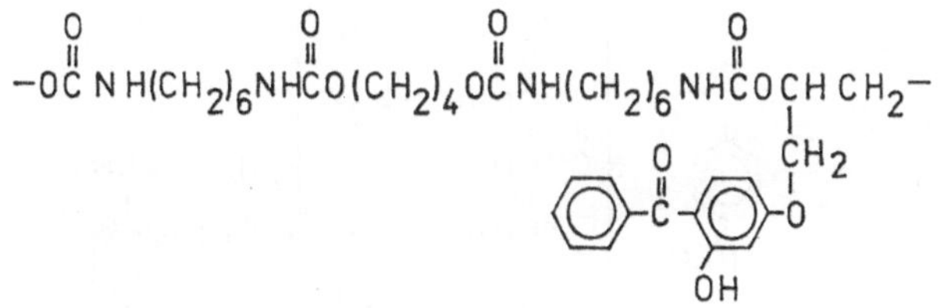

$$-O\overset{O}{\underset{\|}{C}}NH(CH_2)_6NH\overset{O}{\underset{\|}{C}}O(CH_2)_4O\overset{O}{\underset{\|}{C}}NH(CH_2)_6NH\overset{O}{\underset{\|}{C}}O\underset{\underset{O}{\overset{CH_2}{|}}}{CH}CH_2-$$

111

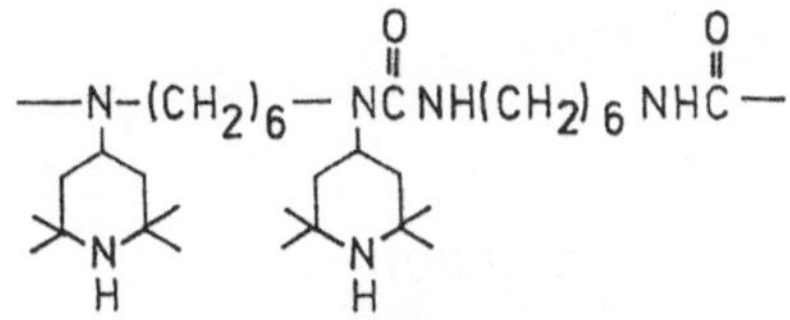

$$-N-(CH_2)_6-\overset{O}{\underset{\|}{N}C}NH(CH_2)_6NH\overset{O}{\underset{\|}{C}}-$$

112

$$-\overset{O}{\underset{\|}{C}}NH(CH_2)_6NH\overset{O}{\underset{\|}{C}}O(CH_2)_2-\underset{\underset{\underset{NH}{(CH_2)_3}}{|}}{N}-(CH_2)_2\ O-$$

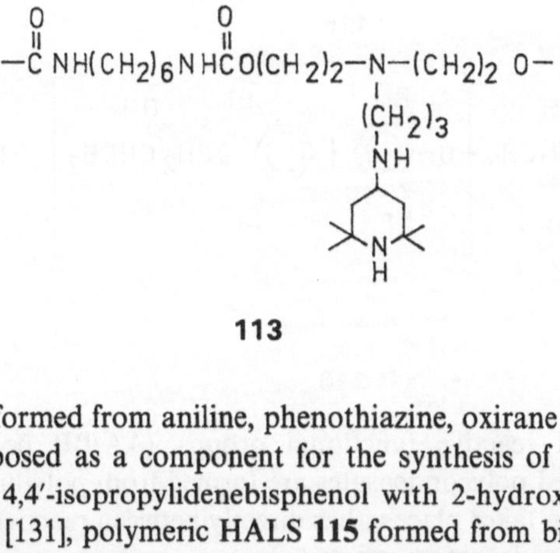

113

A polyadduct formed from aniline, phenothiazine, oxirane and 2,3-epoxypropane [145] was proposed as a component for the synthesis of a light stable PUR. A polyadduct of 4,4′-isopropylidenebisphenol with 2-hydroxy-4-(2,3-epoxypropyl)-benzophenone [131], polymeric HALS **115** formed from bis(2,3-epoxypropyl)aniline and 4-amino-2,2,6,6-tetramethylpiperidine [146] or a brominated oligomer **116** used in combination with antimony trioxide as FR for PET [147] are other examples of polyaddition stabilizers.

3.1.3 Polycondensation

Numerous Ⓢ-functionalized polymers having stabilizing moieties bound in the main chain (type A, Scheme 1), as pending groups (type B) or crosslinks (type E) were synthesized by repeated condensation steps of bi- or polyfunctional reactants (at least one of them must be Ⓢ-functionalized). Polycondensation takes place in solution, melt or solid phase. Some systems were prepared by the phase transfer polycondensation. Low molecular weight compounds (e.g. water, alcohol, hydrogen halide, inorganic salts, etc.) are always split off during the process. Linear polycondensates differing in construction units and character of reactive end groups are formed from a bifunctional monomer bearing two different reactive functional groups (AB-polycondensates) or from two different monomers, each of them

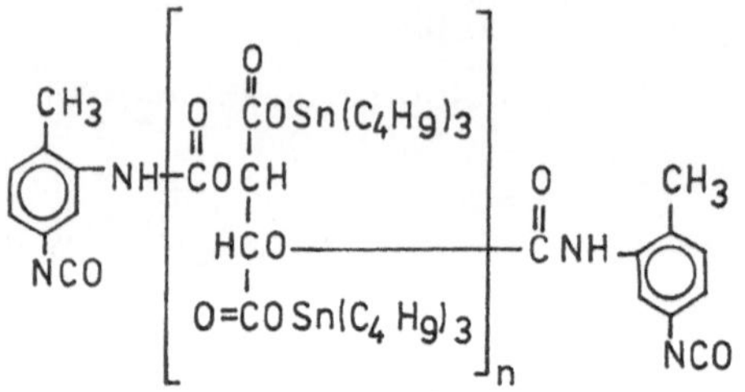

114

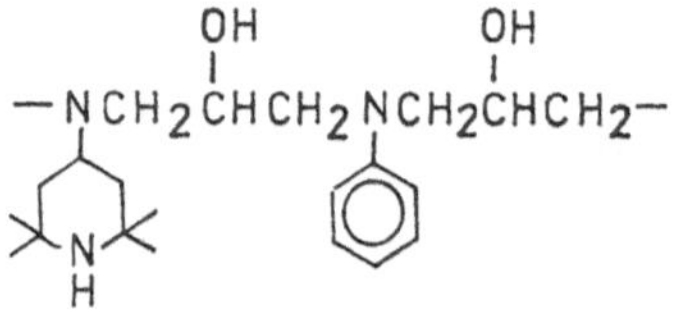

115

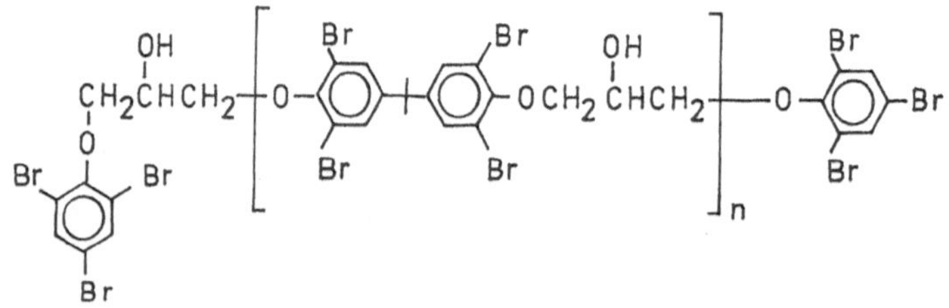

116

bearing two identical reactive functional groups (AA/BB polycondensates). Branched or crosslinked polycondensates are formed from polyfunctional monomers. Polycondensation takes place either directly between reactive monomers or in the presence of polycondensation agents (e.g. an oxidant or water). Depending on reaction conditions, statistical, alternating and block copolycondensates may be prepared. A large number of telechelic oligomers may be used for the synthesis [69]. It is sometimes difficult to distinguish the polycondensation from the polymer modification via reactions on polymers (Sect. 3.2), because the reaction intermediates are mostly not separated from the reaction mixture. Structures of some polycondensates are rather complicated, some products are not exactly defined.

3.1.3.1 Functionalized Products of Autocondensation and Oxidative Coupling

Various oligomeric condensates are formed from phenolic and aromatic amine AO and AOZ as a consequence of interactions of these reactive stabilizers with oxidizing species formed in stabilized polymers, e.g. with alkylperoxyls, alkyl hydroperoxides or ozonides [15–17]. Stabilizing efficiency is preserved in some of these compounds; the latter have been discussed however mostly in connection with mechanism of action of AO or AOZ [6].

Oligomers having molecular weight ~ 400 were prepared by oxidation of o-xylene with CuCl in the presence of AlCl$_3$ [148] and tested in combination with didodecyl 3,3-thiodipropionate as thermal stabilizers for PP.

Anthrone forms in polyphosphoric acid a polycondensate **117** having semi-conducting and paramagnetic properties. Compound **117** was tested as thermo- and photostabilizer of PS. It was concluded [149] that the antioxidant properties are due to the ability of **117** to form CTC with R$^\bullet$ and RO$_2^\bullet$ radicals.

Polycondensates having an irregular arrangement of phenylenediamine and quinone imine units were reported to be formed during oxidation of 1,4-phenylenediamine by anthraquinone or by oxygen or Cu(II) salts in aq. ammonia [150]. Diphenylamine may be oxidized similarly, products containing 5–14% of complex bound CuCl$_2$ were reported to have the properties of thermal stabilizers for PVC.

Various rather undefined polycondensates formed in mixtures of phenols and aromatic amines respectively and proposed as AO for rubbers have been mentioned in the literature.

Oxidative coupling was used for the synthesis of some stabilizers having properties of AO or FR. Products of oxidation of 4,4'-isopropylidenebis(2-methyl-6-tert-butylphenol with potassium ferricyanide [151] or of 4,4'-thiobis(2-tert-butyl-5-methylphenol) with oxygen in the presence of copper salts [152] (**118**) were tested as AO. Thermostable fireproofing additives containing 1 to 4 bromine atoms on a phenolic moiety and designed for the stabilization of thermoplastics, e.g. **119**, were prepared by oxidative coupling of brominated phenols [153].

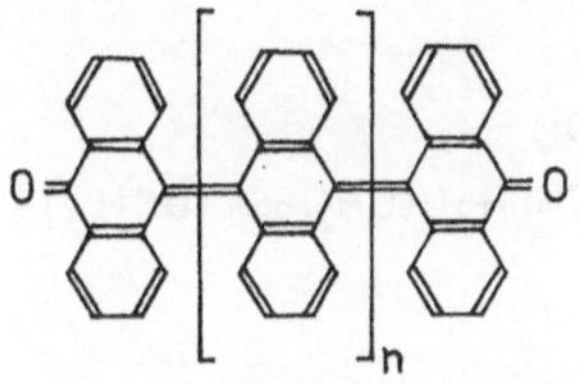

117

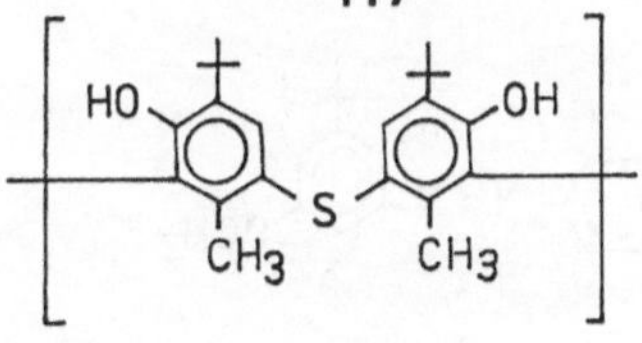

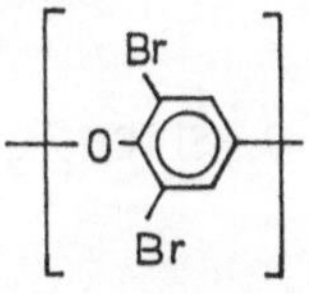

118

119

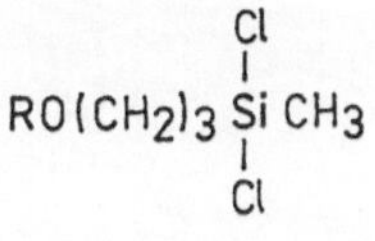

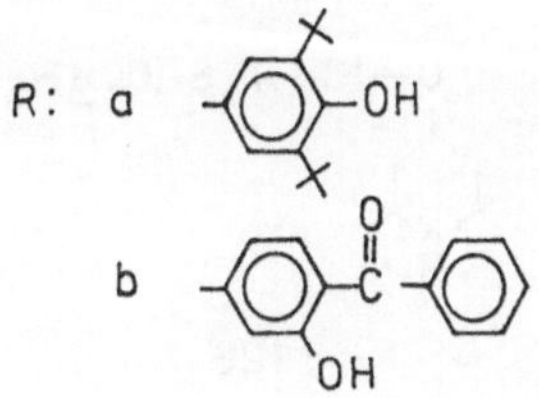

$$RO(CH_2)_3 \underset{\underset{Cl}{|}}{\overset{\overset{Cl}{|}}{Si}} CH_3$$

120

R: a

b

3.1.3.2 Functionalized Polysiloxanes

2,6-Di-*tert*-butyl-4-(dichloromethylsilylpropoxy)phenol (**120a**) and/or 2-hydroxy-4-(dichloromethylsilylpropoxy)benzophenone (**120b**) were bound in the presence of water into siloxane blocks of the butadiene-siloxane rubber. Products formed reduced oxidative crosslinking of butadiene blocks of the latter [154]. Polysiloxanes containing 2,2,6,6-tetramethylpiperidine moieties were prepared too.

Some organofunctional silane coupling agents are well compatible with polyolefins. This phenomenon was exploited for the enhancement of the stabilizer permanency [36]. *N*-Phenyl-3-aminopropyltrimethoxysilane (**121**) or a derivative of phenazine (**122**) were blended with PE. During the long-term weathering, hydrolysis and autopolymerization take place and a network structure **123** bearing

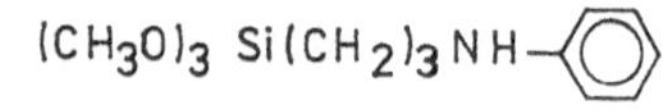

$(CH_3O)_3\,Si(CH_2)_3NH$—◯

121

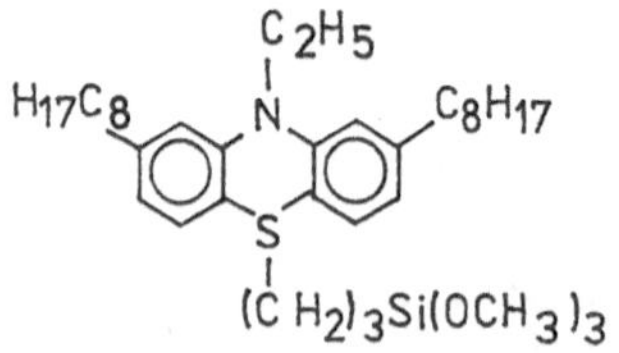

122

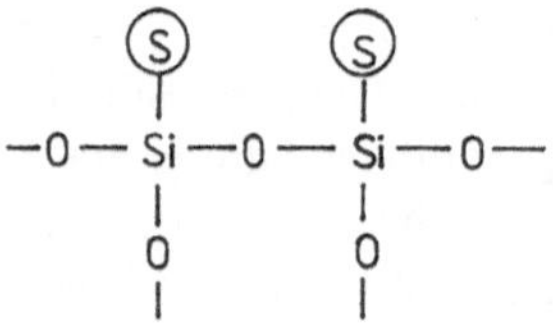

123

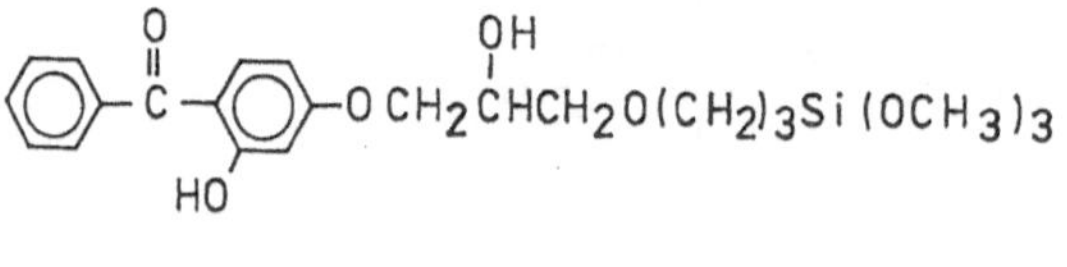

124

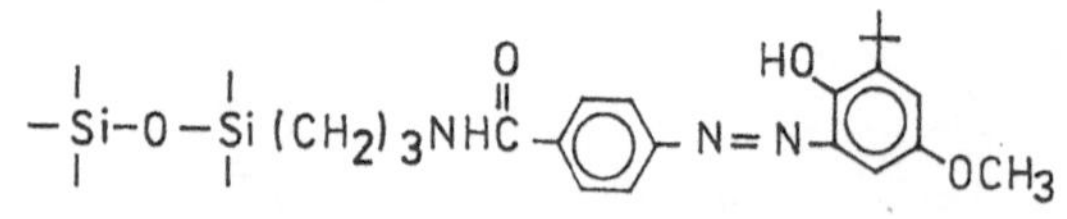

125

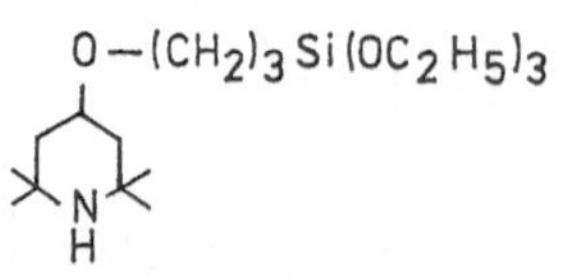

126

the respective functional moieties Ⓢ is created. The permanency of the LS **124** was enhanced in a similar manner [155]. An antioxidant immobilization for application in food stabilization was performed by means of a coupling process of a diazonium salt of the "arylamine glass" with 2-*tert*-butyl-4-methoxyphenol [156]. An active moiety **125** attached to the porous glass was obtained. HALS **126** bearing alkoxysilane groups was contacted with an inorganic filler, e.g. titanium dioxide in an inert solvent. The filler containing immobilized stabilizer was well compatible with LDPE [157].

3.1.3.3 Functionalized Condensates of Oxocompounds

Condensates of various low molecular weight compounds bearing stabilizing moieties with aldehydes and ketones were prepared. Mixtures of products are mostly formed. Condensates of fractions of some natural compounds, like Kraft's lignin, sugar cane bagasse lignin, phenolic fraction of gossypol or creosote bush *(Larix Tridentata)* with aliphatic aldehydes (mostly formaldehyde) do not posses defined structures. This disadvantage was observed also with some polycondensates prepared from various substituted phenols and aromatic amines. All these compounds have properties of antioxidants and have attracted attention since the earliest elucidation of stabilizers. Morawetz [158] was probably the first to perform a mechanistic study with linear phenol-formaldehyde polycondensates in oxidized paraffin white oil. Maximum antioxidant activity was reached in systems having three phenolic nuclei. The optimum activity was reached with systems substituted with bulky alkyls in the *ortho* position to phenolic hydroxyls and with methyls in the *para* position. Most condensates described in the literature have two or three, only exceptionally more than three phenolic nuclei and have been usually listed among high molecular weight AO.

Polycondensates were prepared from pyrocatechol, *o*-cresol, *p*-cresol, mixtures of the latter with 2,4-dimethylphenol,2-*tert*-butyl-4-methylphenol, di-*sec*-butylphenol or other monoalkylphenols and their mixtures with dialkylphenols or alkoxyphenols.

Novolaks **127** formed from *p*-cresol and 2-*tert*-butyl-4-methylphenol were studied in some details [159]. Phenolic groups present in **127** are not equivalent in the reactivity with $RO_2^{\cdot}$. A good antioxidant efficiency was observed in PS and poly(ethylene-*co*-propylene), even with polycondensates having n > 15. Polycondensate **128** is an example of AO which may be used in contact with food [160].

Other aldehydes used for the synthesis of phenolic polycondensates include butyraldehyde, furfural or chloral. Efficient AO, e.g. polycondensate **129**, were prepared with sulfur containing aldehydes [161]. Linear polycondensates of phenols and aldehydes were tested as AO in mineral oils, PE, PP, poly(ethylene-*co*-propylene), PS, PA and/or diene based rubbers [159, 162]. Cyclic phenolic condensates, calixarenes **(130)** posses interesting properties. Calixarenes were synthesized from 4-acyl-, 4-methyl-, 4-*tert*-butyl- or 4-phenylphenol in alkaline catalysed processes. Cycles containing 4 to 7 phenolic units were formed and tested as AO in PE or PP [163, 164]. Nickel(II) salt of **130** (R = *tert*-butyl, n = 1)

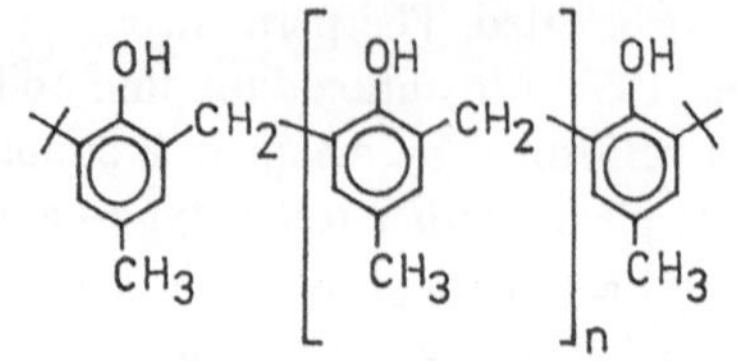

127

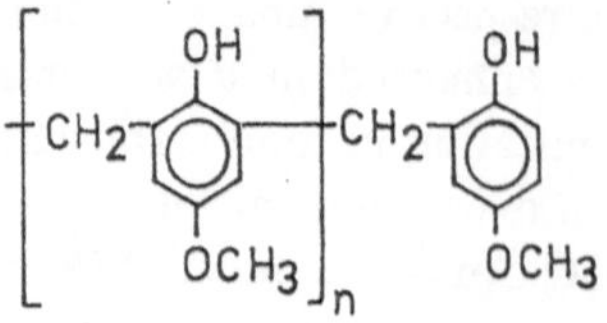

128

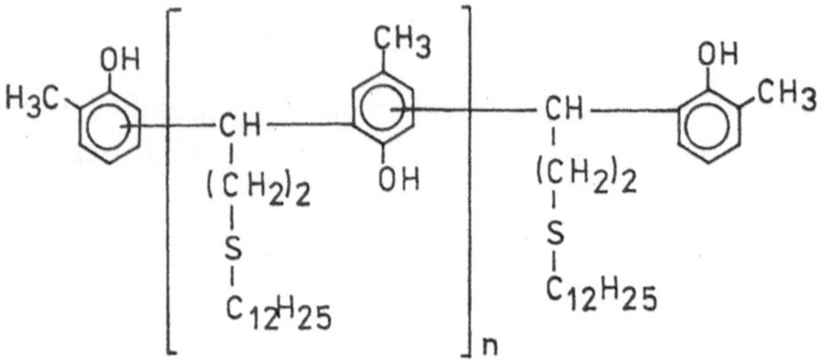

129

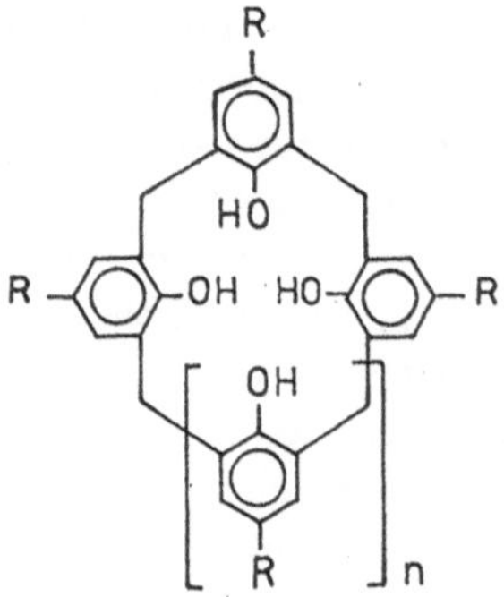

130

and oxidation products of **130** (R = *tert*-butyl, n = 1) with *tert*-butyl hydroperoxide were tested as LS for PE [163].

Aromatic amine antioxidants were prepared by condensation of aniline, aralkylated diphenylamine or 2,2,4-trimethyl-1,2-dihydroquinoline with formaldehyde [165]. The latter was also used for the synthesis of other stabilizers: a LS was formed via condensation with 2,4-dihydroxybenzophenone [166], a MD **131** with 2-hydroxy 4-methoxyacetophenone [167], another MD was formed with 8-hydroxyquinoline [168] and a BIO-S **132** with 4-aminosalicylic acid [169].

Polyfunctional polycondensates were prepared from formaldehyde and a mixture of two different low molecular weight stabilizers, e.g. of 2-hydroxybenzophenone or its derivatives with various 4-alkylphenols [170], e.g. stabilizer **133** (a combination of LS and AO), of a phenol with benzoxazole [171] or 4-nonylphenol and dodecylmercaptan [172] **134** (a combination of CB AO and HD AO). A polycondensate **135** having metal deactivating properties was prepared by condensation of methylenebis(salicyl aldehyde) with methylenebis(salicyl hydrazide) [173].

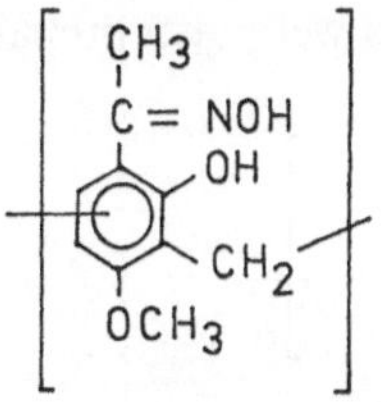

131

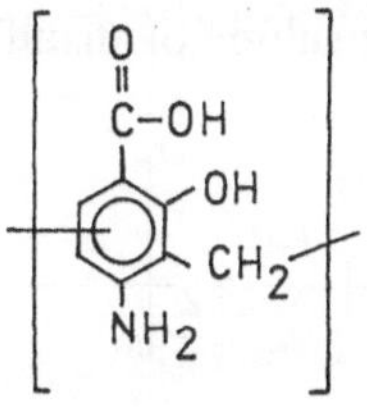

132

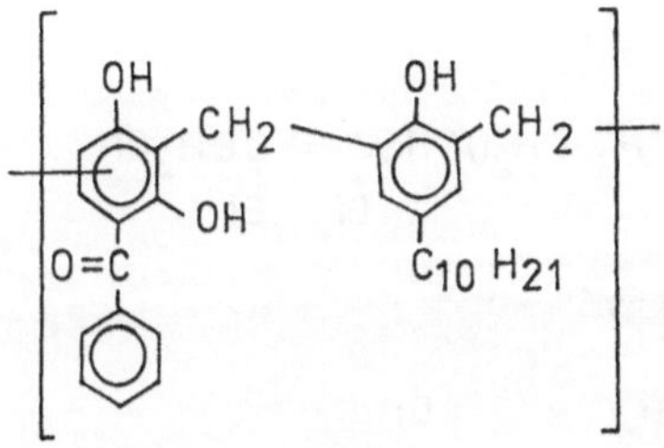

133

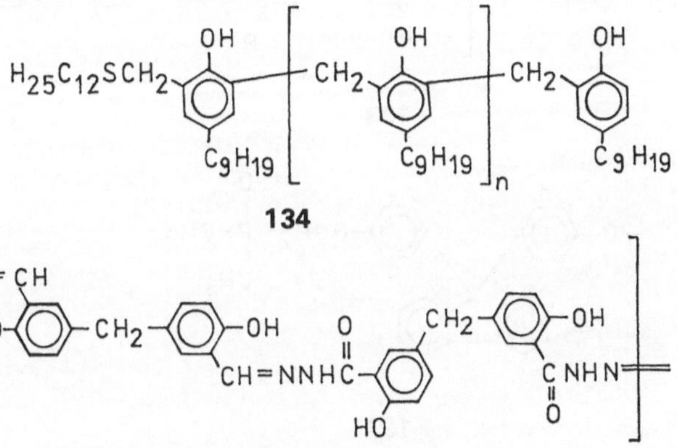

134

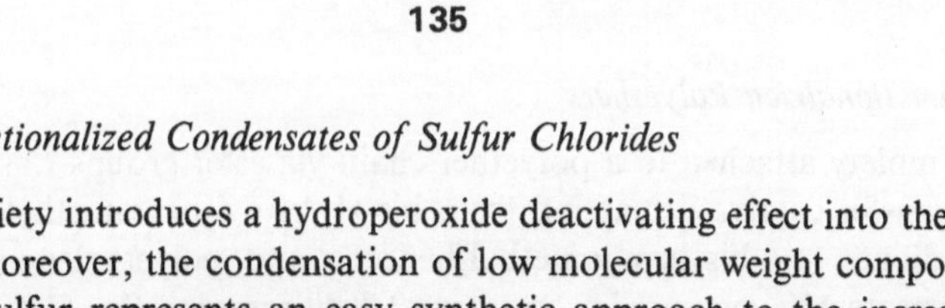

135

3.1.3.4 Functionalized Condensates of Sulfur Chlorides

A sulfur moiety introduces a hydroperoxide deactivating effect into the stabilizer molecule. Moreover, the condensation of low molecular weight compounds with halides of sulfur represents an easy synthetic approach to the increase of the molecular weight and consequently to the improvement of the physical persistency of stabilizers. Efficient AO were prepared by this way from alkylphenols, aromatic amines or phenothiazine [21, 174]. Most of them contain two active units connected with a sulfide/disulfide bridge and should be listed rather among high molecular weight stabilizers. Antioxidant **136** (n > m), prepared from pyrocatechol and sulfurmonochloride [175], increases stability of NR and SBR without affecting tack or viscosity of rubber compounds. Polycondensate of 4,4'-dihydroxy-3,3'-dinitrodiphenyl sulfone with sulfur monochloride was proposed [176] as a thermal stabilizer for radiation crosslinked PE. Some analogous condensates of aromatic diamines containing sulfide or disulfide bridges were also prepared.

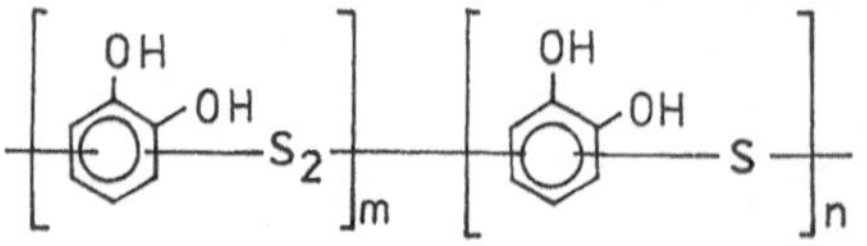

136

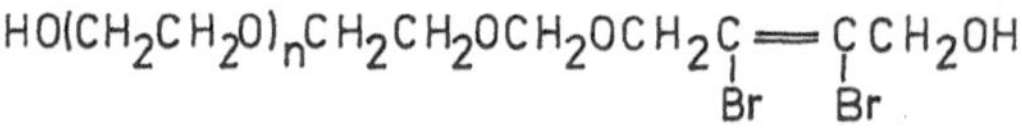

137

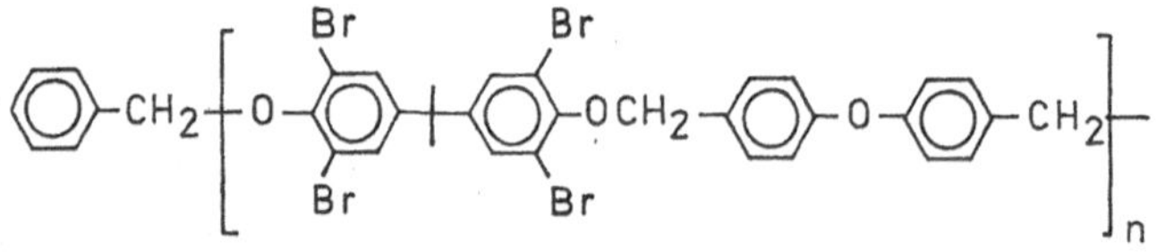

138

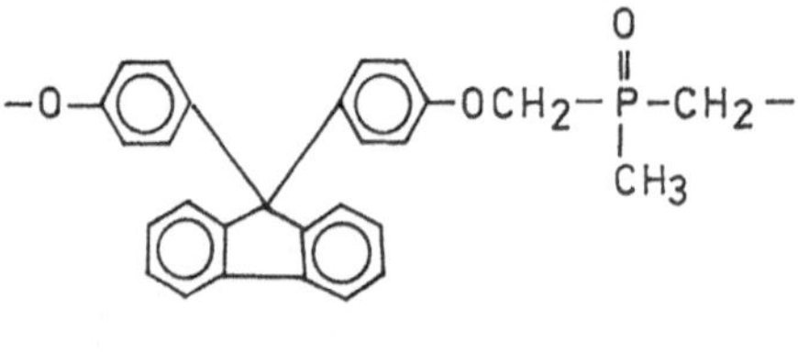

139

3.1.3.5 Functionalized Polyethers

A HALS moiety attached to a polyether chain via ester groups was obtained by a phase-transfer catalysed reaction from bis-(1,2,2,6,6-pentamethyl-4-piperidyl)-propane dioate, 1,4-bis(bromomethyl)benzene and polyethylene glycol [177]. Polycondensate **137**, prepared from *trans*-2,3-dibromobut-2-en-1,4-diol, formaldehyde and poly(ethylene glycol), was used as a reactive component for flame retarded PUR foams [178]. Aromatic functionalized polyethers can be obtained by step growth polymerisation of a substituted diphenoquinone with an oligo-(phenylene oxide) [69]. A brominated polyether **138**, containing approx. 40% of bromine, was prepared from bis(4-chloromethyl)diphenylether, 4,4'-isopropylidenebis(2,6-dibromophenol) and benzylchloride [179] and tested as FR in Novodur PX. Flame resistency of various aromatic polyethers was increased [180] using built-in methyl(bischloromethyl)phosphine oxide derived moiety, e.g. in **139**.

3.1.3.6 Functionalized Polyesters

Esterification and transesterification were used for the synthesis of numerous polymeric stabilizers derived from carboxylic as well as inorganic acids. Systems obtained with poly(alkylene ether)diols contain polyether and polyester links.

AO containing various phenolic moieties were prepared by transesterification in the presence of tetraalkyl titanates. Randomly distributed Ⓢ-active moieties are characteristic of **140** (only the hard polyester segment is given) prepared from dimethyl terephthalate, 1,4-butanediol, poly(tetramethylene oxide)diol and dimethyl 5-(3,5-di-*tert*-butyl-4-hydroxybenzenepropaneamido)isophthalate [181]. The mentioned polymeric AO was used for stabilization of polyether-polyester elastomers. A partial attachement of tetrakis[methylene 3(3,5-di-*tert*-butyl-4-hydroxyphenyl)propionate]methane (**3**) via transesterification reaction was expected in the synthesis of another polyether-polyester elastomer by [182]. A reversible redox polyester was formed from 2,5-bis(2-hydroxyethyl)hydroquinone and dichlorides of aliphatic dicarboxylic acids [137].

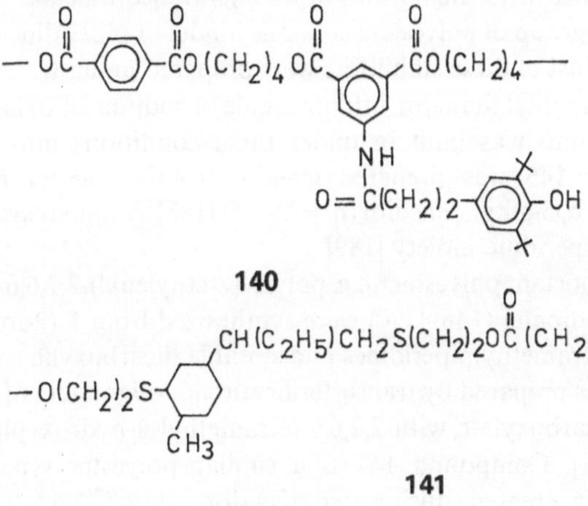

140

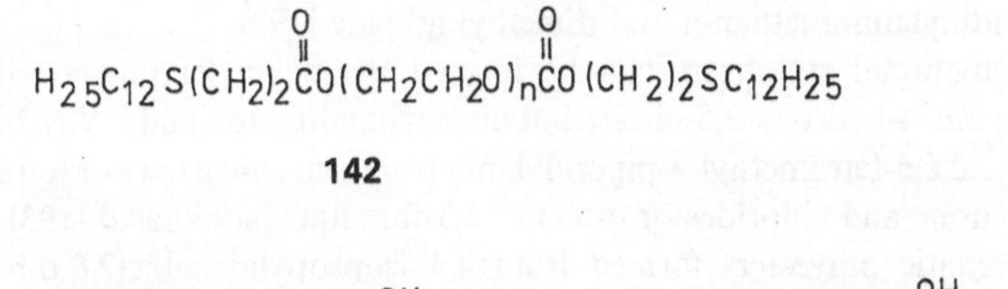

141

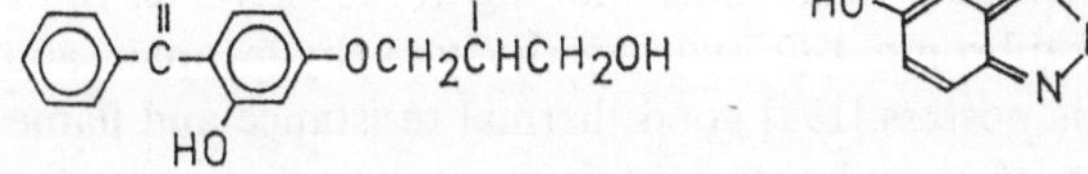

142

143

144

Some sulfur containing polyesters having properties of synergistic AO were described, like a product formed from 2-(2-hydroxyethylthio)-1-methyl-4-[2-(2-hydroxyethylthio)-1-methylethyl]cyclohexane and dimethyl 3,3'-thiodipropionate [183] (141), an esterification product of 3-n-dodecylthiopropionic acid with poly(ethylene glycol) [184] (142) or a polycondensate of sebacoyl chloride with 4,4'-dimercaptodiphenyl sulfide [185].

Various polyester type LS were synthesized and tested mostly in polyesters. 2-Hydroxybenzophenone moiety was fixed as a terminal group during the synthesis of aromatic polyesters (the active reactant was 2,4-dihydroxybenzophenone) [186]. Films or coatings having good UV resistance were prepared by incorporation of 4-(2,3-dihydroxypropoxy)-2-hydroxybenzophenone (143) or analogous reactive benzophenones into a reaction mixture consisting of dimethyl terephthalate and ethylene glycol [139]. Methyl 3-(4-acetoxyphenyl)-2-cyano-2-propenoate was built in during a reaction of dimethyl terephthalate with ethylene glycol [187]. An intensive study was performed with polyesters containing built in 2-(2-hydroxy-phenyl)benzotriazole moiety [89]. 2-(2,4-Dihydroxyphenyl)-2H-5-hydroxybenzo-triazole (144) or its analogue containing the second 2-benzotriazolyl group attached to the phenolic moiety were transformed into linear and soluble polyesters containing 1–3% of benzotriazole moiety using methyl esters of aliphatic or aromatic dicarboxylic acids and appropriate glycols. Crosslinked and branched polyesters were formed with dichlorides of the mentioned diacids.

A comprehensive group of polyesters contains hindered piperidine or piperazine (HALS) moieties. Most of these stabilizers were prepared under transesterification conditions, using tetraalkyl titanates, lithium amide or sodium alkoxide as catalysts. Terminal HALS group was built in under these conditions into a polyether-polyester. Polyester 145 was prepared from a reactive diester derived from piperazinedione and a,ω-alkylidenediol (n = 2–15) [188]. A similar system contains 2,2,6,6-tetramethylpiperidine moiety [189].

Commercially important polyesters, e.g. poly[1-(2-ethylenyl)-2,2,6,6-tetramethyl-4-piperidinylbutane dioate] (146) [190] were synthesized from 1-(2-hydroxyethyl)-4-hydroxy-2,2,6,6-tetramethylpiperidines and suitable dicarboxylic acids. Another polymeric HALS was prepared by transesterification of oligoesters of tetramethyl-butane-1,2,3,4-tetracarboxylate with 2,2,6,6-tetramethyl-4-hydroxypiperidine and 1,10-decanediol [191]. Compound 147 is a similar polyester type HALS. An ester-amide chain is created during esterification of 2-(2,2,6,6-tetramethyl-4-piperidinylamino)ethanol and dimethyl adipate [192].

Bifunctional stabilizers can be formed by means of polyesterification from diethylbis(4-hydroxy-3,5-di-tert-butylbenzyl)malonate and N,N-bis(2-hydroxy-ethyl-2,2,6,6-tetramethyl-4-piperidylamine or from trihydroxy or tetrahydroxyben-zophenone and chlorides or esters of 3,3-thiodipropionic acid [193].

Aromatic polyesters formed from 4,4'-isopropylidenebis(2,6-dibromophenol), from various analogous bisphenols differing in the character of the bridge [94] or from 2,2'-bis[3,5-dibromo-4-(2-hydroxyethoxy)phenyl]propane and terephthalic or isophthalic acids possess [194] good thermal resistance and flame retardance.

Incorporation of phosphorus containing moieties increases thermostability or flame resistance of polyesters. Functionalized 1,4-dihydroxynaphthalene (148) was

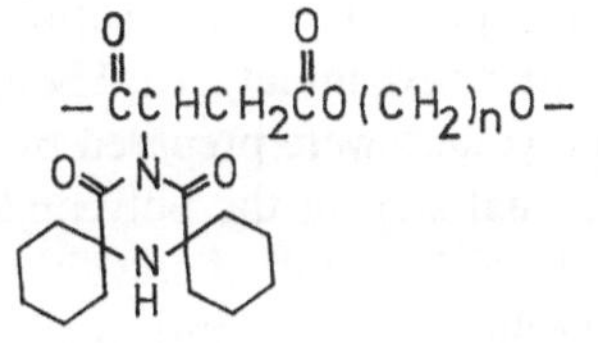

145 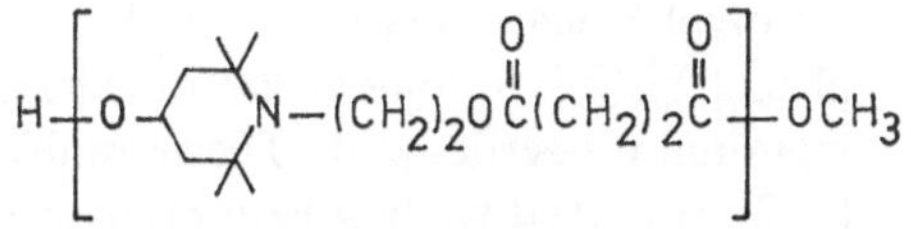   **146**

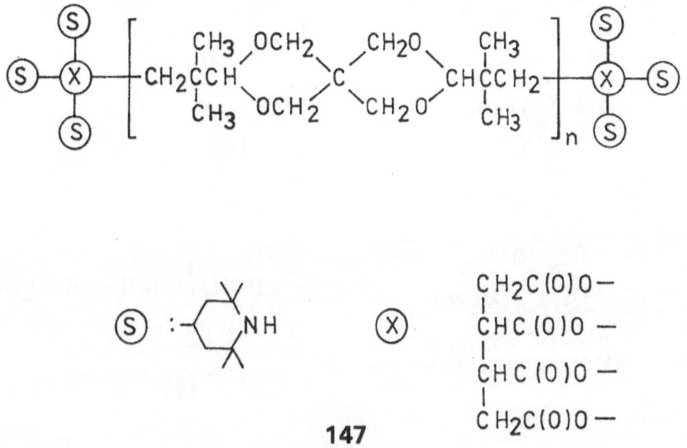

147

esterified with isophthalic acid and a fire resistant aromatic polyester was thus obtained [195]. A dimeric flame retardant additive **149** obtained from phenyl(hydroxymethyl)phosphinic acid was incorporated into a precondensate of ethylene glycol and dimethyl terephthalate [196]. Self-extinguishing and thermally stable polyesters were prepared from 2,8-dichloroformyl-10-phenylphenoxaphosphine-10-oxide (**150**) and diphenyldiols, alkylidenebisphenols or thiobisphenols using an interfacial polycondensation process [197]. 2-Hydroxyethoxyethyl *a*-diethyl phosphonosuccinate (**151**) was treated with bis(2-hydroxyethyl terephthalate); a polyester resin with increased thermostability was obtained [198].

Various polyesters derived from phosphorous or phosphoric acids were prepared. Efficient polyphosphites were synthesised in the early 1960s. Polyphosphite prepared from **152** and 4,4′-isopropylidenebis(cyclohexanol) was tested as a thermal stabilizer for PC [199] or as secondary AO for radiation sterilized EPM [200]. Built-in phosphites obtained by transesterification of trialkylphosphites with 4,4′-isopropylidenebisphenol or 4,4′-thiobisphenol possess antioxidant properties in polyolefins. Stabilizer containing phosphite moiety **153** was prepared from tris(2-hydroxyethyl)isocyanate, decyl alcohol and triphenylphosphite [201]. Various phosphites were derived from polynuclear phenols or dihydric phenols. For example, a polycondensate prepared by reaction of phosphorus trichloride with 2,5-di-*tert*-butylhydroquinone was tested as heat and light stabilizer for PP [202]. A linear polyester with a built-in phenolic moiety was synthesised from (2,6-di-*tert*-butyl-4-methylphenyl)bis(6-hydroxyhexyl)phosphite and dimethyl terephthalate [203].

Polyphosphates derived from 4,4'-dihydric phenols or 4,4'-isopropylidenebis-phenol, bis(2,6-dibromophenol) and having an active moiety like **154** impart flame retardant properties [204]. Thermostable polyesters were prepared by addition of bis(2-hydroxyethyl)phosphate prior to the final step of the polycondensation of ethylene glycol with dimethyl terephthalate [205]. Zinc salt of polydithiophosphate obtained by reaction of phosphorus pentasulfide with 4-substituted anisole was tested as AO in lubricating oils and greases.

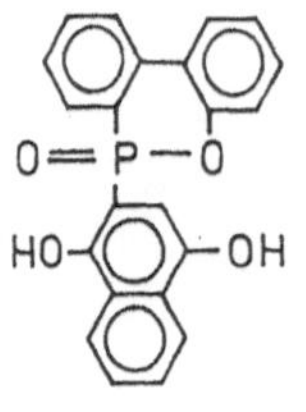

148

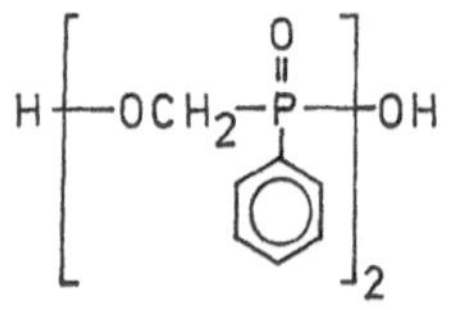

149

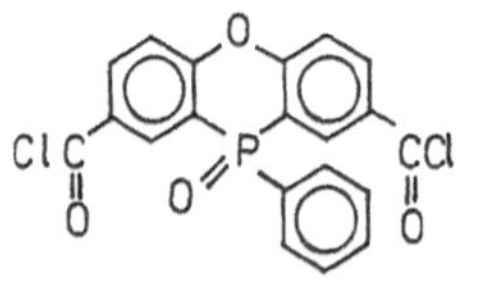

150

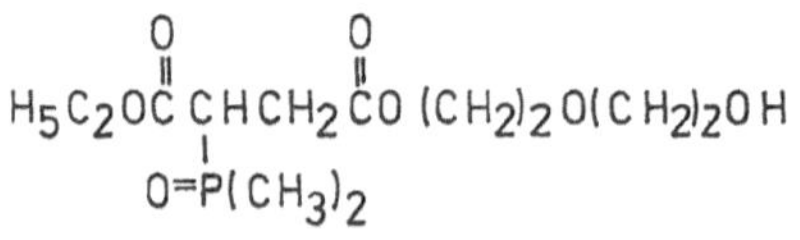

151

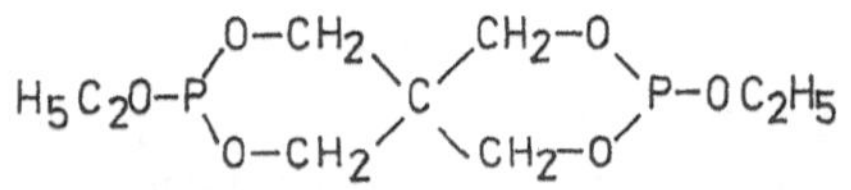

152

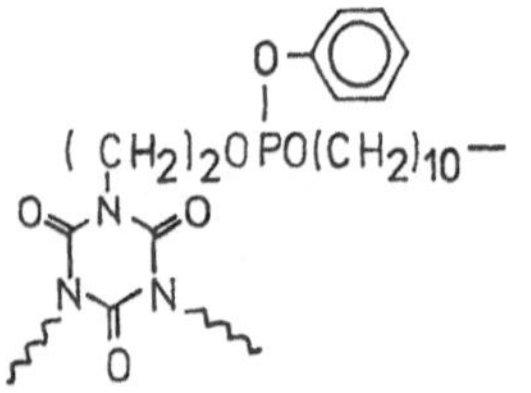

153

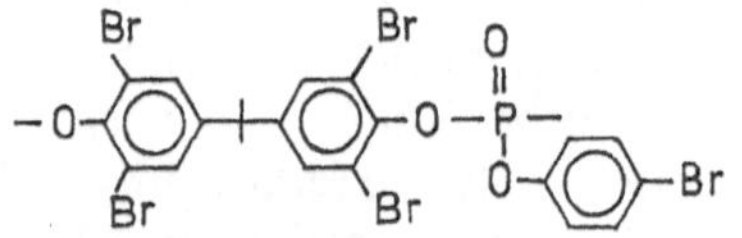

154

Polyesters obtained by interfacial polycondensation of aliphatic and aromatic diacids with dialkyl lead dichloride or dialkyl stannum dichloride possess biocide properties [105]. Ferrocene moieties may be also incorporated.

3.1.3.7 Functionalized Polycarbonates

Various oligomeric and polymeric stabilizers containing PC units were synthesised. Oligomers prepared from phosgene and 4,4'-isopropylidenebis(2-*tert*-butylphen-ol), or 2,5-bis(2-hydroxyethyl)hydroquinone, from diphenyl carbonate and 4,4'-isopropylidenebisphenol, 4,4'-butylidenebis(6-*tert*-butyl-3-methylphenol), or sub-stituted monohydric and dihydric mononuclear phenols, e.g. 1-(3,5-di-*tert*-butyl-4-hydroxyphenyl)-3,3-bis[(3-*tert*-butyl-4-hydroxyphenyl)butane] carbonate (**155**) or more complicated analogue, 1,1,3-tris(2-methyl-4-hydroxy-5-*tert*-butylphenyl) butane carbonate are typical examples of AO for polyolefins [113, 137, 206]. 4,4'-Isopropylidenebisphenol based PC containing terminal or built-in anthraquinone units derived from 1,4-dihydroxyanthraquinone (156) exhibited good antioxidant activity. Trithiocarbonate 157 prepared from 1,9-dichloro-5-oxanonate and sodium tricarbonate by phase transfer catalysis [207] was used for compounding with NR. Permanently incorporated LS, e.g. 158, were prepared from 4,4'-isopropylidenebisphenol, diphenyl carbonate and 2-(2,4-dihydroxyphenyl)-2*H*-4-hydroxybenzotriazole or by transesterification of a polycarbonate precondensate in the melt with the respective benzotriazole derivative [208]. Various PC containing units derived of 4,4'-isopropylidenebis(2,6-dibromophenol), e.g. 159, have properties of FR [209].

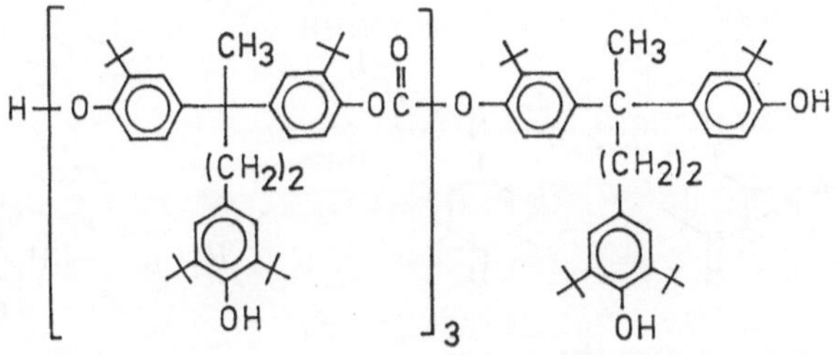

155

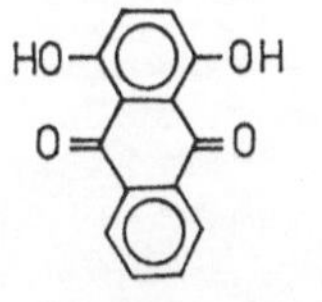

156

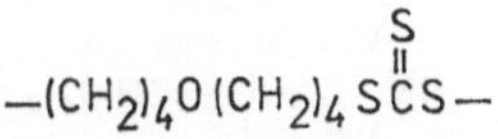

157

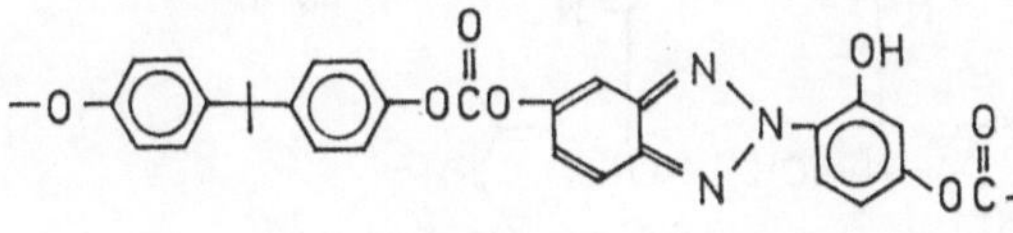

158

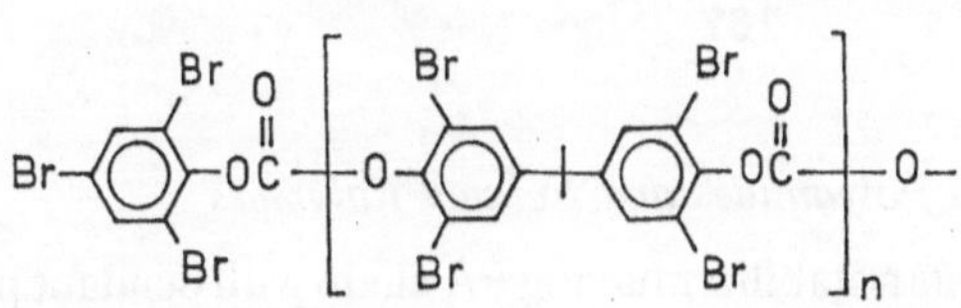

159

3.1.3.8 Functionalized Polyamines

Condensates prepared from various monoalkyl or monoaralkylphenols and hexamethylenetetramine possess phenolic moieties linked with $-CH_2NHCH_2-$ bridges and may be used as AO in rubbers. Commercially very important oligomers designed for stabilization of hydrocarbon polymers and PA contain aliphatic open chain and/or 1,3,5-triazine bridges and 2,2,6,6-tetramethylpiperidine moieties [210], e.g. **160, 161 a, b.** Macrocyclic HALS **162** was used for stabilization of blends of polyphenylene oxide with rubber modified PS [211]. Efficient LS having polyethylene-polyamide chains and pendant substituted 2-oxopiperazine moieties attached via 1,3,5-triazine were synthesised by the Goodrich Co. [212].

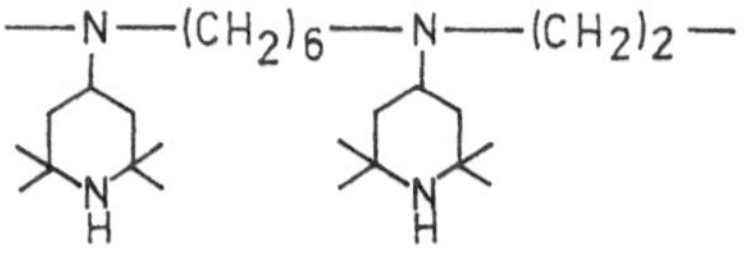

160

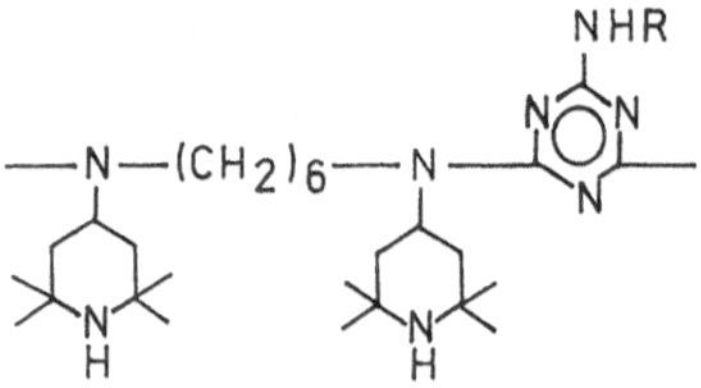

161

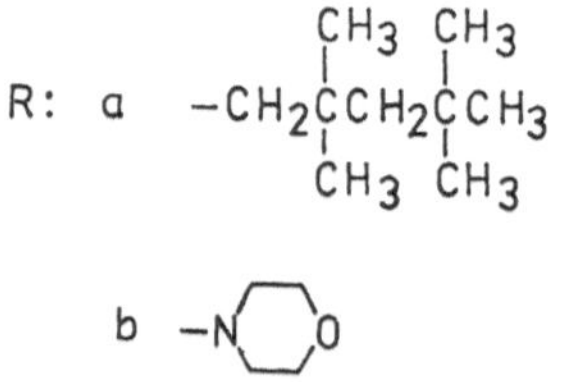

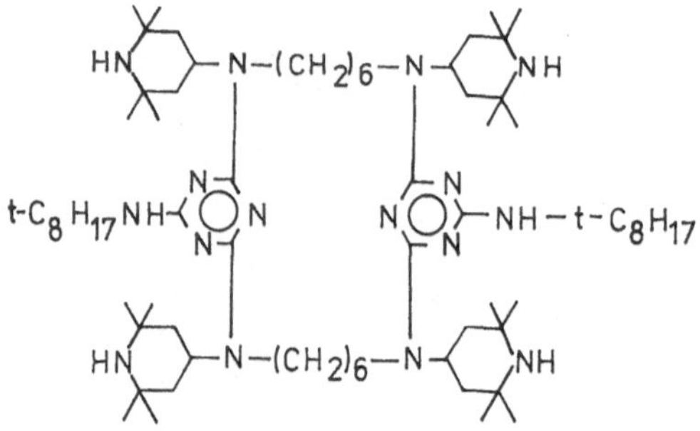

162

3.1.3.9 Functionalized Polyamides and Related Polymers

Oligomer **163** represents a stabilizer having PA chain with pendant phenolic moiety [31]. Pendant hindered piperidine or piperazine moieties were attached to oligomeric stabilizers having polysulfonamide, polyurea (e.g. **164**), or PA (e.g. **165**) unis [213]. Polyhydrazide **166** was tested as HD AO and copper deactivator in PP [214]. Poly(nitrophenylene-carbazide disulfide) **167** was prepared for thermal stabilization of PVC. Phosphorus containing crosslinkable polymers having polyamide, polyimide and polyurea chains were prepared for flame and heat resistant applications [215].

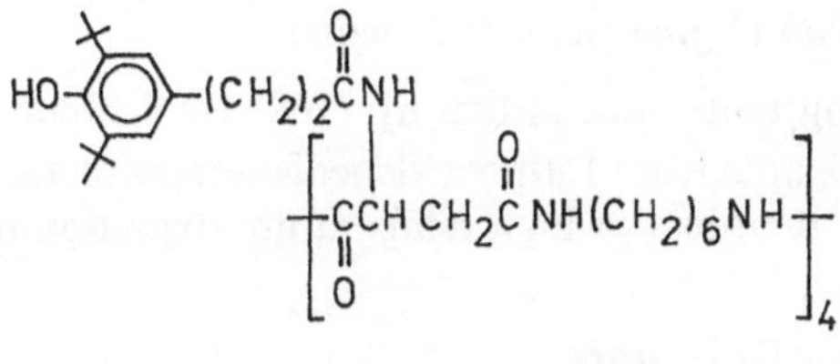

$$HO--(CH_2)_2\overset{\overset{O}{\|}}{C}NH$$

$$\left[\begin{array}{c}\overset{}{C}CH\,CH_2\overset{\overset{O}{\|}}{C}\,NH(CH_2)_6NH\\ \underset{O}{\|}\end{array}\right]_4$$

163

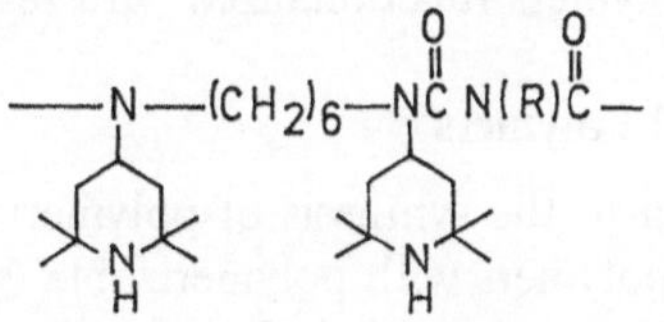

$$-N-(CH_2)_6-\overset{\overset{O}{\|}}{N}C\,N(R)\overset{\overset{O}{\|}}{C}-$$

164

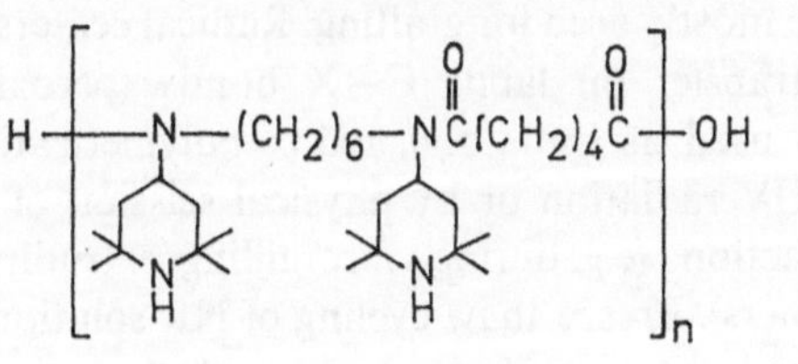

$$H\left[-N-(CH_2)_6-\overset{\overset{O}{\|}}{N}C[CH_2]_4\overset{\overset{O}{\|}}{C}\right]_n OH$$

165

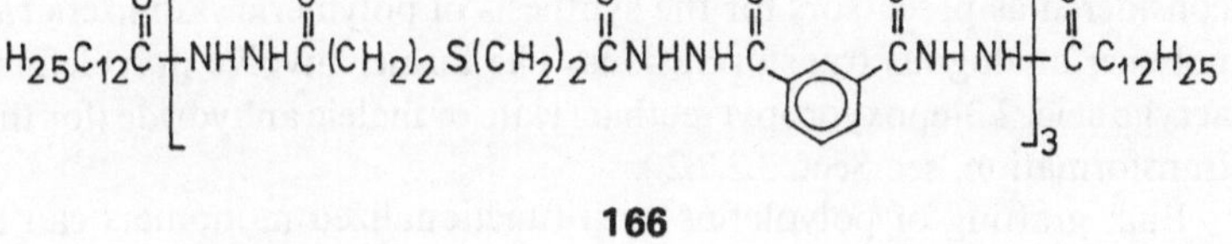

$$H_{25}C_{12}\overset{\overset{O}{\|}}{C}\left[-NHNH\overset{\overset{O}{\|}}{C}(CH_2)_2\,S(CH_2)_2\,\overset{\overset{O}{\|}}{C}NHNH\overset{\overset{O}{\|}}{C}--\overset{\overset{O}{\|}}{C}NHNH\right]_3\overset{\overset{O}{\|}}{C}\,C_{12}H_{25}$$

166

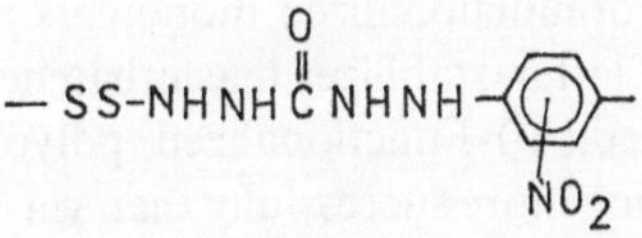

$$-SS-NHNH\overset{\overset{O}{\|}}{C}\,NHNH--NO_2$$

167

$$--Sb-$$

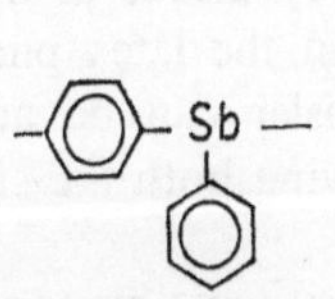

168

3.1.3.10 Systems with Organometallic Moieties

Polymers containing units with antimony (**168**) were prepared by treatment of diazonium salts prepared from 1,4-phenylenediamine with antimonium trichloride [216] and tested as retarders of high temperature oxidation of polymers.

3.2 Reactions on Polymers

This section includes syntheses of stabilizers by means of chemical transformations of conventional polymers or of polymers functionalized with reactive groups.

3.2.1 Grafting Onto Conventional Polymers

Grafting is an important approach to the synthesis of polymeric stabilizers and implies reactions of conventional polymers with polymerizable Ⓢ-functionalized molecules. Depending on the grafting technique, bulk or surface stabilization of polymers can be achieved. Macromolecular stabilizers of the general types D and E (Scheme 1) having attached side chains consisting of Ⓢ-functionalized units are formed. Radical processs are mostly used for grafting. Radical centers on polymers can be formed by chain transfer on labile $C-X$ bonds (peroxides and hydroperoxides or AIBN are used as initiators), redox polymerization, energetic radiation, photosensitized UV radiation or by physical scission of $C=C$ bonds during mechanochemical reactions (e.g. during vibromilling, extruding, high speed stirring or shearing of polymers). Freeze-thaw cycling of NR solutions or swelling of NR by volatile solvents may be also used for generation of free radicals [217, 218].

The number and the length of the grafted side chains are dependent on both the monomer concentration and initiator efficiency. Some grafted polyolefins provide functionalized polymers with pendant reactive groups and may be considered as precursors for the synthesis of polymeric stabilizers *via* consecutive polymer analogous transformations. The former involve polyolefins grafted with acrylic acid, 2,3-epoxypropyl methacrylate or maleic anhydride (for the consecutive transformation, see Sect. 3.2.2.2.).

Bulk grafting of polyolefins with functionalized monomers can be performed by means of conventional processing equipment. Inherently stabilized polyolefins are formed when low concentrations of functionalized monomers are used [218]. Using high concentration levels of the latter, stabilizer masterbatches for blending with undoped polyolefins are created. Ⓢ-Functionalized polyolefins can be prepared by means of the radical grafting more successfully than via copolymerization.

NR and synthetic diene based elastomers are a very versatile material for grafting. Chain transfer grafting may be carried out by means of a functionalized monomer-initiator system in a rubber solution, in the latex phase, in swollen rubber or during vulcanization. The chain transfer can be accompanied in diene-based polymers by addition reactions involving both the starting radicals and growing polymer radicals [217].

Some typical examples of polymeric stabilizer systems prepared by peroxide initiated grafting are mentioned. 4-Vinyl-2,6-di-*tert*-butylphenol was grafted on

HDPE wax, 3,5-di-*tert*-butyl-4-hydroxybenzyl methacrylate on NR or BR, N-(3,5-di-*tert*-butyl-4-hydroxyphenyl)acrylamide on SBR or NBR, an urethane moiety containing phenolic acrylate on poly[butadiene-co-styrene-co-itaconic acid] latex, 3-(4-hydroxy-3,5-di-*tert*-butylphenyl)-1-methylpropyl methacrylate on poly(acrylonitrile-co-butadiene) [218–220] (**169**). A specific attention should be attributed to [2-(2-hydroxy-3-*tert*-butyl-5-methylbenzyl)-4-methyl-6-*tert*-butyl phenyl]acrylate (type **27**), an important elastomer reactive AO (activity mechanism was studied in details by Yachigo et al. [221]. Phenolic and aromatic amine AO moieties can be incorporated into HDPE by means of 2-(subst. arylamino)-4,6-diallyloxy-1,3,5-triazine (stabilizers **170 a, b** are formed) and analogous compounds having Ⓢ-functionalized reactive 1,3,5-triazine moiety, or N-phenyl-N,N'-diallyl-1,4-phenylenediamine [222]. 1-(4-Anilinophenyl)butadiene was built-in into NBR in the presence of N-cyclohexyl-2-benzothiazole during curing [233] at 160 °C. N-(4-Anilinophenyl)methacrylamide, its styrenated derivative or 3-N-(4-anilinophenyl)amino-2-hydroxypropyl methacrylate (**171**) can be grafted onto NR, SBR or NBR [224]. Grafting with aromatic amine monomers was reported to be less efficient than that with phenols. SBR grafted with 2-{[3-(dodecylthio)-2-methylpropionyl]oxy}ethyl methacrylate or its analogues, e.g. (**172**) operates as HD AO [225].

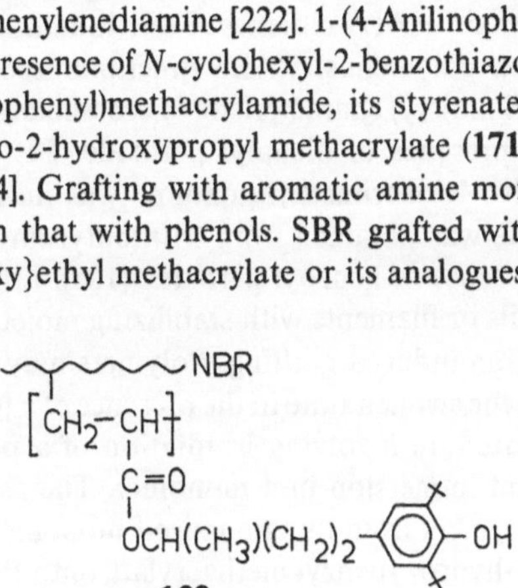

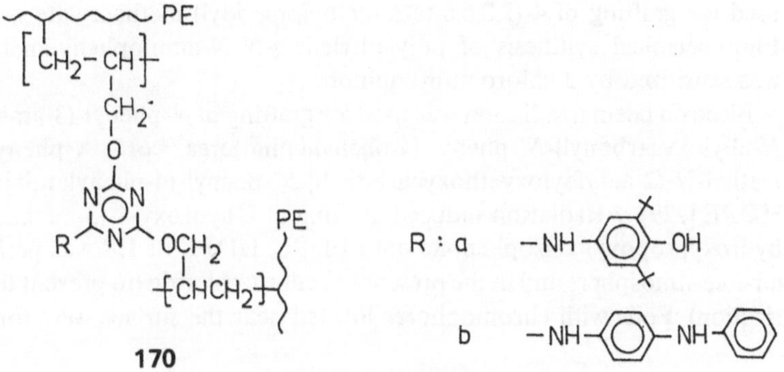

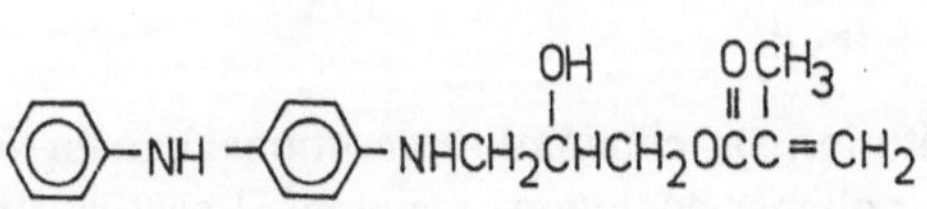

A great attention was paid to grafting of light stabilizers onto polyolefins. 2-Hydroxy-4-methacryloyloxybenzophenone was grafted onto LDPE, HDPE, PP, poly(4-methylpent-1-ene) or ABS in extruders in the presence of peroxides or onto LDPE or PP in the melt without peroxides. Grafting on PS was very efficient,

~~~~~~~~ SBR   ~~~~~~~ EVA , BR

$$\left[CH_2-\underset{\underset{\underset{OCH(CH_3)_2CH_2SC_{12}H_{25}}{|}}{\overset{\overset{C=O}{|}}{C}}}{C}(CH_3) \right]$$

| 172 | 173 |

the methacryloyloxy derivative was more reactive in comparison with the respective acrylate. Grafted copolymer **173** was obtained from 2-(2-hydroxy-5-vinylphenyl)-2*H*-benzotriazole and poly(ethylene-co-vinylacetate), PP or poly(butadiene) [56, 76]. Various grafted systems contain HALS moieties: Functionalized acrylates or maleates were melt-grafted onto polyolefins. HALS functionalized system analogous to **170** (R contains HALS moiety) characteristic of crosslinks, was obtained by melt-grafting of PE in the presence of dicumyl peroxide. PP grafting with 1-methacryloyloxyethyldecahydro-2,2,4-trimethylquinoline was initiated by *tert*-butyl peroxide. A grafted biocide was obtained from tri(*n*-butyl tin) acrylate and unsaturated polyester in the presence of benzoyl peroxide [105].

Modification of surfaces of foils or filaments with stabilizing moieties represents the main application of irradiation induced grafting. Polymers may be irradiated either in solution, emulsion or in the swollen state in the presence of a functionalized monomer or by an indirect procedure involving irradiation of a polymer in an inert atmosphere and subsequent immersion in a monomer. The second process minimize formation of homopolymers from functionalized monomers.

Grafting of 3,5-di-*tert*-butyl-4-hydroxybenzyl methacrylate onto PP films under UV irradiation was sensitized with benzophenone [218]. A similar process was used for grafting of 4-(2,2,6,6-tetramethylpiperidyl) methacrylate onto PP [228]. Photochemical synthesis of poly[ethylene-g-*N*-(4-aminophenyl)methacrylamide] was sensitized by 2-chloroanthraquinone.

Electron beam irradiation was used for grafting of *N*-phenyl-(3-amino-6-methyl-*N*-allyloxycarbonyl)-*N'*-phenyl-(4-phenylamino)urea   or   *N*-phenyl[3-amino-6-methyl-*N*-(2-acryloyloxyethoxycarbonyl)]-*N'*-phenyl-(4-phenylamino)urea   onto HDPE [229]. γ-Radiation induced grafting of 2-hydroxy-4-(3-methacryloyloxy-2-hydroxypropoxy)benzophenone onto LDPE, HDPE or PP was performed in an air-free atmosphere and in the presence of cupric chloride (to prevent homopolymerization). Foils with chromophores located near the surface were formed [218].

3.2.2 General Chemical Methods

Most of addition, substitution or elimination reactions known from organic chemistry can be used in polymer-analogous reactions. Reactive sites along the polymer chain or end reactive groups are exploited. Reactions with low molecular weight compounds carrying stabilizing moieties account for a very effective approach to the synthesis of physically persistent stabilizers. Various systems were prepared by modification of conventional commercialized polymers. The main attention has been paid to processes providing chemical binding of stabilizing moieties during customary operations, e.g. melt-processing, processing of latexes, mastication or vulcanization of rubbers. Much attention has been paid to the modification of polymers bearing reactive moieties. Most of the reactions used are of intermolecular character. Specific differences between chemical transformations of low molecular weight reactants and polymer analogous reactions are due to the macromolecular character of one of the co-reactants taking part in the process.

Numerous synthetical possibilities exist theoretically. However, only some of them are of potential economic interest. Stabilizers of types B, C, E, F and G, respectively, (Scheme 1) can be formed. Some of them are formally analogous to stabilizers prepared by polyreactions.

3.2.2.1 Reactions of Conventional Polymers with Low Molecular Weight Compounds Carrying Stabilizing Moieties

Some post-polymerization treatments of conventional polymers are very attractive. In the ideal case, a commercially available polymer should be modified with a non-hazardous reactant carrying a stabilizing function. The treatment should proceed without any substantial deviation from the conventional technology and with unfavourable influence on the processing operation.

Attachment of Transformation Products of Stabilizers. Up-to-date knowledge dealing with the chemistry of transformation products of phenolic [6, 15, 17, 20] and aromatic aminic [16, 43, 230] antioxidants and photoantioxidants based on hindered piperidines [10] indicates the possibility of attaching compounds having structures of quinone imine or quinone methide, or of radical species like cyclohexadienonyl, phenoxyl, aminyl or nitroxide to polymeric backbones. These reactions proceed mostly via reactivity of macroalkyl radicals derived from stabilized polymers. Various compounds modelling this reactivity have been isolated [19, 230]. These results are of importance mainly for the explanation of mechanisms of antioxidant activity [6, 22, 24].

It was reported [231] that phenolic and aromatic aminic AO are able to form bound-in species in EPM cured with peroxides. This principle was exploited in NR doped with 2,6-di-*tert*-butyl-4-methylphenol and *tert*-butyl peroxide. Other phenolic AO bearing methyl groups may take place in a similar process too [232]. The extent of the coupling of radicals derived from phenolic antioxidants with macroalkyls is influenced by the concentration of phenols. A competitive process, autocoupling of phenol derived radicals, increases with increasing concentration of the phenolic antioxidant [17].

Reactivity With Nitroso Compounds. Functionalization of diene based rubbers with aromatic nitroso compounds bearing aminic or phenolic moieties **174**, like with N,N-diethyl-4-nitrosoaniline, 4-nitrosodiphenylamine, 4-nitrosodiphenylhydroxylamine or 4-nitrosophenol represents an effective way for the synthesis of polymer-bound antioxidants [233]. The respective nitroso compound can be mixed with rubbers during compounding or with concentrated rubber latexes. The chemical attachement of stabilizing active moieties takes place during subsequent operations, e.g. during vulcanization of rubber compounds or drying of latexes. Nitroso compounds can be attached to latexes very slowly even at ambient temperature, before latex coagulation.

Functionalization of rubbers with nitroso compounds (174) involves two reaction steps [233]: formation of a hydroxylamine derivative 175 and its

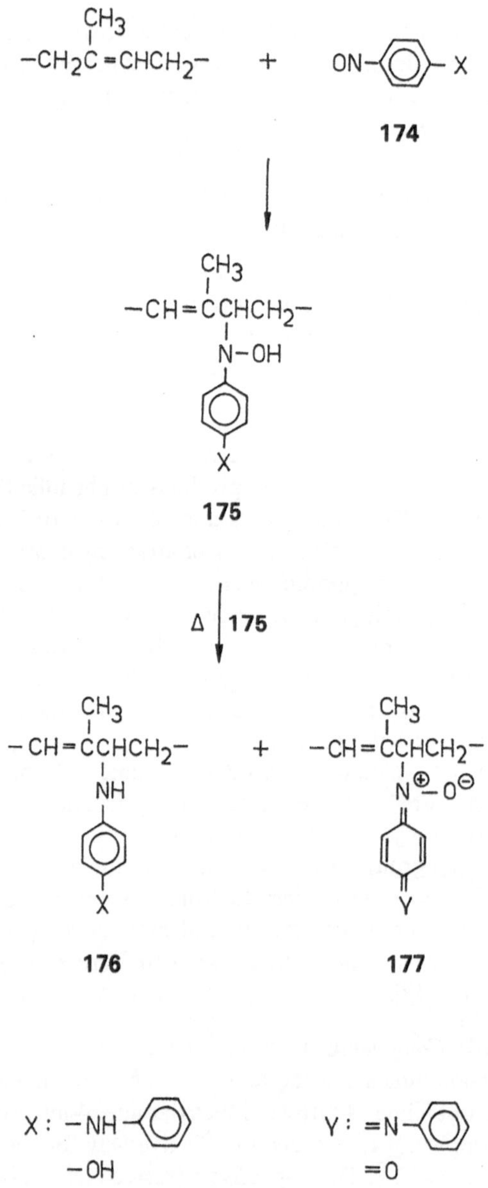

SCHEME 4

thermolysis into a polymer-bound AO **176** and the respective quinone imine-*N*-oxide **177** (Scheme 4). 2-Methylpent-2-ene was used as a model for rubber to confirm the reaction mechanism with diphenylaminic and phenolic derivatives. 40 to 80% binding was confirmed. The procedure enables the synthesis of effective nonvolatile AOs not only from rubbers (NR, IR, BR, SBR, NBR) but also from other naturally occuring or synthetic unsaturated substrates, like vegetable oils or varnishes.

Functionalization with aromatic nitroso compounds can be performed very easily using conventional processing technique without any major modification of technological processes. Nitroso compounds are, however, difficult to handle due to their potential toxicity, they discolor and peptize rubber and reduce its scorch resistance due to the reactivity in the mesomeric oxime form [43]. The latter disadvantage can be removed by means of the blocking of the oxime group by a reaction with isocyanate, alkalies or acyl chloride. Moreover, the acyl derivatives are characterized by having a lower skin irritating effect.

A main chain degradation was observed during the reaction of 4-nitrosodiphenylamine with IR or IIR. Formation of functionalized oligomers containing 13 to 16% of terminal aromatic nitrone-amine moieties of the type **178** and having $\overline{M_n}$ 2506 and 2738 was reported [234].

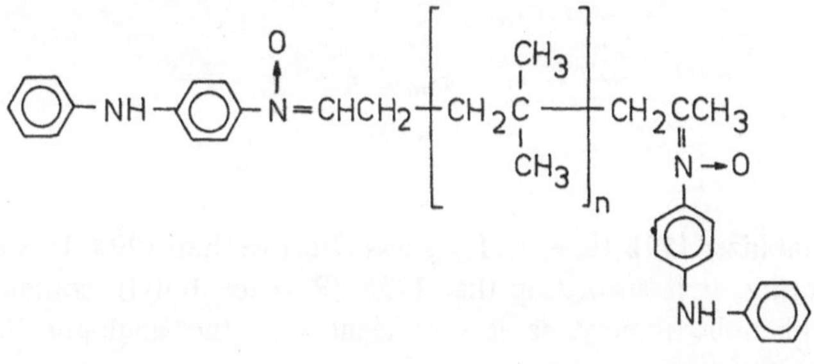

178

Attention was paid to the reactivity of PP with aliphatic nitroso compounds too [235]. It was postulated that the stabilizing effect was based on formation of a PP bound nitroxide and its disproportionation into the respective derivatives of hydroxylamine and nitrone.

Reactivity with Aromatic Nitrones. Rubber bound AO **179** based on 1,3-cycloaddition of aromatic nitrones (**180**) to C=C double bonds of IR or BR were formed during the vulcanization process [236]. In IR, 50–89% of nitrones was bound-in in the dicumyl peroxide vulcanizate, 39–85% in the tetramethylthiuram disulfide vulcanizate. Pendant stabilizing moieties are attached to the rubber chain by means of the isoxazolidine moiety. Crosslinks are formed with bis-nitrones. It should be mentioned that aromatic nitrones unfavourably influence the scorch time of the both conventional and sulfurless IR or BR compounds. Aminic stabilizer **179a** is an effective but discoloring and staining AO. Unhindered

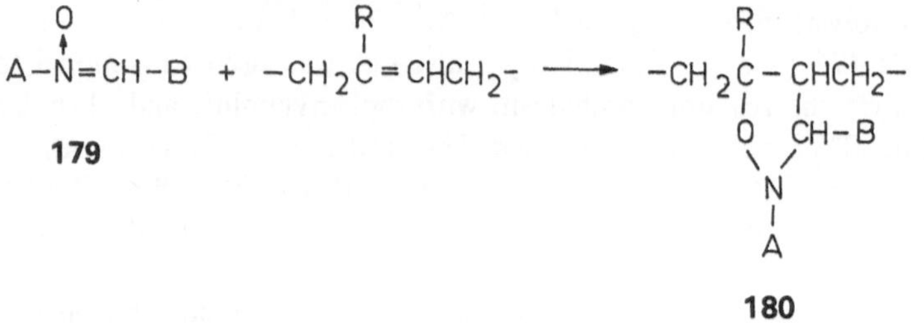

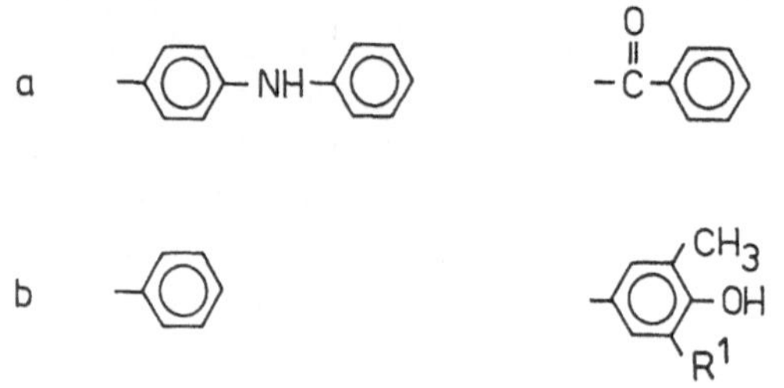

$$
\underset{\textbf{179}}{A-\overset{\uparrow}{\overset{O}{N}}=CH-B} \;+\; -CH_2\overset{R}{\overset{|}{C}}=CHCH_2- \;\longrightarrow\; -CH_2\underset{\underset{\underset{A}{|}}{\overset{|}{N}}}{\overset{R}{\overset{|}{\underset{O}{\overset{|}{C}}}}}-\underset{CH-B}{CHCH_2-}
$$

180

SCHEME 5

phenolic stabilizer **179b** (R = CH$_3$) is less effective than **179a**. However, **179b** is nondiscoloring. It is surprising that **179b** (R = *tert*-butyl), containing a more hindered phenolic moiety, is less efficient that the analogue derived from 2,6-dimethylphenol.

Aromatic nitrones may react also with polymeric alkoxy radicals formed during polymer degradation. Polymeric nitroxides contributing to the AO efficiency are formed [236].

Reactivity with Tetrazoles. Reactive tetrazoles were used for the synthesis of rubbers having an appreciable ageing protection [237]. 2-(Subst.)phenyl -5-(3,5-di-*tert*-butyl-4-hydroxyphenyl)tetrazole (**181**) can be compounded without thermolysis with BR or IR. At vulcanization temperatures, **181** is thermolysed and reactive nitrileimine **182** undergoes 1,3-dipolar addition to C=C double bonds. **183** is formed with yields from 70 to 85%, depending on the character of R in **181** (the activity series: R = CH$_3$ < H < Cl). Mechanistic model experiments were performed with **181** (R = Cl) and styrene, the latter was isued as a model instead of IR [237]. 1-(3-Chlorophenyl) -3-(3,5-di-*tert*-butyl-4-hydroxyphenyl)-5-phenyl-2-pyridazoline was isolated in 44% yield and mechanism of the attachement of **183** to IR was thus confirmed.

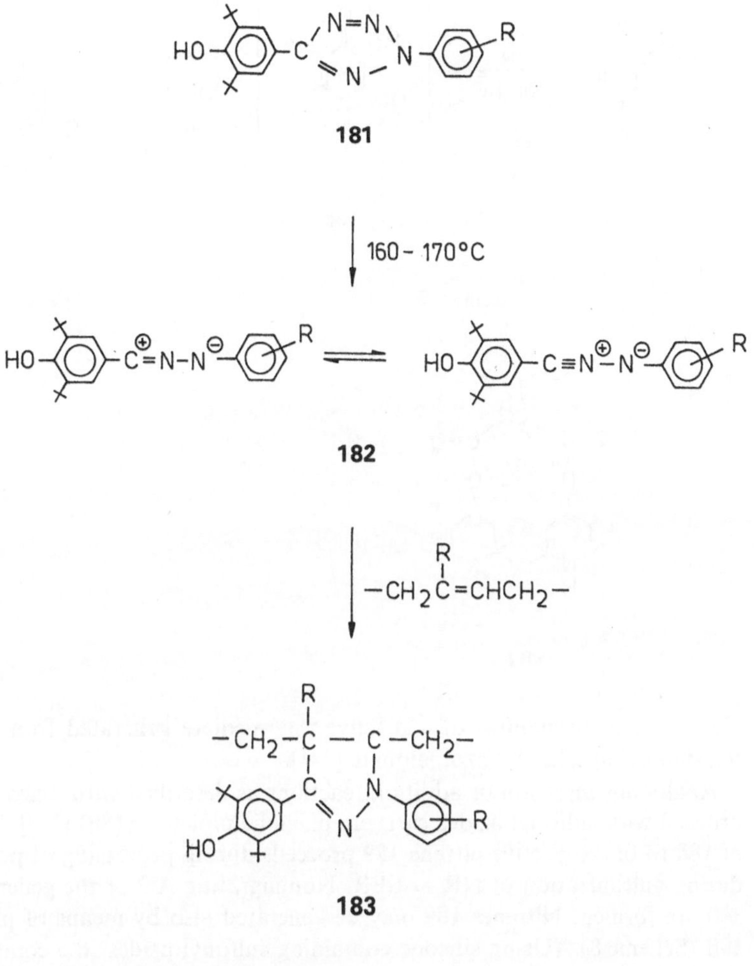

181

160 – 170 °C

182

183

SCHEME 6

Reactivity with Diazo Oxides and Azides. Diazo oxides, e.g. 2,6-di-*tert*-butyl-4-diazo-2,5-cyclohexadiene-1-one (**184**, R = *tert*-butyl) or 2,6-diphenyl-4-diazo-2,5-cyclohexadiene-1-one (**184**, R = phenyl) are thermolabile and photolabile. Carbenes **185** are formed from **184** as reactive intermediates. The process proceeds in a one-step mechanism in which the carbene species reacts in the singlet state [238]. The attachement of **185** to saturated hydrocarbon chains proceeds via insertion in C–H bond during processing. LDPE and PP were modified in this way [239, 240]. Poly(oxymethylene) was modified similarly.

Diene based rubbers are functionalized via addition of **185** to C=C double bonds during vulcanization [240]. Functionalized hydrocarbon chain polymers with pendant phenolic moiety (**186**) are formed. PP bound HALS **187** was formed

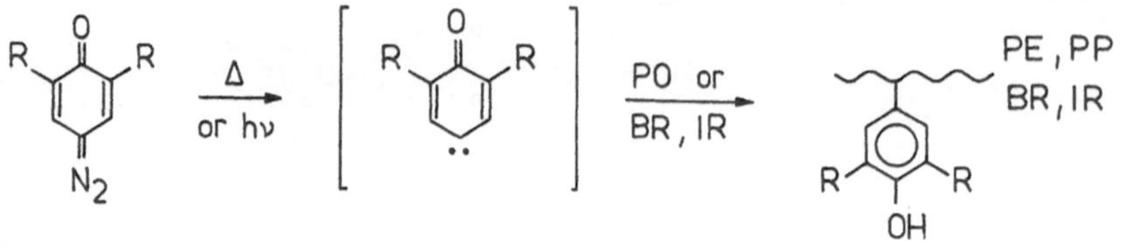

SCHEME 7

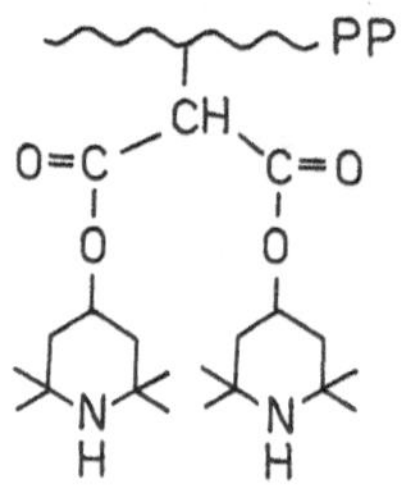

187

via insertion mechanism of a reactive intermediate generated from bis(2,2,6,6-tetramethylpiperidyl) diazomalonate [241].

Analogous insertion or addition reactions as described with liazo compounds proceed with sulfonyl azides carrying phenolic moieties (**188**) [242]. Thermolysis of **188** to the respective nitrene **189** proceeds during processing of polyolefins or during vulcanization of NR or SBR. Nonmigrating AO of the general structure **190** are formed. Nitrenes **189** may be generated also by means of photolysis of **188** (Scheme 8). (Using silicone containing sulfonyl azides, the compatibility of PP with inorganic fillers can be improved by the same mechanism as described in Schema 8). Functionalization of polymers with diazo oxides and sulfonyl azides possess disadvantages due to the high cost, toxicity hazard and explosive character of reactants.

Reactivity with Mercaptans and Disulfides. Thiyl radicals are formed as reactive species during functionalization of polyolefins and diene based rubbers with mercaptans or disulfides substituted with stabilizing moieties. This process exploits experiences earned in mastication of rubber in the presence of reactive sulfur compounds [1, 43, 243] or in curing of rubbers with thiol groups containing 1,3,5-triazines [244] and was successfully used by Weinstein [74] for incorporation of phenol or amino substituted mercaptans or disulfides via radical induced addition to polydiene latexes. The principle was perfected [243] for the functionalization of saturated and unsaturated polymer and rubber modified plastics with antioxidant or light stabilizing moieties. The first data were published in the late 1970s and dealt with the attachment of phenolic and aromatic amine moieties

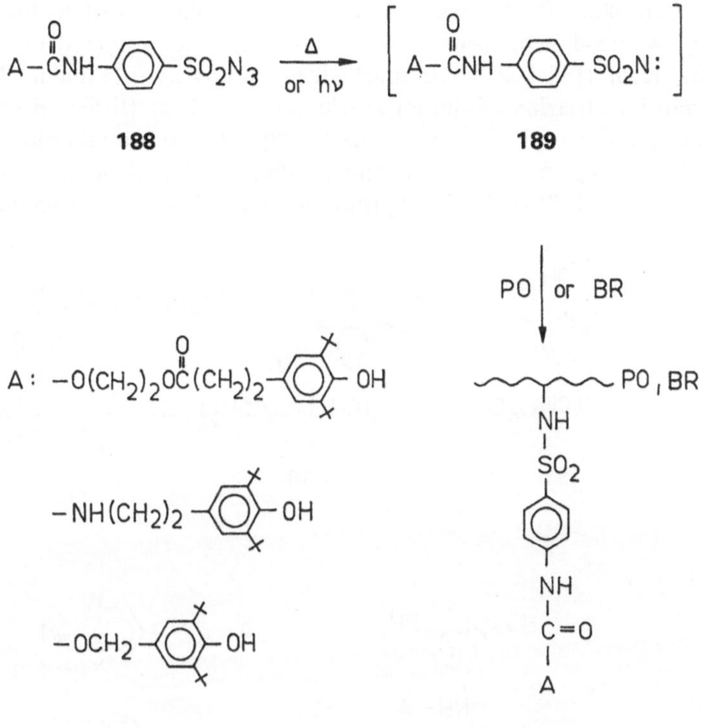

SCHEME 8

[73, 74] or 2-hydroxybenzophenone moieties [245]. Data dealing with the attachment of HD AO moieties were reported in the 1980s [246]. Most effort was concentrated on functionalization of important unsaturated polymers, like NR, BR, IR, SBR or NBR. Two-phase polymers containing an elastomer phase (e.g. ABS), EPDM rubber and PVC dehydrochlorinated in situ during processing were successfully functionalized, too [243, 245]. However, the attachement of stabilizer moieties was effective even with less reactive saturated polymers like with atactic PP, isotactic PP or PE [243].

Various reactive stabilizers were used for the functionalization of polymers: 2,6-di-*tert*-butyl-4-mercaptophenol (**191**, n = 0) [74], 3,5-di-*tert*-butyl-4-hydroxy-benzylmercaptan [243] (**191**, n = 1), 2-mercaptoethyl 3-(3,5-di-*tert*-butyl-4-hydroxyphenyl)propionate [247] (**192**), phenolic 2-mercaptoacetate [243] (**193**), 2-(3,5-di-*tert*-butyl-4-hydroxyanilino)-1,3,5-triazine-4,6-dithiol (**194a**) or an analogous triazine derivative substituted in positions 4 and 6 with trithiomorpholino group (a "vulcanizing" AO) [248], 2,2'-thiobis(4-methyl-6-*tert*-butylphenol) [74] (**195**), N-phenyl-N'-(mercaptoacetyl)-1,4-phenylenediamine (**196**, R = H, n = 1) and an analogous N'-(2-mercaptopropionyl) derivative [74, 243] (**196**, R = H, n = 2), disulfide **197** formed from the latter by oxidation with hydrogen peroxide [249], *N*-(α,α-dimethylbenzylphenyl)-*N'*-(2-mercaptopropionyl)-1,4-phenylenedi-amine [250] (**196** , R = α,α-dimethylbenzyl, n = 2), 2-(4-anilino-*N-iso*propyl-anilino)-1,3,5-triazine-4,6-dithiol (**194b**) and analogous triazine derivatives containing trithiomorpholino groups [244, 248], 2-[(mercaptoacetyl)oxy]ethyl-3-{[4-(phenylamino)phenyl]amino }butanoate (**198**, n = 2) and an analogous stabilizer **198**, n = 6 (ref. 251). 0,0-Di-*iso*propyl (or dibutyl) dithiophosphoric acid [252]

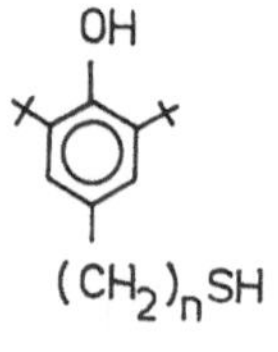

191

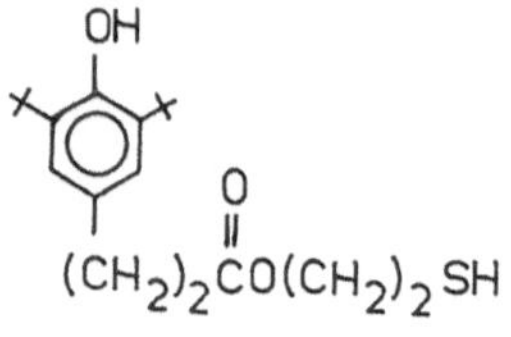

192

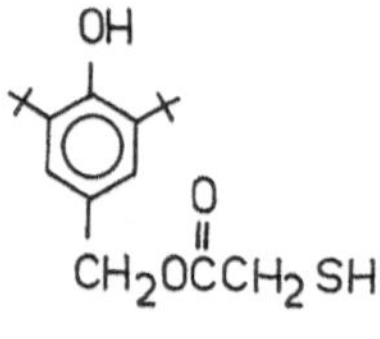

193

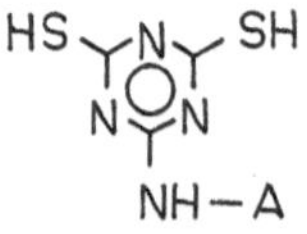

194

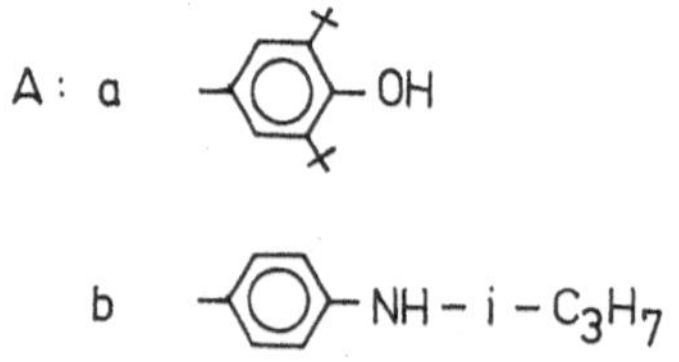

A : a

b

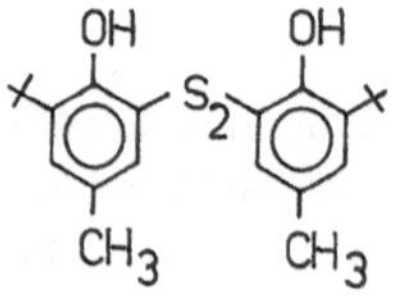

195

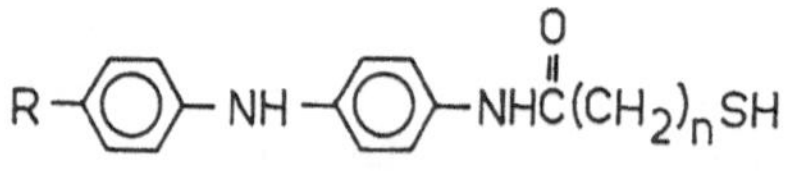

196

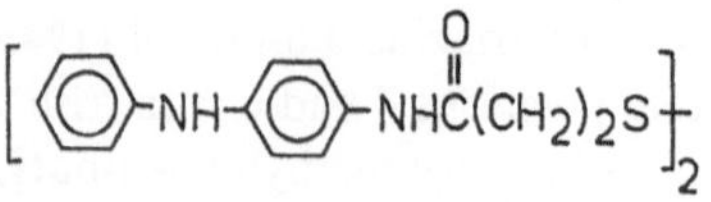

197

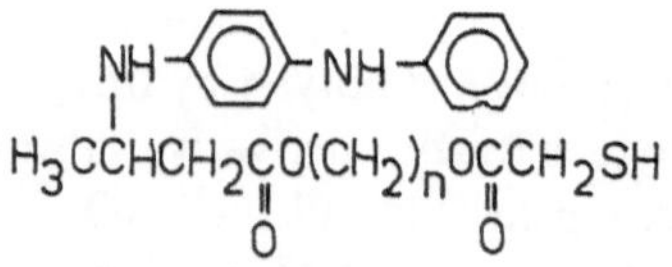

198

(199) or dodecyl 3-mercaptopropionate [246] (200) are bondable HD AO. 2-(4-Benzoyl-3-hydroxyphenyloxy)ethyl-2-mercaptoacetate [243, 245] (201) represents a bondable LS. The binding of functionalized mercaptans or disulfides to polymers by means of the sulfur atom proceeds via thiyl radicals generated by

$$(RO)_2P(S)SH \qquad\qquad HS(CH_2)_2\overset{\overset{\displaystyle O}{\|}}{C}OC_{12}H_{25}$$

199   **200**

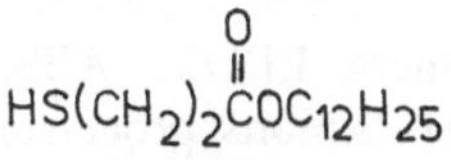

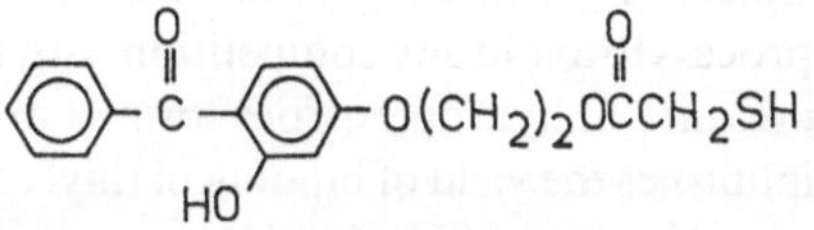

201

radical reactions [74, 243]. Crosslinks can be formed with compounds containing two reactive thiol groups [248], like with 194. Disulfides are formed by dimerization of thiyls as a competitive process. Both the polymer bound stabilizers and the respective disulfides contribute to the integral AO effect resulting of the process. Coloured compounds having quinoid character are formed as by-products.

Mechanism of the thiyl functionalization of polymers involves reactivity of polymeric C-centered radicals with thiyls [243]. The generalized Scheme 9 can be used to explain the process (A = stabilizing active moiety, P, P' = polymer and

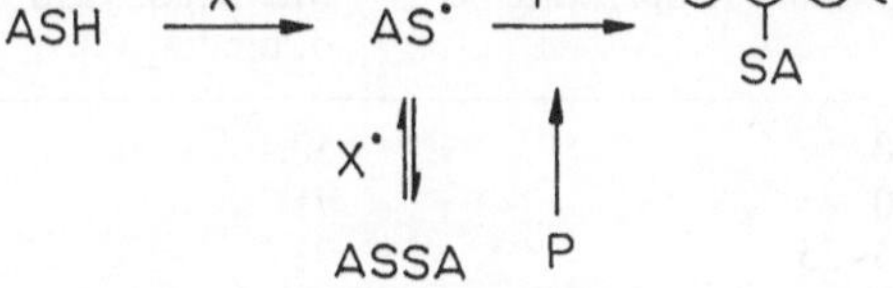

SCHEME 9

the respective polymeric C-centered radical, X' = initiating free radical). It has been considered that radicals P' are generated primarily on carbon atoms forming C=C double bonds in the main chain of diene based polymers, on pendant vinyl groups or on the tertiary C-atom in saturated polymers.

Initiating radicals are formed either from polymers used for functionalization or are generated from initiators. Peroxides and hydroperoxides and AIBN are used as initiators in NR, SBR, NBR or EPDM functionalization [236, 243, 245].

Attachment of active moieties during rubber vulcanization is due to the involved thermal processes generating free radicals [248–251].

An effective formation of X˙ takes place during UV irradiation of latexes, rubber solutions and rubber surfaces or PP films. In the latter polymer, the yield of attachment of compound **191** up to 68% was obtained [253].

A specific attention has been paid to the mechanochemical generation of X˙ in unsaturated elastomers, EPDM, ABS, PVC and polyolefins [243, 252]. This synthetical approach exploits processing operations producing free radicals. In situ chemical reactions can be performed during some important industrial processes like mastication of rubber and extrusion, mixing or reaction injection molding of plastics.

It was confirmed that mechanochemically formed radicals are active enough to initiate polymerization of vinyl monomers [254]. Due to the quick reactivity of the formed carbon centered radicals P˙ with oxygen, an inert atmosphere must be strictly observed during the process to avoid any competition with the generation of thiyls [243]. Moreover, the presence of hydroperoxides formed in polymers due to the reactivity with oxygen diminishes the yield of binding of thiyls to the polymer chain as a result of the HD AO activity of ASH (ASSH).

The yield of the adduct formation is dependent on the efficiency of the mechanochemical formation of radicals, on the viscosity of the mechanochemically treated polymer and the temperature of the process due to the reversibility in the binding of thiyls [243].

$$P^{\cdot} + AS^{\cdot} \rightleftarrows PSA$$

(AS˙ = thiyl radical derived either from a monomeric stabilizer ASH or a polymeric one ASP.) Therefore, an optimum temperature must be maintained for each polymer to reach the maximum yield of the chemical binding of AS˙.

Polymeric substrate	Optimum temperature °C	Maximum yield of binding, %
NR	70	50
SBR	70	75
NBR	45–55	78
IR	70	75
EPDM	150	80

It was reported [243] that the level of binding of **196** into NR or SBR increases further during the subsequent vulcanization.

The attachment of the functionalized mercaptans proceeds with unsaturated polymers more efficiently than with saturated ones. The stabilizing moiety can therefore be targetted toward the most oxidation sensitive component in multiphase systems or blends, e.g. to the BR phase in ABS.

Due to the high level of binding of stabilizing moieties in unsaturated rubbers or EPDM, masterbatches can be formed by this way and used for the stabilization of virgin polymers having the same chemical character. The masterbatches can also be exploited, however, for the stabilization of polymers having different chemical character. For example, functionalized EPDM is suitable for blending with NBR or SBR. This significantly increases the application spectrum of masterbatches. The compounding process is similar to that involving masterbatches consisting of conventional HMW stabilizers and a polymeric binder.

Functional polymers carrying an aminic moiety with AO properties [74, 236] (**202**), a bifunctional intramolecularly cooperating system having properties of AO and LS attached to the elastomeric phase of ABS [255] (**203**) and a functionalized bridge involved in the network system (**204**) formed [244, 248] during BR vulcanization in the presence of **194a** are typical examples of stabilizers prepared by means of mercaptan's activity with conventional polymers.

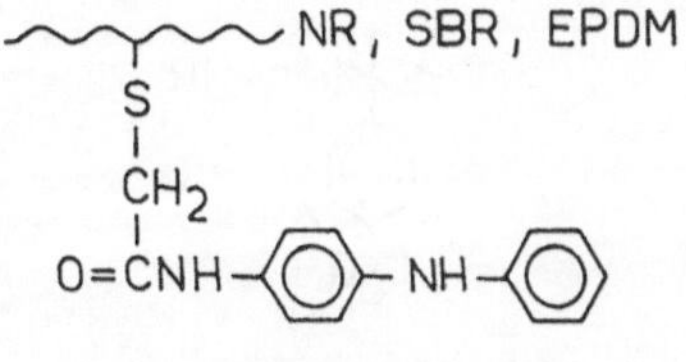

202

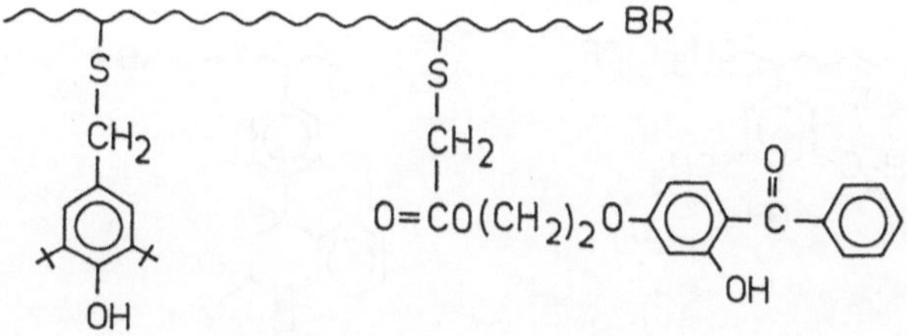

203

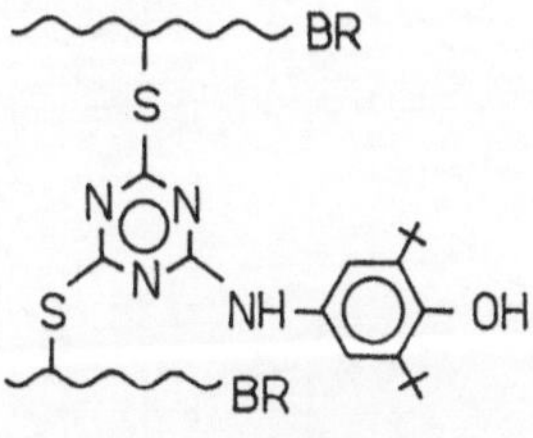

204

Friedel-Crafts Alkylation. Processes catalyzed by Lewis and Bronsted acids offer varied possibilities for the synthesis of polymeric stabilizers. The polymer analogous Friedel-Crafts alkylation involves both the macromolecular and low molecular weight alkylating agents. Polymers containing reactive double bonds like liquid BR containing less than 20% of the structure 1,4-,IR,poly(butadiene-*co*-styrene), EPDM or low molecular weight PP were used for alkylation of low molecular weight CB AO, e.g. of 2-*tert*-butylphenol, 2-*tert*-butyl-4-methylphenol, 2,6-dimethylphenol, 1-naphthol, diphenylamine, *N*-phenyl-2-naphthylamine, *N,N'*-disubstituted-1,4-phenylenediamine or 2,2,4-trimethyl-1,2-dihydroquinoline [43, 256, 257]. Halides of boron, aluminium or zinc, aluminium phenolate or acid clays were used as catalysts. Bound-in stabilizers **205** (formed via reactivity of pendant vinyl groups in liquid PB [258]), **206** (formed via reactivity of C=C

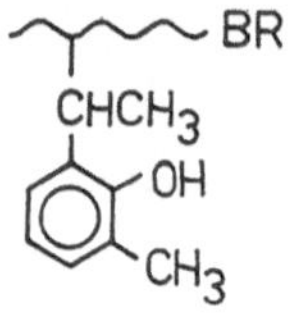

205

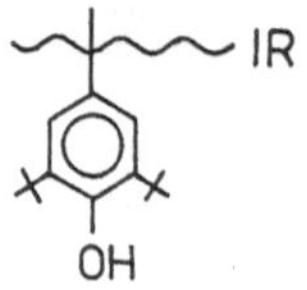

206

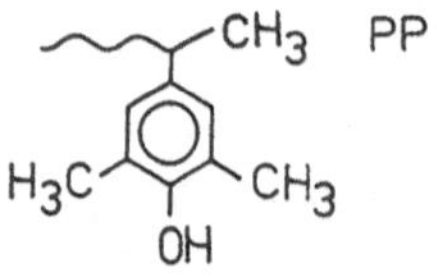

207

208

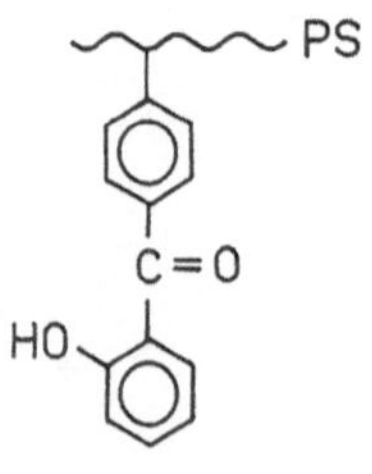

209

double bonds in IR [259]) or **207** (formed via reactivity of C=C double bonds in oligomeric PP [256]) are typical examples. Low molecular weight compounds having reactive C=C double bonds can be used for the functionalization of conventional SBR via the reactivity with pendant phenyl groups [260]. Systems with the attached AO moiety, e.g. **208** can be formed. Pendant 2-hydroxybenzo-

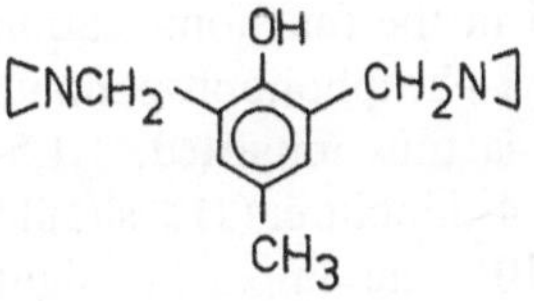

210

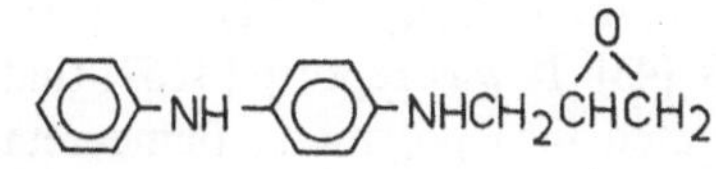

211

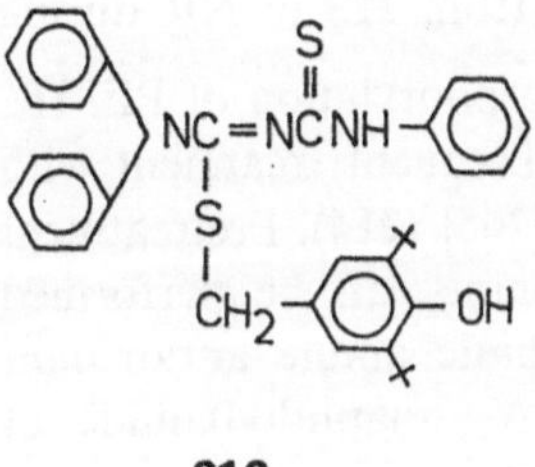

212

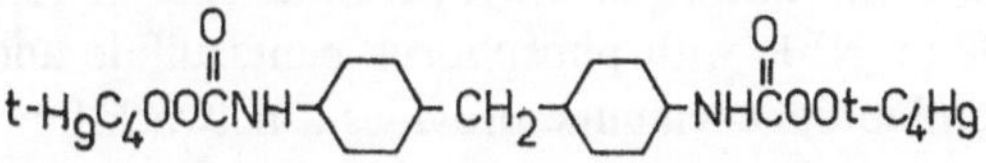

213

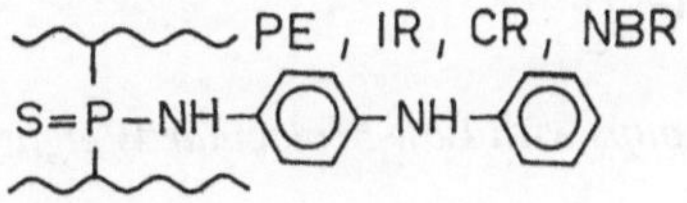

214

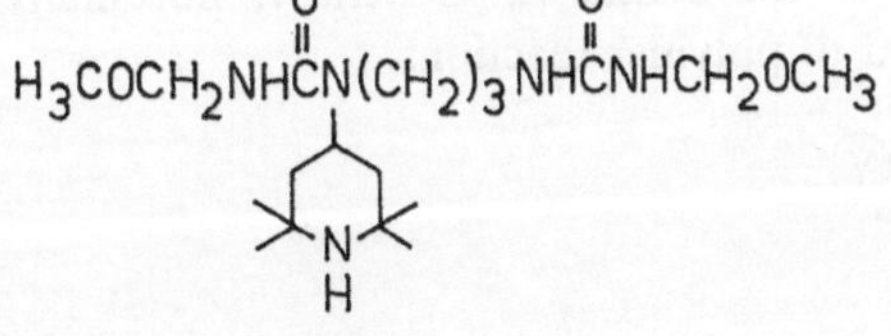

215

phenone moiety **209** was introduced into PS by means of the $AlCl_3$ catalyzed reaction with 2-hydroxybenzylchloride [261].

Reactivity with other Reagents. Thermal treatment of rubber compounds during vulcanization can be exploited for the attachement of stabilizing moieties due to the reactivity of some other reagents than that mentioned in the preceding Sections. Ring opening reactions of phenol bearing aziridine groups (**210**) or 1,4-phenylene-diamine bearing oxirane groups (**211**) can be exploited in the functionalization of NR and SBR [43]. It was reported [262] that glycidyl phosphonopropanoate can be incorporated into polymers; flame retardancy is thus imparted. 1,1,5-Triphenyl-2-thia-(3,5-di-*tert*-butyl-4-hydroxybenzyl)-*iso*-4-thiobiuret (**212**) should be listed among reactive compounds bondable to SBR and imparting light stabilizing and metal deactivating protection [263]. Vulcanizing agents having structure of bis(peroxy carbamates), e.g. **213**, posses properties of "vulcanizing" antioxidants and impart to NR high tensile strength, excellent heat resistance to crack initiation and crack propagation in cyclic deformation fatigue [264]. Network forming moiety $-\text{NH} \hspace{-0.2em}-\hspace{-0.5em}\langle\text{H}\rangle\hspace{-0.5em}-\text{CH}_2-\hspace{-0.5em}\langle\text{H}\rangle\hspace{-0.5em}-\text{NH}-$ is created from **213** in NR during low sulfur and/or dicumyl peroxide cure at 130 °C. Phosphorylation of PE, IR, CR or NBR with phosphorus pentasulfide and the subsequent treatment with 4-aminodiphenylamine provides a network-bound AO [265] (**214**). Pretreatment of PP before functionalization with aminodiphenylamine can be performed similarly with dithiophosphoric acid. Secondary aromatic amine antioxidant moiety was atached to NR by means of *N*-phenyl-*N*-(2-naphthyl)amide of dithiophosphoric acid [34, 97].

Two papers dealing with functionalization of conventional polymers with LS moieties were published. PE was treated with 2-hydroxy-4-(2-chloroethanoyl)ben-zophenone [266] and PUR or PA were functionalized with a reactive methoxyme-thylol urea derivative **215**. Bisnitroxide formed from the latter creates a HALS bearing crosslink between two polymeric chains [267].

3.2.2.2 Reactions of Polymers Bearing Reactive Groups with Low Molecular Weight Compounds Carrying Stabilizing Moieties

Exploitation of polymers bearing reactive groups for the attachment of stabiliz-ing moieties represents another way for the synthesis of polymeric stabilizers. A suitable prepolymer having right molecular weight and molecular weight distribution must be used. The synthetic approach to stabilizers via polymer analogous reactions is more expensive than synthesis from conventional polymers (see Sect. 3.2.2.1) and can be used only for the synthesis of stabilizers having an appreciable value justifying the price of the synthesis. Generally, not many polymers are available for the right kind of polymer reactions.

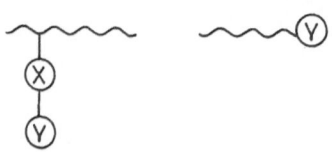

Reactive groups Ⓨ can form a pendant or terminal part of the polymer chain to be functionalized (Ⓧ represents a spacer). The most frequently exploited reactive groups Ⓨ are listed in Scheme 10. The chemical modification of Ⓨ with suitable

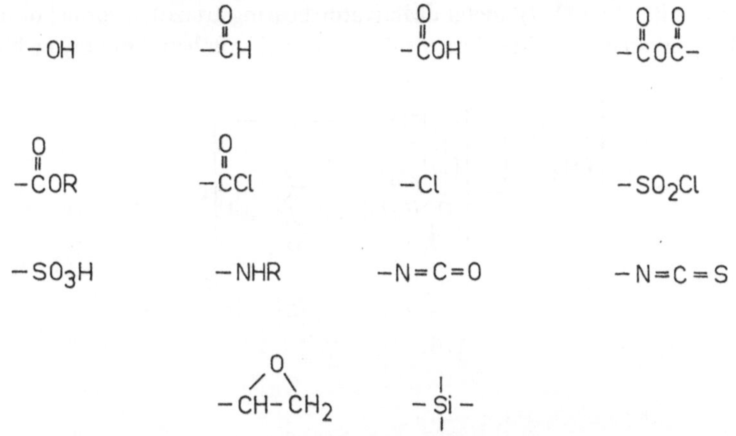

$$-OH \qquad -\overset{\overset{\displaystyle O}{\|}}{C}H \qquad -\overset{\overset{\displaystyle O}{\|}}{C}OH \qquad -\overset{\overset{\displaystyle O\ \ O}{\| \ \ \|}}{C}O\overset{}{C}-$$

$$-\overset{\overset{\displaystyle O}{\|}}{C}OR \qquad -\overset{\overset{\displaystyle O}{\|}}{C}Cl \qquad -Cl \qquad -SO_2Cl$$

$$-SO_3H \qquad -NHR \qquad -N=C=O \qquad -N=C=S$$

$$-\overset{\displaystyle O}{\overset{\displaystyle /\ \ \backslash}{CH-CH_2}} \qquad -\overset{|}{\underset{|}{Si}}-$$

SCHEME 10

low molecular weight compounds Ⓢ considered as precursors of stabilizing moieties Ⓢ proceeds generally only to a limited degree of completion due to the macromolecular character of the substrate [46]. Formation of polymer bound by-products Ⓩ introducing side effects into the formed functionalized polymer must be considered. Therefore, functional groups Ⓨ, Ⓩ and Ⓢ may be attached to the formed macromolecular stabilizer.

$$\left[\ \underset{Ⓨ}{\ }\ \right] \xrightarrow{\ \ Ⓢ'\ \ } \left[\ \underset{Ⓨ}{\ }\right]_m \left[\ \underset{Ⓩ}{\ }\right]_n \left[\ \underset{Ⓢ}{\ }\right]_o$$

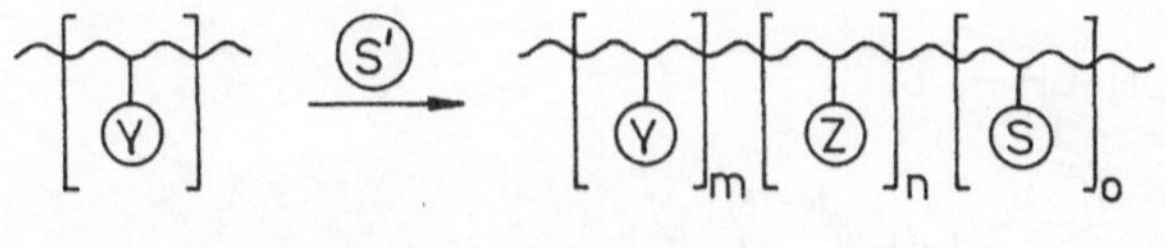

Exploitation of Pendant Reactive Groups. Pendant and linkage forming stabilizing moieties can be created in this way. Macromolecular stabilizers thus formed have general structures B, C, and E (Scheme 1) and are formally structurally related to stabilizers prepared by other polyreactions. The distribution mode of stabilizing moieties and the presence of foreign structures Ⓨ and Ⓩ mentioned above represent the main difference between stabilizers prepared via different synthetic approaches.

Chlorinated, sulfonated, chlorosulfonated or epoxidized polymers, homopolymers and copolymers of functionalized monomers, e.g. poly(methacryl aldehyde), poly(2,3-epoxypropyl acrylate), poly(4-vinylphenol), poly(propylene-*co*-10-undecene-1-ol), poly(butadiene-*co*-methacryl aldehyde), poly(butadiene-*co*-acrylic acid), poly(ethylene-*co*-alkyl acrylate), poly(alkyl acrylate-*co*-2,3-epoxypropyl acrylate), poly(alkyl acrylate-*co*-maleic anhydride), poly(styrene-*co*-4-vinylbenzyl chloride) or poly(styrene-*co*-4-vinylphenyl isothiocyanate) are examples of polymers used successfully for the synthesis of various polymeric stabilizers. Examples of some systems prepared by the polymer analogous reactions are given (only functionalities Ⓢ are exemplified, functions Ⓨ and Ⓩ are not given). Phenolic AO [268] (**216**), phosphite [269] (**217**), metal deactivator bearing triazolylazophenol moiety [270] (**218**), an analogous MD bearing 4-(3-triazolylazo)phenol moiety or biocides [271]

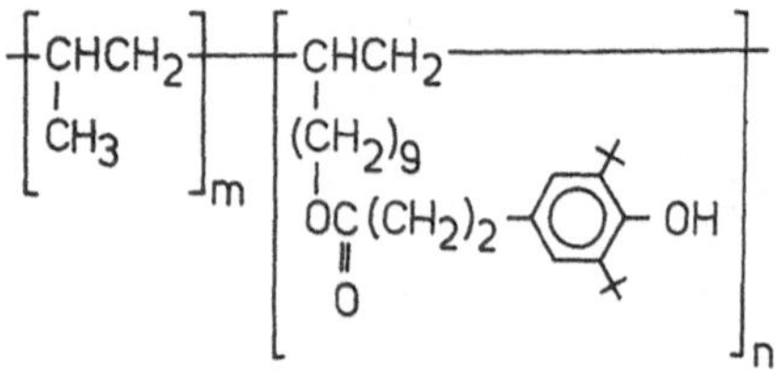

216

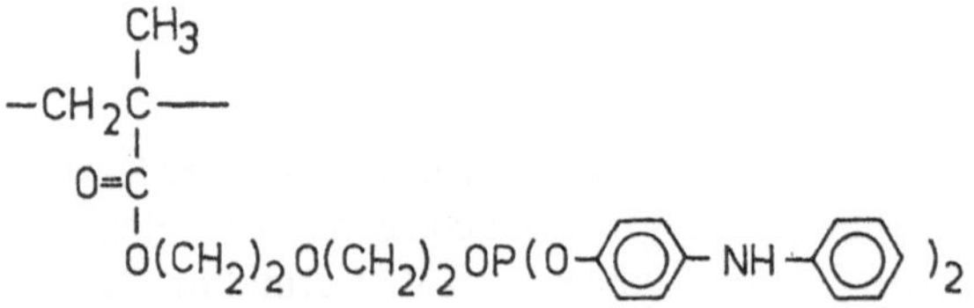

217

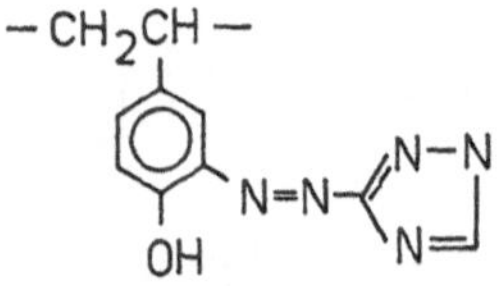

218

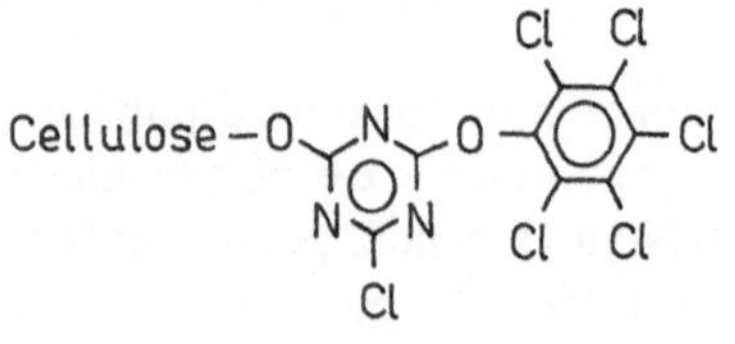

219

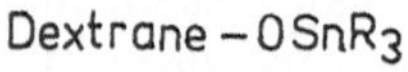

Dextrane $-$ OSnR$_3$

220

219 and **220** were prepared via reactivity of pendant hydroxy groups. Interesting phenolic (e.g. **221**) or aminic AO were prepared from polymers or copolymers with pendant −CHO [272] group. Various stabilizers are based on functionalization of polymers with pendant carboxylic acid groups or their derivatives. Various CB AO, e.g. **222a** or analogous HD AO **222b** were prepared from PP grafted with methacrylic acid. Poly(propylene-g-maleic anhydride) can be functionalized similarly with a thiobisphenol, 4-amino or 4-hydroxy-2,2,6,6-tetramethylpiperidine. A polymeric AO obtained by functionalization of poly(methyl vinyl ether-co-maleic anhydride) with 3,5-di-*tert*-butyl-4-hydroxybenzylamine is acceptable for the stabilization of foods and pharmaceuticals [272], an analogous system obtained from poly(octadec-1-ene-*co*-maleic anhydride) and primary aromatic amines is designed for lubricating oils and polyester oils [273]. Light stabilizer **223** or antioxidant [274] **224** were prepared by functionalization of suitable copolymers. Copolymerized or grafted maleic anhydride can be functionalized with β-(3,5-

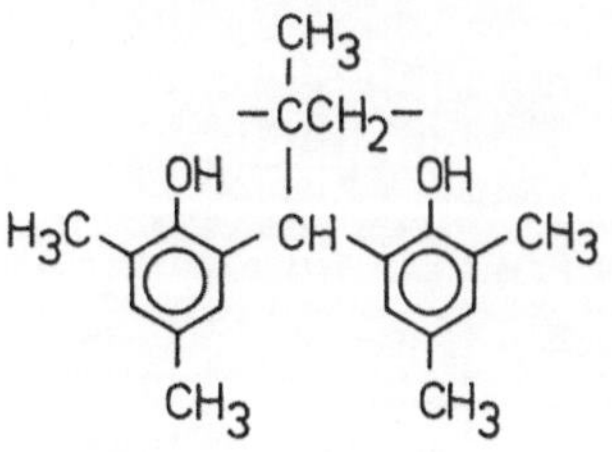

221

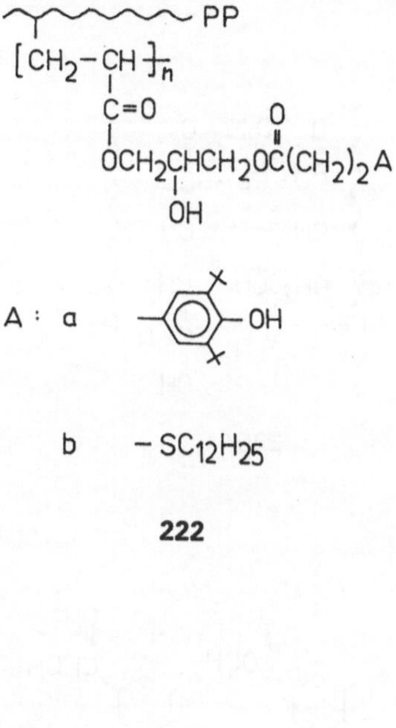

222

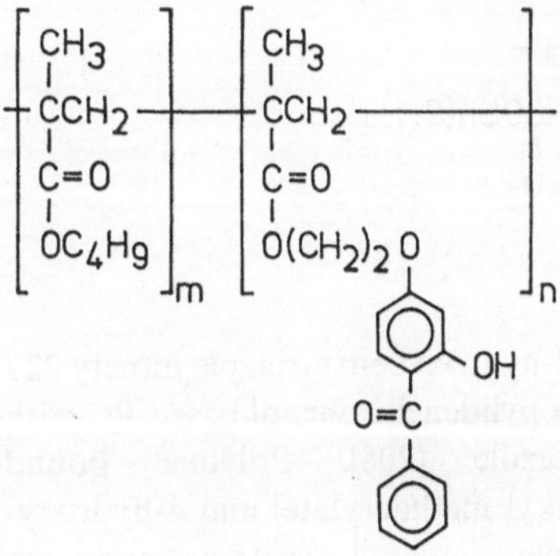

223

di-*tert*-butyl-4-hydroxyphenyl)propionhydrazide, α-(4-benzoyl-3-hydroxyphenyl) acetylhydrazide, 4-amino-2,2,6,6-tetramethylpiperidine, dialkyl lead dichloride [in the latter case, interchain (225) and intrachain carboxylates acting as synergistic flame retardants are formed [275]) or with trialkyl tin oxide or trialkyl tin chloride (biocides like 226 can be formed [105]). Transesterification was used for the synthesis

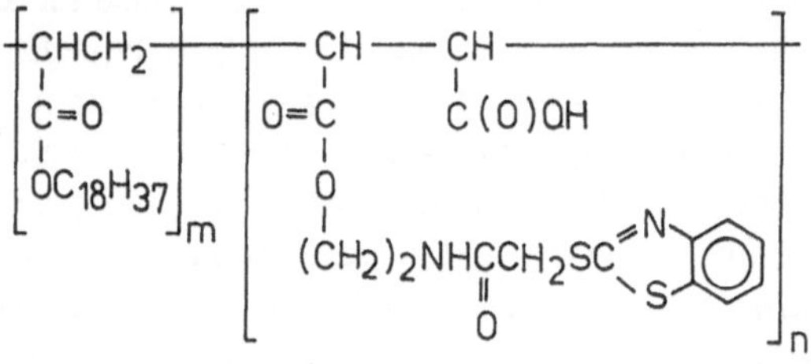

224

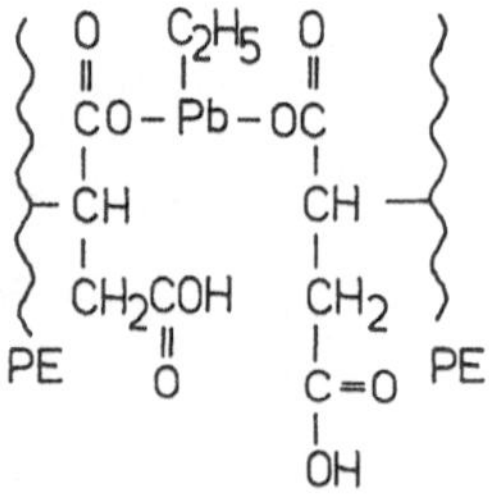

225

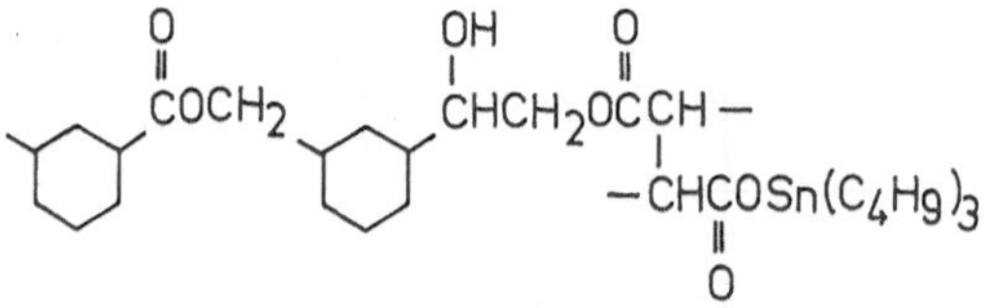

226

of some LS. For example, a system containing 2-hydroxybenzotriazole moiety 227 was obtained by transesterification of 4,4′-isopropylidenebisphenol based PC with 2-(2,4-dihydroxyphenyl)-2*H*-4-hydroxybenzotriazole [208]. Polymer bound HALS were prepared from poly(ethylene-*co*-alkyl methacrylate) and 4-hydroxy-2,2,6,6-tetramethylpiperidine from poly(ethylene-*co*-vinyl acetate) and the respective esters of HALS [276].

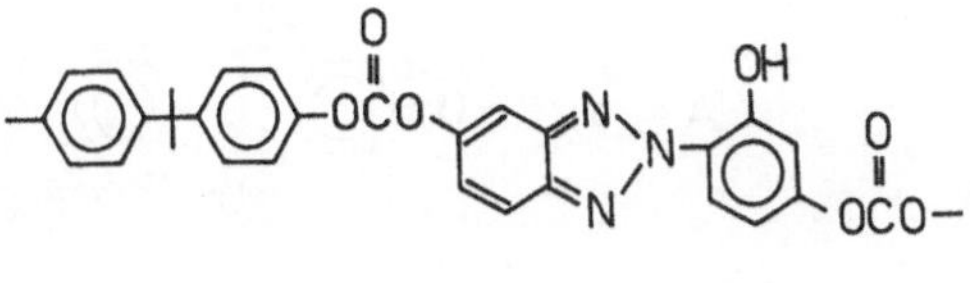

227

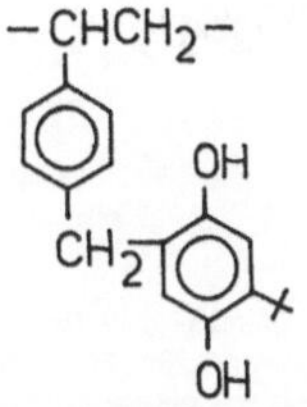

228

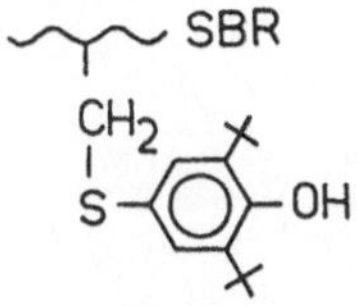

229

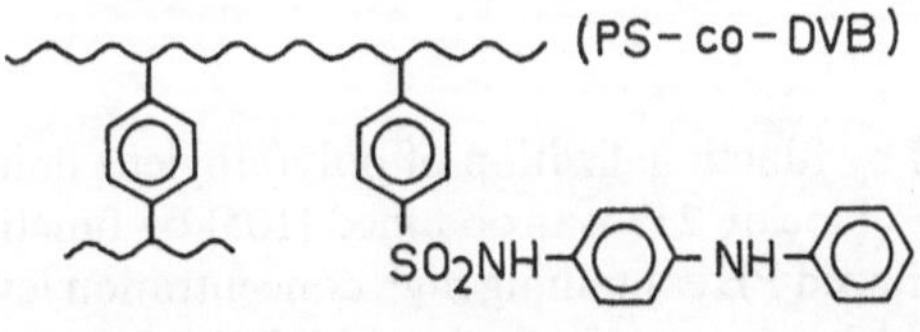

230

Various polymeric stabilizers were synthesised from polymers bearing chlorine, bromine, chloromethyl, benzyl chloride or sulfonyl chloride groups. Some examples are given: Phenolic AO [277] (**228**), phenolic sulfide [278] **229** prepared from chloromethylated SBR latex or an analogous phenolic sulfide based on partially chloromethylated PS, aromatic amine AO for NBR [279] (**230**) prepared from poly(styrene-*co*-divinylbenzene), systems with benzophenone [280] (**231a**), benzotriazole [281] (**231b**) and phosphorus containing pendant moieties [282] (**231c**) formed by functionalization of brominated poly(2,6-dimethylphenylene oxide) designed as LS for PPO or blends PPO/PS, polymeric MD [283] **232a** or FR [284] **232b** for PE, PP, and PS synthesized from chloromethylated PS. The attachment of *N-sec*-alkyl-*N*′-aryl-1,4-phenylenediamine to CR via nitrogen bearing secondary alkyl groups proceeds during high shear mixing or vulcanization at 160 °C [285]. Metal deactivator **233** containing thiosemicarbazide moiety was prepared from poly(styrene-*co*-4-vinylphenyl isothiocyanate) [286]. Light stabilizer **234** containing 2,2,6,6-tetramethylpiperidine moieties attached via 1,3,5-tria-

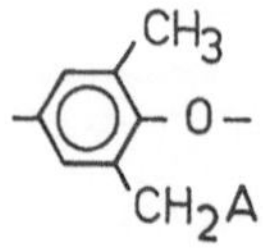

231

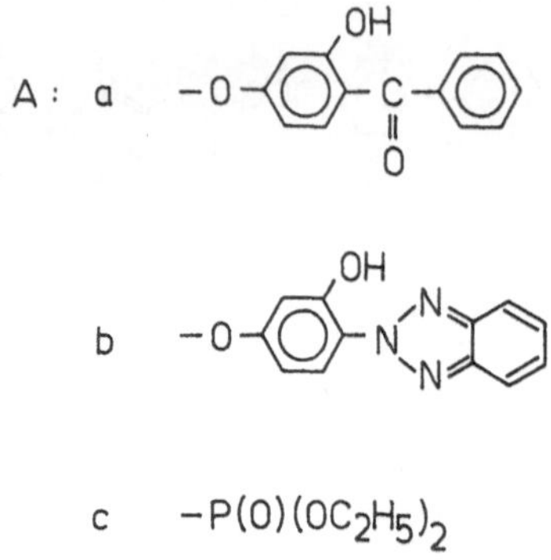

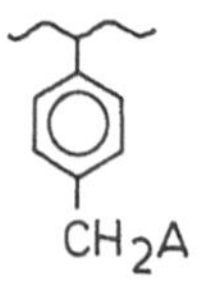

A: a $-O-$ ... $-C(=O)-$ (OH) phenyl

b $-O-$ ... (OH) $-N=N-N$ benzotriazole

c $-P(O)(OC_2H_5)_2$

232

A: a $-N(-\text{pyridyl})_2$

b $-P(O)(OCH_3)_2$

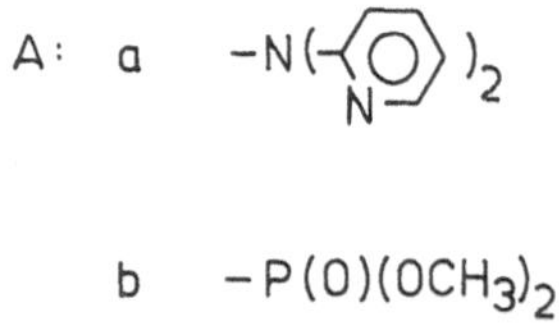

zine spacer was prepared by functionalization of poly(ethylene imine) [287]. Film forming and hydrophobic biocide **235** was obtained [105] by functionalization of the same polymer. Preoxidized PE containing high concentration level of carbonyl groups was functionalized be means of a hydrazide derived from 3-(3,5-di-*tert*-butyl-4-hydroxyphenyl)propinic acid [288]. The formed hydrazone augmented the thermal stability of PE.

Friedel-Crafts alkylation was used for introduction of *tert*-butyl groups into poly(4-vinylphenol) [55].

Many authors elucidated functionalization of polymers containing reactive oxirane moieties. Epoxidized NR, BR, IR and/or the respective model hydrocarbons, poly(butadiene-*co*-isoprene, various epoxy resins, poly(2,3-epoxypropyl methacrylate) and its copolymers or grafted systems were mostly exploited. Stabilizers based on epoxidized unsaturated rubbers are of the top interest. The mechanism of the functionalization process was studied in details by means of 3,4-epoxy-4-methylheptane and 1,2-epoxy-3-ethyl-2-methylpentane as model compounds [289]. The ring opening of the asymmetric oxirane is regiospecific. Aliphatic primary amines attack the least substituted carbon atom and can be involved in crosslink formation. Aromatic primary and secondary amines are less reactive than aliphatic ones because of their lower basicity; the attack on the least substituted carbon atom is however preferred too.

Systems with attached phenolic moiety [272] (**236 a, b**) aromatic amine moiety [289] (**237**) or a combination of moieties forming a homosynergistic system [290]

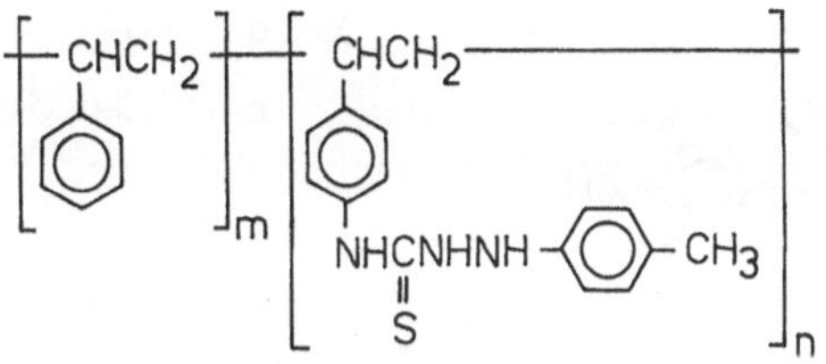

233

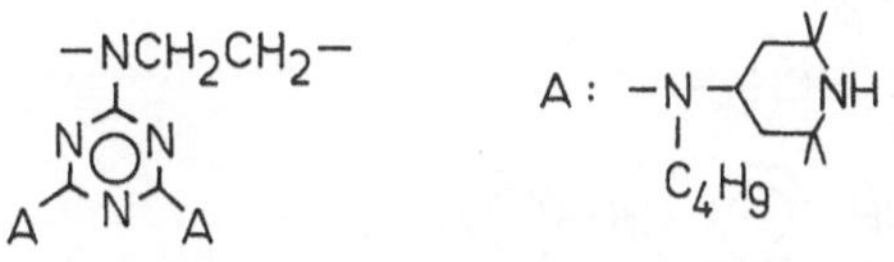

234

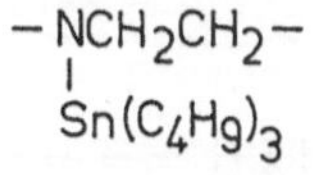

$$-NCH_2CH_2-$$
$$|$$
$$Sn(C_4H_9)_3$$

235

(238) are examples of stabilizers based on epoxidized rubbers. All of them were designed for rubber stabilization. 4-Aminodiphenylamine was used for functionalization of low moecular weight epoxide resins. Interesting LS were prepared by means of functionalization of copolymers of 2,3-epoxypropyl methacrylate: as examples, a terpolymer with ethyl acrylate and methyl methacrylate [291], a copolymer with divinylbenzene functionalized with 2,4-dihydroxybenzophenone [292] (239) or a copolymer with methyl methacrylate functionalized with 2-(2,4-dihydroxyphenyl)-2H-benzotriazole [293] (240) are given. All these LS were tested in PP. Numerous systems were prepared via functionalization of poly(propylene-g-2,3-epoxypropyl methacrylate) or of similarly grafted PE and PS, respectively. The reactive prepolymer was prepared by means of melt or surface grafting. The latter process is diffusion controlled and is very favourable mainly for the synthesis of LS because a high surface concentration of stabilizing moieties can be reached [294]. A considerable number of epoxy groups was, however, consumed (depleted) in side reactions. Polymers containing CB phenolic moiety [218], e.g. 241a, HD sulfidic moiety [218], e.g. 241b, light stabilizing benzophenone [295] (241c) or 2,2,6,6-tetramethylpiperidine moieties [294, 295], e.g. 241d, flame retardant moieties [296], e.g. 241e, or metal deactivating moiety formed via functionalization with ethylenediamine tetraacetic acid are representative examples of prepared stabili-

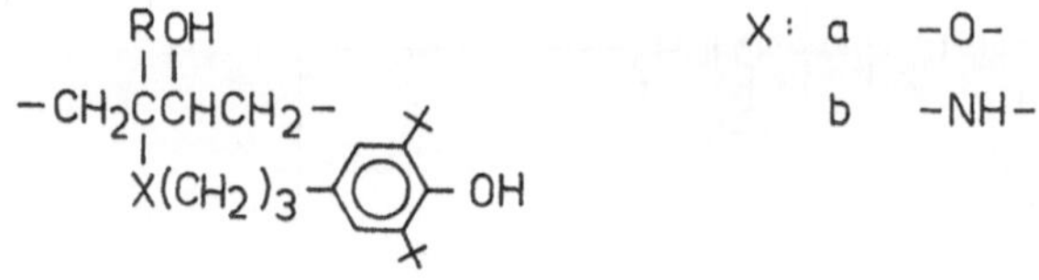

$$-CH_2\underset{X(CH_2)_3}{\overset{R}{\underset{|}{C}}} \overset{OH}{\underset{|}{C}} HCH_2-$$

X: a   -O-
    b   -NH-

236

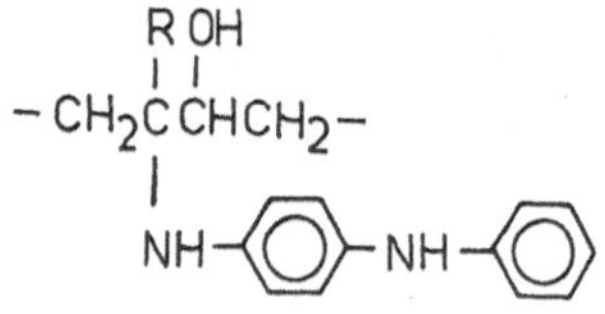

237

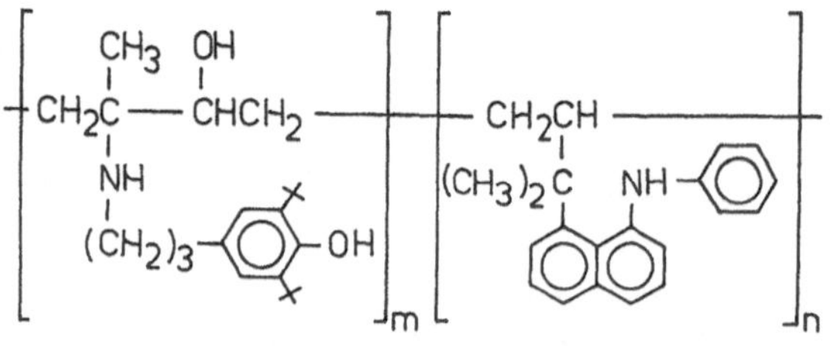

238

zers. In this section, also an oligomeric AO **242** should be mentioned. **242** was formed by ozonolysis of adducts of 3-(3,5-di-*tert*-butyl-4-hydroxyphenyl)propionic acid with 1,4-*cis*-polyisoprene [297].

Exploitation of Terminal Reactive Groups. Various groups introduced as a result of initiation, termination and/or chain transfer reactions during polymerization processes or remaining in polycondensates as a result of residual functionality of bifunctional monomers may be chemically modified. Stabilizers of types F and G (Scheme 1) can be synthesized from endreactive (telechelic) polymers and oligomers or from endreactive species formed in living polymerization by means of end-capping [69]. Living polymers and copolymers of butadiene or isoprene, styrene or methyl methacrylate, prepared in lithium catalyzed polymerization or PMMA prepared by group-transfer polymerization were end-capped with stabilizing moieties. Antioxidant **243** for stabilization of elastomers [298], AO **244** for

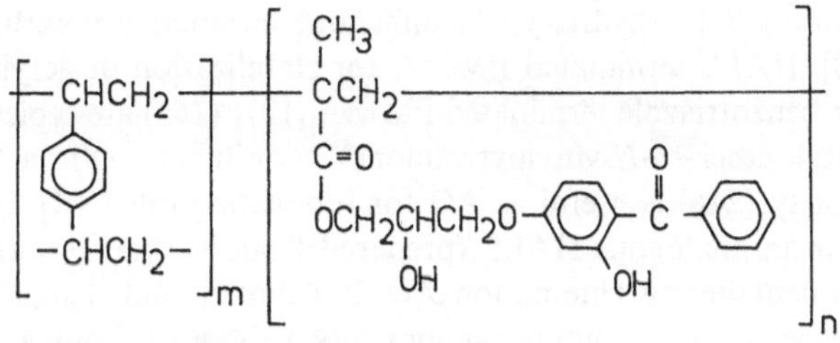

239

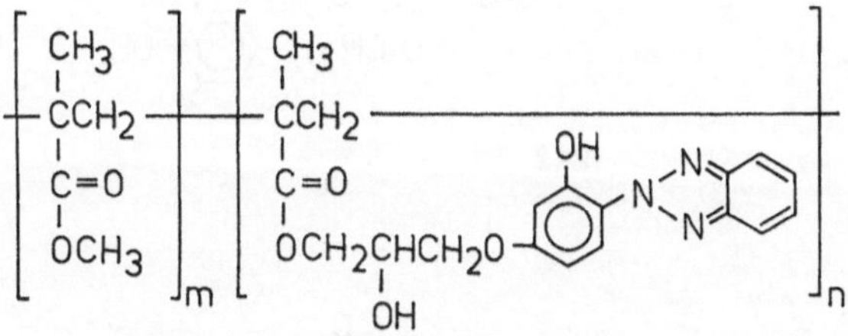

240

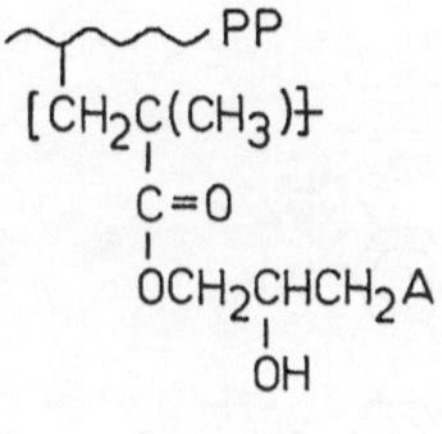

241

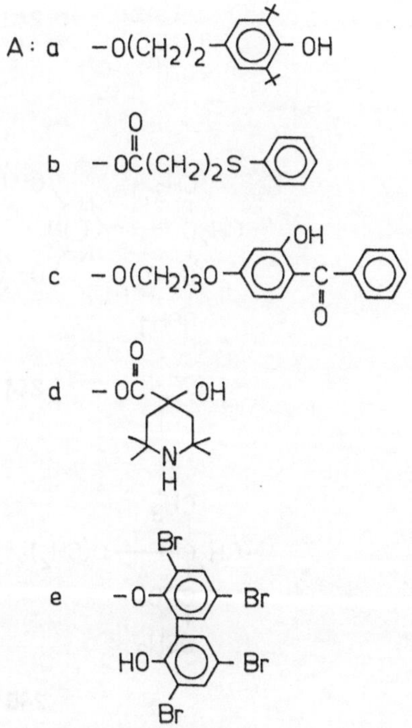

A : a $-O(CH_2)_2$ —(ring with X, X)— OH

b $-O\overset{O}{\overset{\|}{C}}(CH_2)_2S$ —(ring)

c $-O(CH_2)_3O$ —(ring with OH)— $\overset{}{\underset{O}{C}}$ —(ring)

d $-O\overset{O}{\overset{\|}{C}}$ —(piperidine ring with OH, N—H)

e $-O$ —(ring with Br, Br)— HO —(ring with Br, Br)

space application [299], 2-hydroxy-benzophenone terminated poly(butadiene-*co*-isoprene) [300], HALS terminated PMMA for stabilization of acrylic-melamine paints [126] or benzotriazole terminated PMMA [127] (**245**) are typical examples. Poly(vinyl cetyl ether-*co*-N-vinylpyrrolidone) terminated with a phosphorus containing moiety (**246**) is useful as AO for lubricating oils [301]. Phenolic AO **246** and **247** or an analogous HALS (prepared from isocyanate terminated BR with the subsequent thermal elimination of carbon dioxide and of aminoterminated oligodienes respectively), aromatic amines **249** (obtained from a hydroxyter-

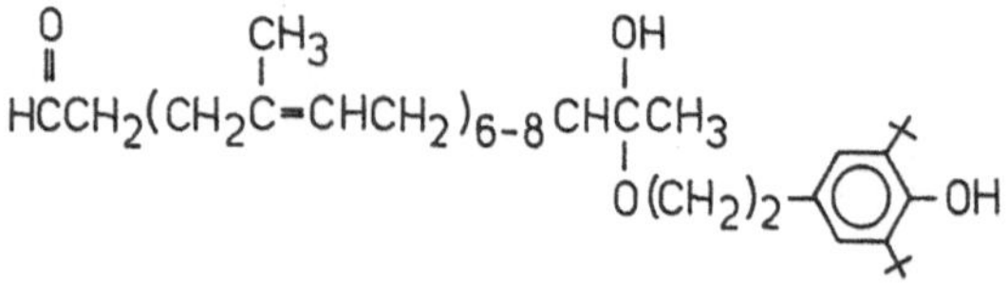

$$\text{O} \atop \text{HCCH}_2(\text{CH}_2\text{C}=\text{CHCH}_2)_{6-8} \overset{\text{CH}_3}{\underset{\text{O(CH}_2)_2}{\text{CHCCH}_3}} \text{—OH}$$

242

$$\left[\text{CHCH}_2\right]_m \left[\text{CH}_2\text{CH}=\text{CHCH}_2\right]_n \overset{\text{O}}{\text{C}}(\text{CH}_2)_2\text{—OH}$$

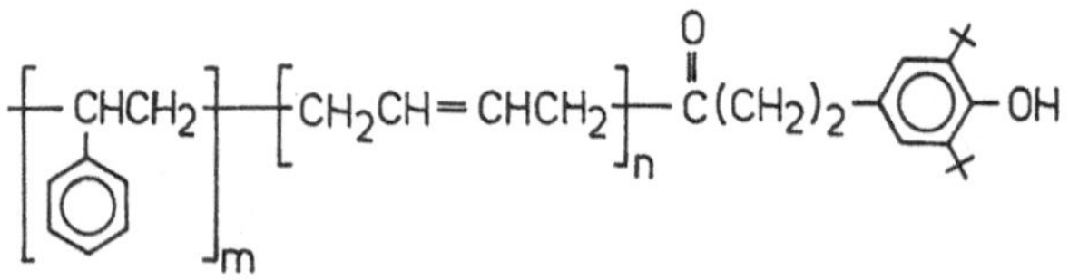

243

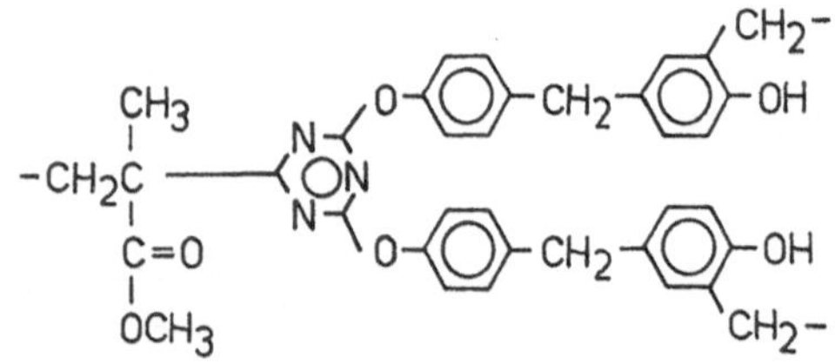

244

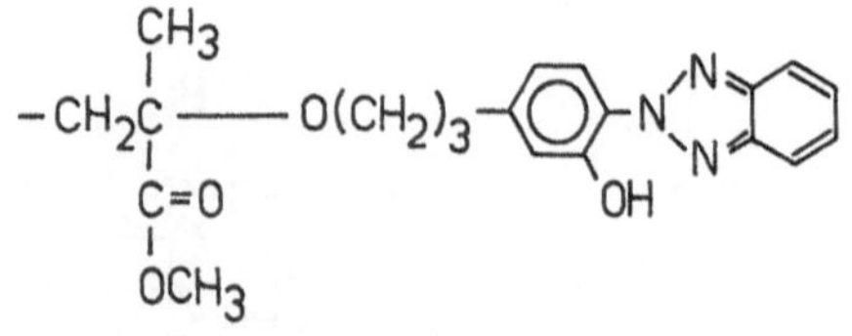

245

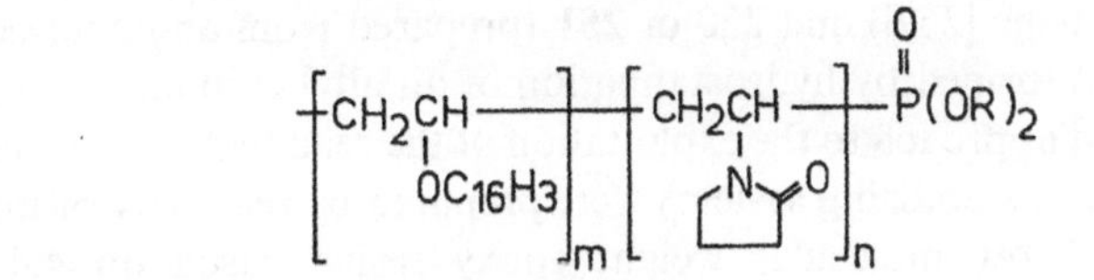

$$\left[CH_2CH \underset{OC_{16}H_3}{\mid} \right]_m \left[CH_2CH \underset{N \diagdown_O}{\mid} \right]_n P(OR)_2 \overset{O}{\parallel}$$

246

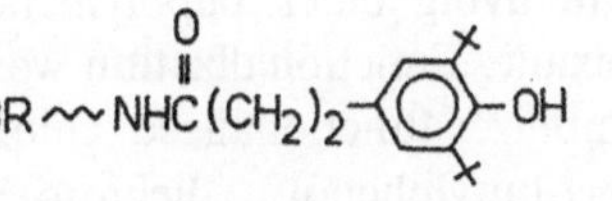

$$BR \sim NH\overset{O}{\overset{\parallel}{C}}(CH_2)_2 - \underset{X}{\overset{X}{\bigcirc}} - OH \qquad BR \sim NHCH_2 - \underset{X}{\overset{X}{\bigcirc}} - OH$$

247 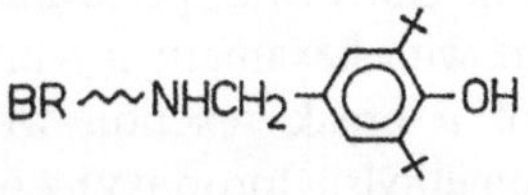   **248**

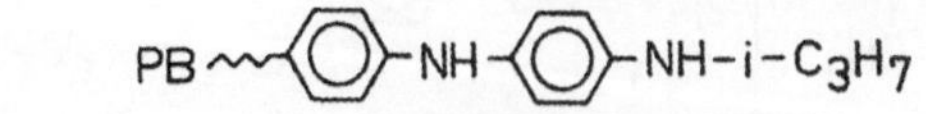

$$PB \sim \bigcirc - NH - \bigcirc - NH - i - C_3H_7$$

249

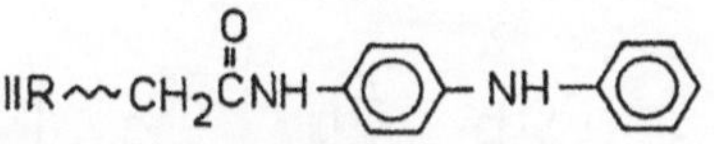

$$IIR \sim CH_2\overset{O}{\overset{\parallel}{C}}NH - \bigcirc - NH - \bigcirc \qquad IIR \sim CH_2\overset{O}{\overset{\parallel}{C}}S\overset{N}{\underset{S}{C}}\bigcirc$$

250 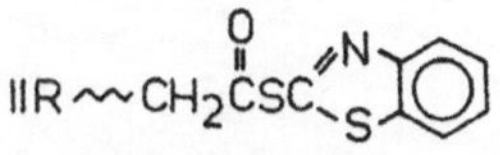   **251**

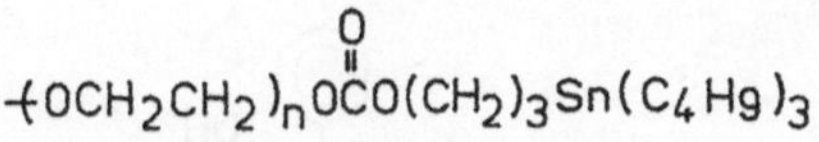

$$\left. OCH_2CH_2 \right)_n O\overset{O}{\overset{\parallel}{C}}O(CH_2)_3Sn(C_4H_9)_3$$

252

$$HO - CH\overset{O}{\overset{\parallel}{C}}OSn(C_4H_9)_3$$
$$\sim CHCH_2O\overset{\mid}{C}H$$
$$\underset{OH}{\mid} \qquad \underset{\overset{\parallel}{O}}{COSn(C_4H_9)_3}$$

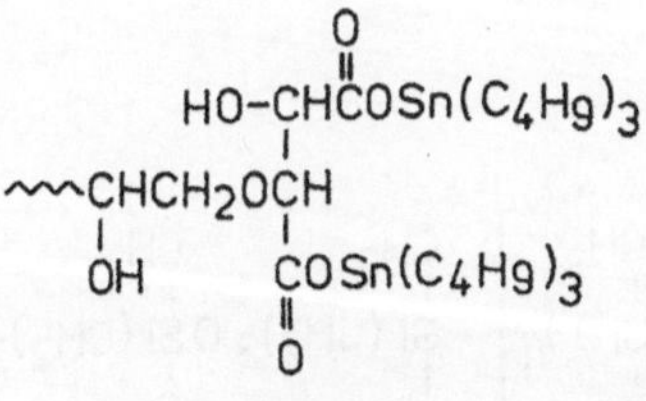

253

minated liquid polybutadiene [257]) and **250** or **251** (prepared from an ozonized IIR [303]), or biocide **252** formed by hydrostannation of an allyl carbonate [304] are examples of the varied approach to the exploitation of the reactivity of various terminal groups. Numerous stabilizing systems were prepared by reactivity of the terminal epoxy groups. Low molecular weight epoxy resins based on 4,4'-isopropylidenebisphenol were mostly used as polymeric reactants for the attache-ment of antioxidant phenolic or 1,4-phenylenediamine functions, light stabilizing benzophenone or HALS functions, PVC thermostabilizing 4-amino-1,2,3-triazole or benzimidazole functions or biocide tri-*n*-butyl tin bearing functions [105] **253**. A detailed study was performed with living block copolymers of styrene or butadiene with hexamethylcyclotrisiloxane. Functionalization was performed by means of a break reaction with stabilizer-functionalized chlorosilanes. 4-(3-Dichloromethylsilylpropoxy)-2,6-di-*tert*-butylphenol,   dichloro(3-chloropropyl)

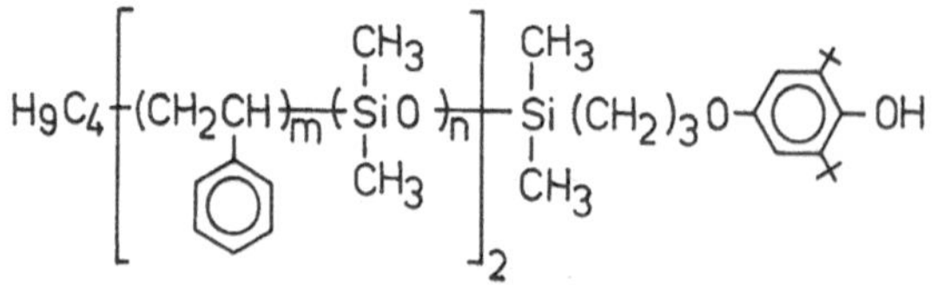

254

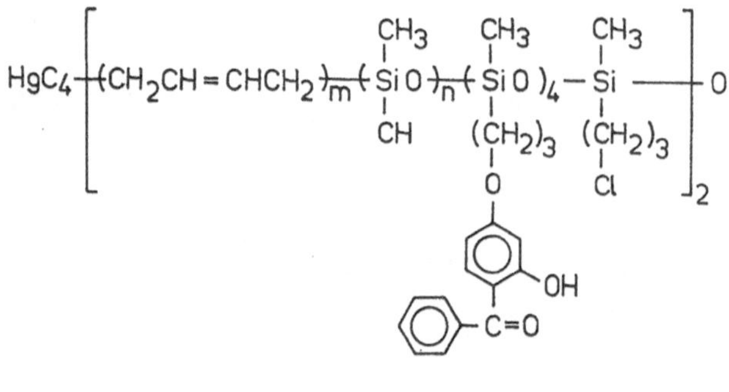

255

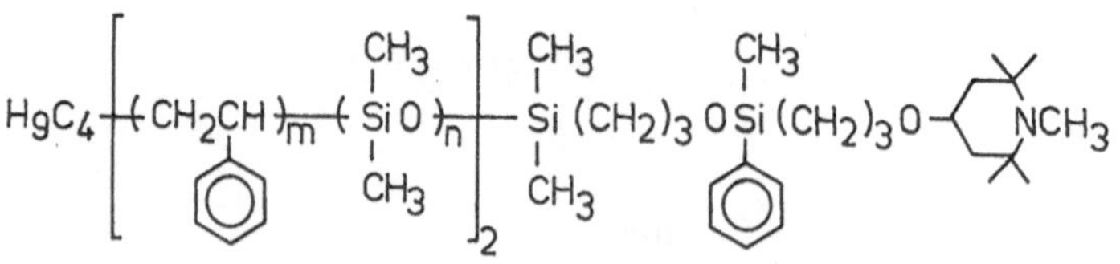

256

methylsilane (stopping agent) with 2,4-dihydroxybenzophenone and 4-[3-(ethoxymethylphenylsilyl)propoxy]-1,2,2,6,6-pentamethylpiperidine were used [305] for the synthesis of hydrolysis resistant stabilizers 254–256.

3.2.3 Intramolecular Rearrangements
of Polymers Bearing Precursors of Stabilizing Moieties

Acetophenone or benzophenone moieties forming a part of a polymer chain or of pendant groups can be created by means of the Fries rearrangement of suitable aromatic ester moieties. Aromatic polyesters should be, therefore, considered as

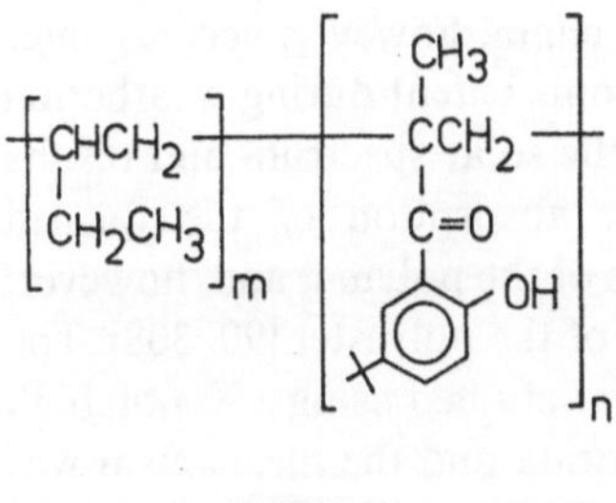

257

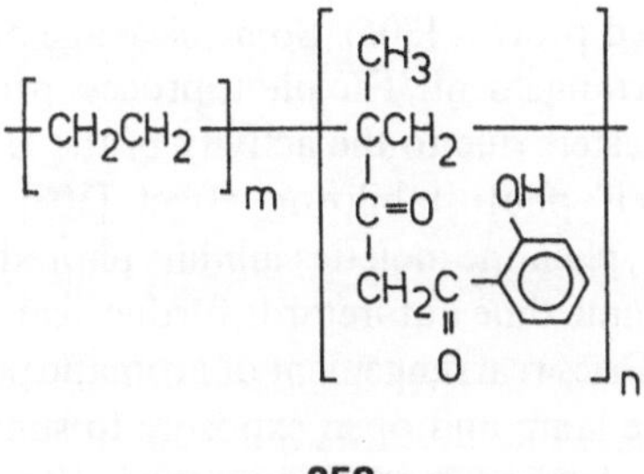

258

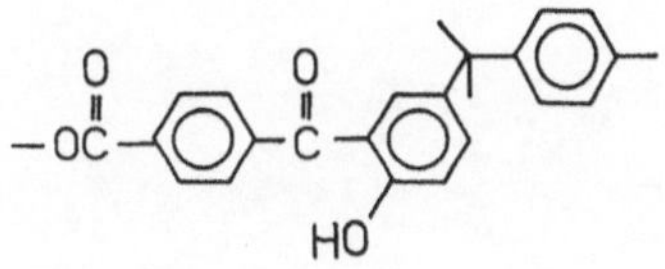

259

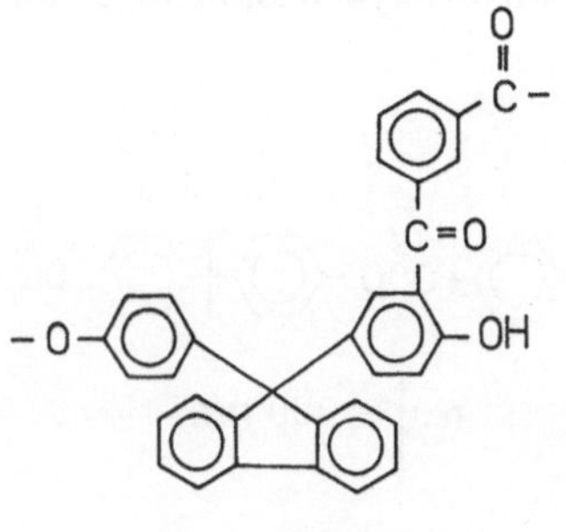

260

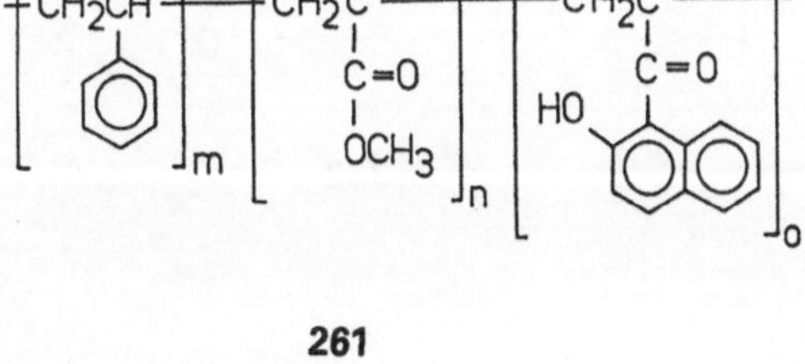

261

latent UV absorbers (UV absorber progenistors). Stabilizers of types A, B and C (Scheme 1) can be formed using this technique. Aluminium chloride catalyzed thermal rearrangement of poly(4-methylphenyl acrylate), poly[4-*tert*-butylphenyl methacrylate-*co*-(1-butene)] and of a polyester formed from terephthalic acid and 4,4′-isopropylidenebisphenol, or alkali catalyzed rearrangement of poly(ethylene-*co*-2-acetylphenyl methacrylate) are typical examples. Light stabilizers with pendant 2-hydroxyphenone [306] (**257**) or 1-(2-hydroxyphenyl)-1,3-propanedione moieties [307] (**258**) and with chain built-in 2-hydroxybenzophenone moiety [306] (**259**) are formed. Formation of dark products and aluminium containing polymers are undesired by-products of the AlCl$_3$ catalyzed rearrangement. The dark products are not formed during the photo-Fries rearrangement of aromatic esters. The structure of the rearranged products is however similar to that formed by the catalyzed process [306]. Some cleavage of the main chain, however, accompanies the rearrangement. The photoprocess proceeds to some extent during weathering of polyesters due to the activity of the UV part of the solar spectrum and results in a self photostabilizing effect [308]. The light absorption of the formed hydroxyphenone moieties inhibits photodegradation of the polymer and, however, at the same time autoretards further rearrangement of the polyester [90, 308]. The photo-Fries rearrangement of aromatic polyesters was studied using a Xenon high pressure lamp and open exposure to sunlight in Florida and the mechanism was explained using model compounds. The internal photostabilization effect was exploited in various systems, e.g. in polyacrylates based on fluorene (a system **260** containing 2-hydroxybenzophenone moiety was formed [308] after the photo-Fries

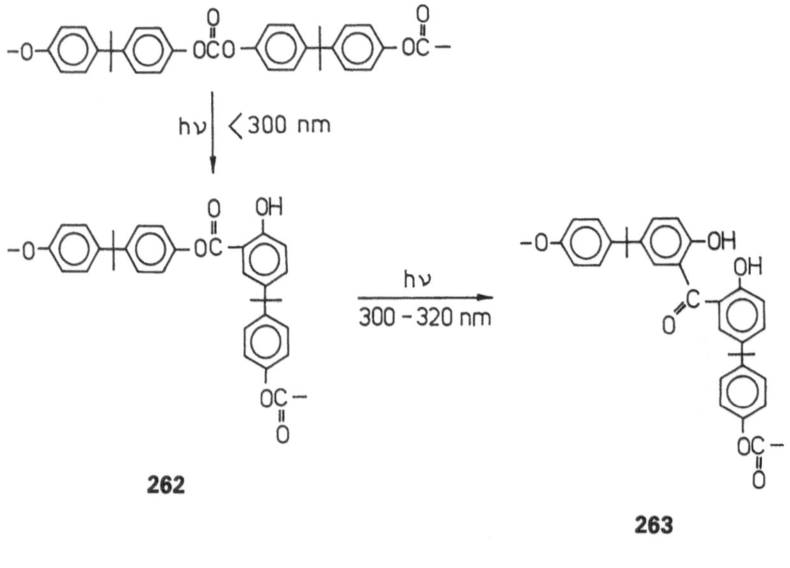

SCHEME 11

rearrangement), in a multilayer system formed from a PC film coated with a copolymer of methyl methacrylate with a 3-benzoyloxyphenyl urethane moiety bearing ethyl methacrylate and with a alkoxysilane/SiO$_2$ blend [309], in solid poly(styrene-*co*-methyl methacrylate) containing minor amounts of 2-naphthyl methacrylate [310] (moiety **261** formed during rearrangement is responsible for the photostabilization effect), in a polycarbonate resin, using a two stage radiation process [311]: phenyl salicylate moiety **262** was formed in the first stage; the 2,2′-dihydroxybenzophenone moiety **263**, formed in the second stage, improves the weathering resistance of PC against yellowing. Recently, the mechanism of the photo-Fries rearrangement of [2-(2*H*-benztriazol-2-yl)-4-*tert*-octylphenylbenzenesulfonate] creating a 2-(2-hydroxyphenyl)benzotriazole moiety was explained [312]. The data can be exploited also for polymer bound moieties.

4 Properties of Oligomeric and Polymeric Stabilizers

The optimum exploitation of stabilizers having both the high inherent chemical efficiency and physical persistence may be achieved only if the stabilizers are soluble and evenly distributed in the polymer matrix or localized in the areas of concentrated attack by deteriogens. This involves e.g. localization of AO in amorphous parts of semicrystalline polymers or of LS and AOZ in surface layers of a polymer. Moreover, good compatibility must endure long term application of a doped polymer. All these properties have been required from the both conventional HMW and macromolecular stabilizers. The latter have been involved in the innovative programme of stabilizer synthesis. Oligomers and polymers prepared from reactive functionalized low molecular weight compounds by means of methods described in Sect. 3 and having properties of AO, MD, LS, FR and BIO-S, respectively, belong generally to specialty polymers. The latter attained specific importance in the last decade [46]. Their price is generally much higher than that of conventional polymers. This is due to the necessary modification in the polymer synthesis via polyreactions. A specific performance can be however achieved using these systems. Economically favourable synthetical methods exploiting conventional processes for polymer functionalization, e.g. processing of polyolefins or vulcanization of rubbers, are, therefore, very much appreciated. The high cost/performance ratio allows the exploitation of macromolecular stabilizers only in very demanding applications.

It has been generally accepted that macromolecular stabilizers protect polymers by the same physical or chemical mechanism as conventional stabilizers containing a comparable functional moiety. (For details dealing with activity mechanisms of macromolecular stabilizers see Sect. 4.2.) Some differences must be however taken into consideration due to the macromolecular character of polymeric stabilizers.

Improvement of physical persistence was the main aim of the development of the latter. The problem of stabilizer volatilization was solved practically by the application of HMW stabilizers. The extraction problem remains however open even with the macromolecular stabilizers. Oligomeric stabilizers are slowly lost, mainly by a repeated solvent leaching [243]. Compatibility problems have been involved in the aplication of the fully extraction resistent macromolecular stabilizers. It seems to be difficult to construct one general-purpose polymer AO which would be compatible with several important classes of polymers [313]. The compatibility of the former with some conventional polymers drops with increasing molecular weight of the functionalized polymer (polymeric stabilizer). In contrast to HMW stabilizers, a more precise selection of individual components of the couple host polymer-polymeric stabilizer must be done. This limits a more general exploitation of polymeric stabilizers, each of them should be almost a tailor-made system for a definite polymer to be stabilized. The compatiblity problem is more serious in crystalline than in amorphous polymers. A slow separation of phases and formation of domains oversaturated with polymeric stabilizers take place in incompatible blends. This makes the proper exploitation of the efficacy of these domains for stabilization questionable. It was not experimentally determined if domains consisting of polymeric AO or AOZ and having a high intrachain concentration of functionalized groups are able to protect the residual bulk or the surfaces of the polymer matrix. Moreover, the crystalline domains of semi-crystalline polymers can be depleted after blending with polymeric stabilizers. It was reported [218] that grafted stabilizing moieties do not disrupt the crystallites if the graft content is lower than 20%. The permanence of a stabilizer is not only related to its molecular weight (the latter may increase only the physical persistency) but also to its specific molecular architecture. For example, a "dilution" of polar stabilizing moieties by nonpolar conventional units (like in copolymers or copolycondensates) increases the solubility in a nonpolar polymer [27]. The volume of the stabilizer molecule is of greater importance than its length. To improve the compatibility, the chemical structure of polymeric stabilizers must be therefore properly selected. For example, homopolymers of polymerizable 2-hydroxybenzo-phenones and 2-hydroxyphenylbenzotriazoles have very low compatibility with acrylic coatings. Therefore, the physical deterioration of the latter cannot be slowed down [26]. However, copolymers of the same monomers with methyl methacrylate can be used efficiently because of the improved compatibility.

An intramolecular incompatibility takes place also between functionalized grafts and the main polymer chain [314]. A tendency for the separation of phases is increased with the increasing molecular weight of grafts. Each component of the system, i.e. the functionalized graft and the main chain are looking for the maximum "homogeneous" contact with the same component ("simili similia solvuntur"). The incompatibility with the "foreign" component is thus increased. A surface accumulation of functionalized moieties takes place in systems where a low amount of the grafted stabilizer has been blended with a polymer having the same chemical character as the main chain of the grafted stabilizer. A favourable state may be reached using a functionalized polymer having an optimum number of grafts with a proper length. This, in principle, has not been determined experimentally up to now for stabilizer-grafted polymers.

The heterogeneity in the distribution of polymeric stabilizers and a limited solubility of the latter in the mass of the solid host polymer may deplete a substantial part of advantages arising from the high physical persistency of stabilizers.

It seems that stabilizers prepared by copolymerization of functionalized monomers with conventional monomers and some polycondensates having $\overline{M}_n < 5000$ are the most acceptable from the point of view of the compatibility with the semicrystalline host polymer. Systems prepared by polymer analogous reactions may have higher values of $\overline{M}_n$. A relatively homogeneous distribution of functionalized groups can be achieved after blending with the host polymer. Segmental mobility of functionalized chains should assure interactions with severed sites in the polymers to an extent similar to that of HMW stabilizers [313]. It seems that the limitation of the mobility of functionalized segments in polymer bound AO is not so important in heat aging tests [243]. The chain mobility in amorphous areas of the host polymer is probably sufficient to reach the attacked polymer sites and scavenge $RO_2^\bullet$ radicals. It was mentioned that the restricted chain mobility of oligomeric AO does not fully limit their antifatigue activity. The mobility of the functionalized chain can be restricted by other segments or construction units present in the stabilizer functionalized polymer [315]. This may be due to the formation of relatively strong van der Waals bonds between various groups. Functions formed via side reactions during the polymer analogous synthesis of stabilizers can be involved in these interactions. The influence of the character of the polymer chain bearing a functional moiety or a mutual interaction of various polymer attached functional groups on the final stabilizing effect have been considered up to now only rarely. In stabilizers of the general type A (Scheme 1), the functional moieties are immobilized. Only a long spacer (type C, Scheme 1) allows to reach the attacked sites localized relatively far from the polymer chain carrying stabilizing moiety [315].

The transport of polymeric stabilizers via diffusion through the host polymer drops with the increasing molecular weight of the former. This fact may represent the main kinetic limitation of the efficient protection in processes where a high local concentration of stabilizers is a necessary condition. A limited migration ability of oligomeric AO, AOZ or LS was considered as one of reasons of their lower stabilization effect observed in various experiments [27]. Important experimental data were obtained with aromatic amines [251] and some conflicting data dealing with AO/AOZ efficiency of polymer bound species were explained. It was shown using properly selected model compounds that the AOZ effect was lost after attachment of the stabilizing moiety to the polymeric chain. The long-term antioxidant effect of bound species remained, however, comparable with that of migratable amines. The limited migration was, therefore, the main restriction of the antiozonant − and most probably even of antifatigue − activity in articles where protection against dynamic stressing during weathering requires a renewable surface concentration by migration. An optimum molecular weight of stabilizers must be, therefore, maintained.

The analysis of information available dealing with chemical efficiency and physical persistence of AO and with important physical relations between the stabilizer and the host polymer reveals [313] that polymeric antioxidants having molecular weights in broad limits of $\overline{M}_n$ 3000 to 20,000 are acceptable according to the character of the host polymer. Minagawa [38] reports $\overline{M}_n \sim 2000$ as an optimum for HMW phenolic AO. It was found that the antioxidant effect decreases in multifilament yarns with increasing molecular weight of AO [313]. Because of the longer physical persistence, however, the chemical loss of AO becomes more significant in accelerated tests.

For stabilizers from the group of HALS, the value of $\overline{M_n}$ in the range of 3000 to 5000 was reported as an optimum [316]. The data were obtained with PP tapes (50 µm) doped with poly(1,2,2,6,6-petamethyl-4-piperidyl acrylate) of different $\overline{M_n}$ and the validity of data thus obtained was extended for all polymeric HALS. It should be mentioned that the application of polymeric stabilizers having lower $\overline{M_n}$ assures a greater flexibility in application in various polymers.

$\overline{M_n}$	Hours of UV irradiation in Xeno-test 1200 until 50% tensile strength	Days in circulating air oven at 120 °C until brittle
1800	6420	125
2720	6720	127
6800	4140	98
13200	3350	88
23400	2710	51
Control	690	39

Thermal and hydrolytic stabilities of polymeric and polymer bound stabilizers under conditions of polymer processing and/or application were reported only exceptionally. The advantage of the physical persistence of macromolecular stabilizers can be lost due to the low thermal stability of the latter. This case was reported with poly(3,5-di-*tert*-butyl-hydroxyphenyl acrylate) [317]. Differences were found in the thermostability of various polymer bound 2-hydroxybenzo-phenones [318]: systems attached to a polyester chain were less stable than that formed via reaction of terminal epoxy groups ($\overline{M_n}$ of the both systems was in the range 1700 to 2300). A three stage thermal degradation was observed with poly(2,2,6,6-tetramethylpiperidinyl methacrylate-*co*-methyl methacrylate) [319]. The copolymer suffers a greater weight loss than PMMA below 650 K. It seems that an initial chain scission takes place between functionalized and nonfunctional-ized methacrylate units. A lower thermal stability was also observed with some copolymers of styrene or acrylonitrile with brominated phenyl acrylates (the latter have properties of FR) in comparison with nonbrominated systems [76].

Studies of hydrolytic stability were performed with LS containing benzotriazole moieties. It was reported [293] that copolymers prepared from various vinyl or isopropenyl derivatives of 2-(2-hydroxyphenyl)-2*H*-benzotriazole were hydrolyti-cally stable. However, acryloyloxy and methacryloyloxy groups linked directly to the 2-hydroxyphenyl ring of the benzotriazole moiety were suspected of lower hydrolytic stability. Therefore, systems having the benzotriazole moiety connected with the main chain via an aliphatic spacer were prepared, e.g. copolymers of 2-[2-hydroxy-4-(2-hydroxy-3-methacryloyloxy)propoxy]-2*H*-benzotriazole with methyl methacrylate [293].

Due to the above-mentioned physical facts, some differences in mechanistic details in the effectivity between conventional and macromolecular stabilizers should be expected.

Many macromolecular stabilizer-functionalized systems and reactive stabilizer-functionalized monomers have been described in the literature. Only some of them have been however commercialized. Monomeric AO [2-(2-hydroxy-3-*tert*-butyl-5-methylbenzyl)-4-methyl-6-*tert*-butylphenyl] acrylate (**97**, Irganox 3052, Ciba-Geigy; Sumilizer GM, Sumitomo) is a reactive stabilizer for unsaturated rubbers [221, 320]. Mechanism of its activity was studied in details [221]. A polymerisable 4-(2-acryloyloxy)-2-hydroxybenzophenone (Cyasorb UV 2098) has been commercialized by American Cyanamide. Some macromolecular AO were available on the market only for a limited period. These include Poly AO-79 (**97**, Dynapol), a nondigestable AO for food application, Modanox 2000 (**118**, Monsanto Corp.), a processing AO for polyolefins, and Poly TDP-2000 (**141**, Eastman Chemical Products), a HD AO for polyolefins. Some very important antidegradants have been commercialized for the stabilization of rubbers. Chemigum HR (**86**, Goodyear Tire and Rubber) has been available as a terpolymer containing 22, 27, 32, 39 and 45% acrylonitrile, respectively, (the various grades of Chemigum HR have been specified with number indexes 967, 765, 665, 365 and 266, respectively). This grade range covers the available spread over which conventional NBR is used. Polymerization stabilized NBR (Chemigum HR) imparts the stability normally obtained by the use of the same concentration level of efficient conventional HMW antioxidants [50–53, 77, 94]. Minimum losses in physical properties of NBR were observed however when the stabilized rubber was exposed to sequence aging in hot oil. The Chemigum HR/NBR blend is compatible also with PVC and ABS. Wingstay K (**134**) and Wingstay L (**99**) are other oligomeric AO of Goodyear designed for rubber stabilization. Stabilizer concentrates of the type **196** (R = H, n = 1, 2) are an AO development by Robinson Brothers Ltd. [321] and designed for rubbers. Chemantox AO-49 (**207**, Chemopetrol) has been used for polyolefin stabilization. Oligomeric 2,2,4-trimethyl-1,2-dihydroquinoline (**100**) has been commercialized under various trade names, such as Agerite Resin D (R. T. Vanderbilt), Antigene RD-6 (Sumitomo), Flectol H (Monsanto), Good Rite 3140 (B. F. Goodrich), Naugard Q (Uniroyal), Nocrack 224 (Ouichi Shinko), Nonflex RD (Seiko Chemicals), Nonox TQ (ICI), Permanax TX (AKZO) or Vulkanox HS (Bayer). An oligomeric polycondensate of acetone and diphenylamine, Naugard A (Uniroyal) was designed for PA stabilization. A more complicated polycondensate of diphenylamine, acetone and formaldehyde has been commercialized under the trade name Naugard BG (Uniroyal).

National Starch 78-6121, poly[methyl acrylate-*co*-(3-hydroxy-4-benzoylphenyl) acrylate] [82] and Cyasorb UV 2126, poly[4-(2-acryloyloxy)-2-hydroxybenzo-phenone] (**71**, American Cyanamide) are representatives of polymeric benzophenones designed for polyolefins. Some very important oligomeric LS from the HALS group are available on the market: Chimassorb 944 (**161a**, Ciba-Geigy), Cyasorb UV 3346 (**161b**, American Cyanamide), MARK LA-63 (**167**, Asahi Denka), Spinuvex A-36 (**160**, Montefluos) and Tinuvin 622 (**166**, Ciba-Geigy).

Trade names of two polymeric flame retardants have been reported in the technical literature: Fire Guard 7500 (**159**, Sankyo Organic Chemicals, Co), a functionalized oligocarbonate for ABS prepared from 4,4'-isopropylidenebis(dibromophenol) and FR 1025, poly(pentabromobenzyl acrylate) (**76**, Technion) [68].

Technical reports give some information about the average molecular weights ($\overline{M_n}$) of commercial stabilizers: Poly AO-79 (**97**) ~5500, Modanox 2000 (**118**) ~1100, Tinuvin 622 (**146**) ~3600, Chimassorb 944 (**161a**) >2500, Cyasorb UV 2126 (**71**) ~50000.

It can be expected that some other functionalized polymers will be commercialized in the future due to an increased pressure of legislative requirements against physically non-persistent additives used for stabilization of polymeric materials in contact with food or regulations for handling toxic rubber chemicals [322, 323]. The main interest has been given to systems prepared by current technologies.

Using polyreactions and polymer analogous reactions, polymers with two different kinds of stabilizing moieties bound in one molecule in a proper concentration ration can be prepared. The experiences of organic chemists have been exploited to find out ways to change the molecular architecture of stabilizers and to target the stabilizing properties to the most vulnerable segments of multiphase systems.

4.1 Areas of Use

Modern technology is making ever-increasing demands on the durability of plastics, rubbers and coatings. Highly demanding operating conditions together with a more discriminating consuming public have contributed to the increasing demand for higher persistence of polymer stabilizers. Permanent stabilizers assuring retention in the polymers are therefore necessary. In spite of the high price of functionalized polymers, the extent of their consumption as stabilizers of commercial polymers is growing. This is mainly true for LS. The most appreciated fields of application are in articles where long-term troubleless performance under extreme conditions is required, in household, machinery, automotive industry, solar energetics, space or durable medical replacements application. Ecological and legislation rules require a high safety performance in packaging materials.

Engine seals, diaphragms, O-rings; coolant, lubricant and transmission fluid seals; components of fuel pumps, gaskets and hoses in contact with extraction fluids; rubber parts of engines working under hoods at high temperatures; oil field equipment (packers, drill pipe protectors, swab cups); heavy duty truck tires operating at high speeds and in ozone and acid rain rich areas; hose and cable jackets, wire insulation for hot air environment; conveyors used in mines; shoe soles; and/or printing rolls are examples of rubber articles which must have extreme persistence during technical application in aggressive environments. Various plastic articles are expected to operate without failure when exposed to UV light, heat or solvents. Textile fibres, packaging materials for foods, pharmaceuticals and cosmetics, parts of washing machines, hot water pipes, durable, protective, transparent or pigmented coatings, various articles applied in solar energetics, medical devices, implants and replacements in contact with body fluids, toys and articles where dermatologic complications or digestion problems may arise, may serve as examples.

Oligomeric and polymeric stabilizers which have high physical persistence can be used as masterbatches and blended with unstabilized polymers using conventional techniques [243]. The concentration level of polymeric additives should be kept to limits ensuring a sufficient molar concentration of stabilizing moieties, i.e. an efficient protection of the host polymer.

Reports dealing with the technical application of polymeric AO are concerned mostly with aromatic amines. Excellent results in the stabilization of SBR or NBR articles in engineering applications were obtained with Chemigum HR (86) [77]. Oligomeric 2,2,4-trimethyl-1,2-dihydroquinoline (100) has been used successfully for many years in stabilization of vulcanized rubbers. Both 86 and 100 are efficient AO. The imporance of the application of polymeric phenolic AO for the stabilization of packaging materials for foods may be exemplified [114]: The absorption of Poly AO-79 (97) in the gastrointestinal tract was studied with relationship to the $\overline{M_n}$ of the AO:

97, $\overline{M_n}$	Absorption, %
760	1.5
7300	0.44
67000	0.34

This indicates that polymeric stabilizers having $\overline{M_n} > 800$ have high resistance against extraction by components of foods, and, moreover, they do not enter into the metabolic cycle. Public concern over the contamination of the human environment can be thus satisfied. As a consequence, stabilizers like Poly AO-79 (97), Chimassorb 944 (161a) or Tinuvin 622 (146) may be legally used in stabilization of polymers in contact with food. Possibilities for application of other polymeric stabilizers for articles used in the home are open. Extractability problems may be thus overcome. Articles made from weather resistant butadiene based multiphase systems like ABS, MBS or MABS [84] stabilized in the most sensitive BR phase with polymer bound stabilizers may serve as an example.

Polymer supported 2-hydroxybenzophenones and 2-(2-hydroxyphenyl)-2H-benzotriazoles have useful properties in articles used under high exposure of UV light, examples are transparent durable acrylic films, outdoor paints, transparent coatings for optical lenses, transparent protective layers of silver mirrors in solar concentrators or films used in encapsulation systems for photovoltaic modules in solar conversion devices [26, 54, 82, 92, 293]. Polymeric LS can also be used in cosmetics as sunscreening components of emulsions and ointments for protection of sensitive human skin against sunburn [324] or in medical aids, as nonleachable UV absorbing components of silicone rubber for vision aids (contact lenses) [99].

Excellent results have been obtained with commercial oligomeric HALS (e.g. **146, 147, 160, 161a**) in stabilizing multifilaments, fibers, and thin foils from PP [9].

Polymers of benzoquinone, some tin-containing polymers, like poly(dialkyl tin maleate), zinc or cadmium containing polycondensates of 8-hydroxyquinoline with formaldehyde, polymers containing phosphorus, N,N'-diphenylguanidine, 4-amino-1,2,3-triazole, benzothiazole, benzimidazole, benzothiazolone or dithiocarbamate moieties were reported as thermostabilizers for PVC [67, 325–327].

Brominated styrene grafted onto PET is capable of acting as a nonvolatile FR for PET fabrics [328]. Polyphosphate esters can be used as flame retardant plasticizers [204], phosphorus containing moieties act as color improvers and thermostabilizers in polyesters [198, 205]. Antistatic properties of copolymers containing moieties with sulfonic acids have been reported [107]. Some polymers, like poly(N-vinylimidazole) impart anticorrosive and antirusting properties to protective coatings of metals [329].

There are various applications of polymers under conditions of a potential growth of microorganisms. This involves polymers applied as interlinings of carpets, floor coverings containing softeners, coverings of electrical or electronic equipments or coatings of many articles used in a moist and warm environment. The application of macromolecular BIO-S has been concentrated mainly on preservation of the both decorative aspects and protective efficiency of coatings against attack by bacteria, lichens, fungi, sponges and protozoans [105]. This has been exploited very specifically for protection of ships or pipelines immersed in the sea with paints containing antifouling agents. A specific application involves exploitation of relative toxic but nonleachable and nonvolatile biostabilizers in wall paints to help prevent the spread of infections in hospitals and other social institutions [105]. Other applications involve wood preservation and bioprotection of water dispersion coatings.

4.2 Stabilizing Activity and Mechanisms

Most data available on stabilizing activity deal with antioxidants and/or the light stabilizing efficiency of macromolecular stabilizers before and after extraction under specific working conditions of the respective article made from a stabilized polymer. Cyclic testing, such as cycles involving dry cleaning (or laundering) and drying used for polyolefins, or cyclic contact with hydraulic fluids in "contamination" tests for rubbers [321], are probably more relevant models for practical application of persistent stabilizers under severe conditions than hot air oven tests or continuous total immersion in a leaching medium. These cyclical tests can differentiate the physical persistence of stabilizers under demanding conditions very efficiently.

Some data deal with the stabilizing efficiency of polymers and copolymers bearing a stabilizing functionality. Three isomeric poly(vinylphenols) provided a better resistance to oxidation than the corresponding monomers (the o-hydroxy isomer provided the best protection). All tested poly(vinylphenols) were however slightly inferior to 2,6-di-*tert*-butyl-4-methylphenol [330]. The presence of the phenolic moiety in poly[styrene-*co*-(2,6-di-*tert*-butyl-4-vinylphenol)] had almost no protective effect on the mechanical properties of photooxidized hydrocarbon polymers [331]. Formation of phenoxyls in photooxidized systems was detected. Copolymers of 4-vinyl-2,6-di-*tert*-butylphenol with butadiene or isoprene were weaker AO in BR, IR or SBR than the conventional HMW phenols at 0.1% level in short-term tests. The copolymers were however more effective in long-term protection [56]. Copolymers containing aromatic amine moieties do not affect rubber vulcanization process, cause comparatively little discoloration of rubbers during ageing and no migratory stain while providing antioxidant protection equivalent to that of conventional amines [50, 52]. ABS, EPDM or NBR functionalized by means of polymer analogous reactions and bearing phenolic and aromatic amine moieties were found to be more degradation resistant than the same polymers doped with conventional AO. Tests performed with bound-in stabilizers prepared via reactivity of aromatic nitroso compounds (see Sect. 3.2.2.1, Scheme 4) revealed a much higher antioxidant efficiency of the bound aromatic amine species in comparison with that of phenols. No antiozonant properties were detected with these systems [233].

Grafted polyolefins (type D, Scheme 1) are of interest due to a relatively good compatibility with the host polymer. In the case of the graft content not exceeding 20%, crystallites in the host polyolefin are not disrupted [218]. Melt grafting seems to be more suitable for the synthesis of AO. Radiation induced processes provide surface grafting (see Sect. 3.2.1): LS, AOZ or BIO-S may be thus enriched at the sites of concentrated attacks of deteriogens. Radiation-initiated grafting can be performed only with finished articles. The practical importance of radiation grafting is thus limited.

It must be considered that a part of the reactive monomer used for grafting remains in the system as a monomer or homopolymer [218]. This changes the integral physical permanency and may influence the final chemical efficiency too. For example, PP surface grafted with 3,5-di-*tert*-butyl-4-hydroxybenzyl acrylate was a more efficient AO than the respective monomer or homopolymer. Similar results were obtained in experiments with monomeric, homopolymeric and PP grafted 2,2,6,6-tetramethylpiperidyl methacrylate [228]. Longer lifetime was observed with LDPE or PP grafted with 2-hydroxy-4-(3-methacryloyloxy-2-hydroxy-propoxy)benzophenone in comparison with the respective monomer [218]. A comparable level of photostability was obtained with LDPE films either surface grafted with 0.5% 2-hydroxy-4-(3-methacryloyloxy-2-hydroxypropoxy)benzophenone or bulk grafted with 1% of the same monomer [332]. Both grafted systems were more efficient LS in LDPE films that the relevant but only physically dispersed monomer. None of these systems reached however the efficiency of 2-hydroxy-4-octyloxybenzophenone.

More precise data are lacking dealing with the influence of the length and frequency of grafts on the stabilizing effect. It has been considered [236] that more frequent low chain grafts exert a more effective antioxidant protection than few but long chain grafts.

The efficiency of grafting is important for the final stabilizing effect and is influenced by the structure of the polymer matrix as well as of the monomer. For

example, 56% of 2-(2-hydroxy-5-vinylphenyl)-2H-benzotriazole was grafted onto
cis-1,4-polybutadiene contrary to 14% bound to poly-1,2-butadiene [56]. The
efficiency of grafting of 2-hydroxy-4-acryloyl or methacryloyloxybenzophenone
decreased in the order PS > LDPE > PP. The methacryloyloxy derivative was
always more reactive [226]. The light stabilizing activity of melt grafted bis(2,2,6,6-
tetramethylpiperidyl) maleate onto PP was linearly related to the concentration
of HALS moieties up to 70% binding. The formed grafted stabilizer was as
effective as HMW HALS Tinuvin 770 (**10**) when compared on the molar basis
[333].

A very limited amount of information deals with the efficiency of polycondensate
antioxidants. It was reported [158] that the efficiency of linear phenol-formaldehyde condensates drops in systems containing more than three phenolic nuclei.
Similarly, an optimum AO effect was observed with 4-methoxyphenol-formal-
dehyde condensates having a molecular weight of 1000 as a maximum [160]. Both
reported data [158, 160] deal more or less with the inherent chemical efficiency
of stabilizers. The physical persistency was not particularly considered in this case.

Important data were published on activity of light stabilizers having structures
of 2-hydroxybenzophenones, salicylates, 2-(2-hydroxyphenyl)-2H-benzotriazoles
and hindered amines. Some contradictory results can be found in the literature
dealing with the data comparing efficiency of polymeric and HMW systems. This
is mostly due to the differences in testing conditions. A comparable light stabilizing
efficiency in PE was observed with poly-(2-hydroxy-4-acryloyloxybenzophenone)
and a HMW 2-hydroxy-4-alkoxybenzophenone. Both these derivatives of benzophenone were however inferior to a HMW 2-(2-hydroxyphenyl)-2H-benzo-
triazole [61]. This reveals that the importance of the inherent chemical efficiency
can surpass that of the physical persistency in the case when systems differing in
chemical composition are compared.

Poly[styrene-*co*-(2-hydroxy-4'-vinylbenzophenone)] was less efficient in PS than
2-hydroxy-4-methoxybenzophenone [334]. Similarly, PE films doped with 4-
dodecyloxy-2-hydroxybenzophenone (0.1 mol%) were more stable than PE doped
with copolymers of ethylene with polymerisable benzophenones having a comparable content of chromophores [54]. The efficiency of a SAN type LS, a
terpolymer of 2-hydroxy-4-(4-vinylbenzyloxy)benzophenone with acrylonitrile
and styrene did not exceed that of conventional LS [84]. No efficiency loss of
2-hydroxy-4-methacryloyloxybenzophenone in ABS was observed after bonding
into a terpolymer with styrene and acrylonitrile. The homopolymer was slightly
inferior to both the monomer and terpolymer [84]. A better protection of PP was
provided by poly[(2-hydroxy-3-allyl-4-methoxyphenylbenzophenone)-*co*-dibutyl
maleate] than with 2-hydroxy-3-allyl-4-methoxybenzophenone [335] (stabilization
tests were performed in the presence of phenolic antioxidants). A comparable or
better light stabilizing efficiency of poly[vinyl acetate-*co*-(5-methylacryloyloxy
salicylate)] or poly(2-allylphenyl salicylate-*co*-dioctyl maleate) than that of alkyl-
phenyl salicylates was observed in polyolefins [335].

Polymeric stabilizers were prepared from benzotriazoles having polymerizable
functions attached in position 5 or in positions 3, 4 and 5, respectively, of the
2-(2-hydroxyphenyl) moiety [336]. Free hydroxy groups of the latter must remain

in the system formed after polymerization to preserve the light stabilizing activity. High activity was mostly reported with polymeric benzotriazoles. Poly[2-(2-hydroxy-3-allyl-5-methylphenyl)-2H-benzotriazole-co-dimethyl maleate] or analogous systems imparted a better light stability to PP than 2-(2-hydroxy-5-methylphenyl)-2H-benzotriazole [87]. Poly[2-(2-hydroxy-4-acryloyloxyphenyl)-2H-benzotriazole-co-styrene (or methyl methacrylate)] was a more efficient LS in poly-cis-1,4-butadiene than the respective monomer or homopolymer [337]. Various experimental data are available dealing with the physical persistence and light stabilizing efficiency of oligomeric HALS after extraction of stabilized PP fibers or films with hot water (90 °C), aqueous solutions of detergents or solvents used in dry cleaning. Activity losses of HMW HALS are mostly five times greater than that of oligomeric HALS. The high persistence of the latter therefore assures good protection also when the efficiency of HMW HALS before extraction surpasses that of oligomers [338]. The compatibility of stabilizers with the host polymer matrix, however, exerts, an important influence. This was evidenced by comparison of the activity measured with various types of HALS in PP. An activity series homopolymer < monomer < copolymers with long-chain alkyl methacrylates (alkyls C_{12} or C_{18}) was found [338]. The molecular architecture of the polymeric stabilizer may be expressed too: e.g. a copolymer of vinyl (2-ethylbenzyl) ether with a mixture of bis(1,2,2,6,6-pentamethyl-4-piperidyl) itaconate and mesaconate was – at the same concentration level – a stronger LS in PP sheets than a relevant copolymer with ethylenebis(1,2,2,6,6-pentamethyl-4-piperidyl) fumarate [339]. It is not surprising that only minor activity differences were observed between structurally related oligocondensates Cyasorb UV 3346 (161b) and Chimassorb 944 (161a) in polyolefin stabilization.

Contrary to many data dealing with the synthesis of macromolecular stabilizers, the number of mechanistic studies is rather poor. Most data available deal with polymeric LS. The earliest studies were published by Tocker at the end of the 1960s [54] and were foccused on polymer systems containing 2-hydroxybenzo-phenone, 2-hydroxyacetophenone or salicylate units. The benzophenone derivatives were the most photostable species at the Delaware and Florida test stations or in the Atlas Weatherometer. Excellent mechanistic studies were published in the 1980s with various copolymers containing light stabilizing units by Vogl and Gupta [26, 293, 336, 337, 340-2], Munro [85, 86] and Neidlinger [343]. The outdoor photochemistry of these copolymers is expected to be initiated by an energy absorption of pending chromophores. Time resolved spectroscopy, electron spectroscopy chemical analysis (ESCA) and laser photoacoustic techniques were used for monitoring the photochemical processes [26, 85, 86, 341].

UV spectra of copolymers containing benzophenone moieties were studied in detail and compared with that of monomers. Characteristic maxima were found at 270 and 332 nm [343]. Some differences from the spectra of monomers were observed in the π-π^* transition region. UV spectra of monomeric and polymeric salicylates were essentially similar.

Optically clear films of poly[(2-hydroxy-3-allyl-4,4'-dimethoxybenzophenone)-co-methyl methacrylate] having $\overline{M}_n$ 60000 and 83000 respectively were irradiated and both the character and rate of the decay of the excited state of attached benzophenone moieties were investigated by picosecond laser flash spectroscopy [341]. The loss of bound-in chromophores from protective films formed from blends of PMMA with poly[(2-hydroxy-3-allyl-4,4'-dimethoxybenzophenone)-co-methyl methacrylate] during photodegradation is negligible [340]. A detailed mechanistic study revealed chain scission, hydrogen abstraction and crosslinking as results of photooxidation of the functionalized polymer bearing a benzophenone moiety. A gradual increase in branching and crosslinking due to hydrogen abstraction by photosensitization at branch points was observed also with poly[methyl methacrylate-co-(4-methacryloyloxy-2-hydroxybenzophenone)] in photodegradation studies performed below T_g [342].

The physical loss of poly[(4-methacryloyloxy-2-hydroxybenzophenone)-co-methyl methacrylate], $\overline{M}_n$ 7000 to 40000 in blends with PMMA was negligible during accelerated weathering [343]. The chemical consumption was manifested by a steady fall in the concentration of the UV stabilizing moieties without any noticeable change in the molecular weight. This indicates a cleavage and/or modification of the benzophenone pendant group involved in the chromophore consumption process. The photodegradation of tested films proceeded from the surface into the bulk. The phenomenon of the chemical consumption of benzophenone chromophores observed with the copolymeric stabilizers was similar to that observed in PMMA doped with 4-n-octyloxy-2-hydroxybenzophenone, in spite of the high inherent UV stability of the latter LS [343].

UV spectra of monomeric polymerizable derivatives of benzotriazole show typical absorption at λ_{max} 335 to 340 nm. When the benzotriazole moiety was incorporated into polymers, λ_{max} was somewhat higher (between 340 and 350 nm) depending on the polymer [89, 337]. The introduction of the second benzotriazole unit into the system causes a small bathochrome shift and has a significant effect on the fluorescent spectrum. The extinction coefficient increases dramatically when more than one hydroxy group, which is capable of forming H-bonds, is present in the system [344]. A detailed photophysical study was performed with poly[2-(2-hydroxy-5-vinylphenyl)-2H-benzotriazole-co-styrene] and poly[2-(2-hydroxy-4--methacryloyloxyphenyl)-2H-benzotriazole-co-methyl methacrylate] films and solutions [345]. Emissions at λ_{max} 400 and 500 nm respectively were observed. The longer one can be assigned to an excited state following the excitation of the ground state planar form of the benzotriazole chromophore. The 400 nm fluorescence arises from the excitation of stabiler molecules having a nonplanar ground state configuration. It was suggested [345] that this conformer may be less efficient in stabilizing polymers.

Copolymers of styrene or methyl methacrylate with monomers containing 2-(2-hydroxyphenyl)-2H-benzotriazole moieties or polymers grafted with benzotriazole moieties, e.g. poly(ethylene-co-vinyl acetate) grafted with benzotriazole functionalized groups have a superior screening capacity [346]. A photochemical study of poly[2-(2-hydroxy-5-vinylphenyl)-2H-benzotriazole-co-methyl methacrylate] (**95**, R = H, $\overline{M}_n$ 90000), and of its blends with PMMA was performed using FT-IR spectra and ATR FT-IR difference spectra. It was shown using ESCA that the pendant chromophores were excluded from the polymer surface [346]. Photooxidative crosslinking resulting in gel formation was observed. The rate of the crosslinking was dependent on the chromophore concentration. The mechanisms of the crosslinking as well as of chain scission processes were explained as a result of the reactivity of tertiary C—H bonds. Crosslinking can be avoided by application of 2-(2-hydroxy-5-isopropenyl) derivative (95, R = CH₃), having a $-C(CH_3)-$ group in the backbone instead of the respective vinyl derivative (95, R = H). An accelerated photochemical study revealed that 95 (R = H) was considerably more photostable than poly[methyl methacrylate-co-(2-hydroxy-4-methacryloyloxyben-zophenone)] [342].

UV spectra of 95 (R = H) films did not changed after 400 h of accelerated aging at 300–400 nm. Some surface photooxidation observed after 500 h was attributed to the presence of tertiary C—H bonds in the copolymer. This consideration was confirmed by testing the relevant 5-isopropenyl derivative 95 (R = CH₃): the respective copolymer remained photostable for 1000 h.

A detailed long-term study of the photochemical behaviour of 95 (R = H) revealed that after several thousands of hours of accelerated ageing, surface deterioration becomes noticeable [342]. This deterioration was due to some photochemical oxidation of the methyl methacrylate units. The benzotriazole chromophore moiety remained unchanged.

ESCA was used to monitor the surface changes of polystyrene and poly[2-(2-hydroxy-3-vinyl-5-methylphenyl)-2H-benzotriazole-co-styrene] (94) during photo-oxidation at λ_{max} 300 nm. A surface photosensitisation effect, dependent on the content of benzotriazole moieties, was observed [85, 86]. With a 0.5% concentration of the latter in 94, the rate of the surface oxidation appears to be similar to that of virgin PS or PS doped with 0.5% 2-(2-hydroxy-5-methylphenyl)-2H-benzo-triazole. For a concentration of chromophores greater than 1%, the surface oxidation is rather enhanced. For example, copolymer 94 with 2% of benzotriazole moieties exhibited a clearly greater degree of photooxidation after 4 weeks than all the other above-mentioned lower-doped systems. The photosensitizing effect of benzotriazole moieties was experimentally confirmed by monitoring oxygen uptake in the surface of poly[2-(2-hydroxy-3-vinyl-5-methoxyphenyl)-2H-benzo-triazole] by ESCA [85, 94].

Oxidation of the styrene moiety and changes in the nitrogen environment in benzotriazole took place in 94. Available data confirm that the surface photooxida-tion of polymers containing aromatic moieties can differ from that of the bulk material [86]. This is due to the higher partial pressure of oxygen and incident photon flux in the surface area. Transmission IR data indicate that the quenching capabilities of benzotriazole moieties in the copolymer 94 are not active in inhibiting surface oxidation.

UV absorption spectra, fluorescence emission spectra and photostabilization effect of 2-(2-hydroxy-4-acryloyloxyphenyl)-2H-benzotriazole and of its polymer bound forms were studied in poly-cis-1,4-polybutadiene [337]. The following activity series was found: copolymer with methyl methacrylate > homopolymer > monomer. It seems that chromophoric units incorporated into a macromolecule behave cooperatively (causing self-absorbance of the emitted radiation).

Some mechanistic studies were performed with polymeric HALS. It was determined using SEM [347] that 0.1 to 1.0% of poly[styrene-co-(2,2,6,6-

tetramethylpiperidyl methacrylate)] was dispersed in PP matrix in the form of spherical globules having size of 10^{-1} μ. The phase separation lowered the light stabilizing effect. If the same copolymer having pendant HALS moieties was used for stabilization of poly-*cis*-1,4-butadiene in the presence of Rhodamin 6G, some quenching of $^1O_2$ was observed [348].

Mechanism of the light stabilizing effect of calix[n]arenes (**130**, R = *tert*-butyl, n = 3) was studied [163]. Carbonyl containing compounds formed in situ from **130** (R = *tert*-butyl, n = 3) via reactivity with alkyl hydroperoxides were considered as the actual light stabilizers.

The activity mechanism of macromolecular biocides has not been fully explained [105]. Most macromolecular biocide systems studied contain Bu_3Sn-groups, fixed by hydrolyzable (amide or ester) or nonhydrolyzable bonds. The former are of advantage in sea applications, where the biocidal moiety Bu_3SnOH can be released.

Using a proper selection of functionalized monomers, systems containing two different stabilizing moietis can be prepared, e.g. **203**, **204**. An intramolecular cooperative effect between the phenolic and the amine moiety in **238** was reported in rubber stabilization [290]. Intramolecular synergism was found between polymer bound phenolic and benzophenone moieties (**203**) in ABS and PVC photostabilization [243, 245, 255]. Two cooperating flame retardant moieties are available in poly[methyl methacrylate-*co*-(4-methacryloyloxy-2,3,5,6-tetrabromobenzyl phosphonite)] [104].

4.2.1 Combinations of Polymeric and High Molecular Weight Stabilizers

Technically and economically acceptable stabilization of polymeric articles can be achieved with mixtures of HMW and polymeric stabilizers. Because of differences in the contribution of the diffusion to the stabilization mechanism between AO and LS, application of polymeric (oligomeric) LS in combinations with HMW AO can be of advantage. Various efficient systems important for the practical stabilization were described by Gugumus [9]. A better efficiency of a combination of poly(2-alkylphenyl salicylate-*co*-dibutyl maleate) having $\overline{M}_n$ 2500 and/or poly[(2-hydroxy-3-allyl-4-methoxybenzophenone-*co*-dibutyl maleate] (0.2 parts) with 0.1 part of each of Irganox 1010 (**3**) and tris(2,4-di-*tert*-butylphosphite) was reached in PP stabilization in comparison with a stabilizing system containing 2-alkylphenyl salicylate as LS [335]. Addition of *N*-lauroyl-4-aminophenol increased the weather resistance of poly[ethylene-*co*-(acryloyloxy-2-hydroxybenzophenone)] [54]. An effective protection of hydrocarbon polymers was achieved with combinations of HMW and oligomeric HALS [9].

The migration principle was suggested as the reason of the enhancement of the antioxidant activity of polyester-polyether elastomer-bound hindered phenol by the addition of 0.25% of an easier migrating AO, 4,4'-bis(α,α-dimethylbenzyl)diphenylamine (**5**) [181]: this easier migrating amine is regenerated by the immobilized phenolic moiety, by means of the principle of homosynergism [5]. Similarly, a blend of polymeric redox hydroquinone-benzoquinone AO with equal amounts of *N*-phenyl-*N*'-(1,3-dimethylbutyl)-1,4-phenylenediamine exerted a pronounced increase of antioxidant efficiency in SBR [124]. A synergistic combination based on aminic stabilizers differing in migration ability consists of 65% diphenylamine-acetone condensate (Naugard A) and 35% N,N'-diphenyl-1,4-phenylenediamine (Naugard J) and has been commercialized under a trade name Flexamine (Uniroyal).

An effective heterosynergism was observed [240] in PP stabilization between PP functionalized with 2,6-diphenyl-4-hydroxyphenyl moiety (**186**, R = phenyl) and a mobile thioester (secondary AO), having molecular weight of about 400. The importance of the migration of the latter was confirmed in experiments performed with immobilized thioesters. The synergism was only weak in the latter case. It is surprising that analogous immobilized 2,6-di-*tert*-butylphenol moiety in **186** (R = *tert*-butyl) did not participate in synergistic phenomena with the same thioesters. Synergism was observed however in PP doped with other polymeric stabilizers bearing phenolic moieties and didodecyl 3,3-thiodipropionate (DLTP) or a polyfunctional AO 2-(3,5-di-*tert*-butyl-4-hydroxyanilino)-4,6-dioctylthio-1,3,5-triazine (Irganox 565, Ciba-Geigy) [55]. Similar cooperative phenomena were determined in poly-*cis*-1,4-butadiene doped with a combination of DLTP with BR functionalized with the 3-(3,5-di-*tert*-butyl-4-hydroxyphenyl)propyloxy moiety [272] (**236a**).

Effective synergistic combinations were formed with bound N-(4-anilinophenyl)methacrylamide (Chemigum HR, **86**) and 1,11-(3,6,9-trioxaundecyl)bis-(3-*n*-dodecylthio) propionate (**6**) in SBR stabilization [77, 184].

In the case that the secondary AO moiety is immobilized, like in [183] or in NBR grafted with 2-{[3-(dodecylthio)-2-methylpropionyl]oxy}ethyl methacrylate [225], at least some migration ability must be exhibited by the primary AO: synergism stronger than that with DLTP was observed [183] in PP stabilized with **141** and Irganox 1010 (**3**) and in SBR stabilized with the NBR grafted with 2,6-di-*tert*-butyl-4-hydroxyphenyl methacrylate [225].

5 Conclusions

The use of polymer-bound stabilizers as a tool for solving problems of the physical loss of stabilizers from polymers has been considered since the 1960s. Various systems have been synthesised in the meantime. Commercial interest in polymeric stabilizers has been restricted by a high cost/performance ratio, by a relatively low application spectrum as well as migration and compatibility problems of individual stabilizers. Up to now, the knowledge dealing with the relations among the supermolecular structure of polymer stabilizers, concentration of stabilizing moieties, the mode of the distribution and/or attachment of the latter in the functionalized macromolecule and the efficiency in a particular host polymer has been very limited. Most data available deal with LS. The latter are also the most frequently commercialized functionalized oligomeric stabilizers and the only stabilizers where data dealing with the activity mechanisms are available. An increase in practical application of other oligomeric (polymeric) stabilizers is to be expected in the future. The future potential use of polymeric AO will be in the automotive and aerospace industries, in biocompatible replacements in human medicine and environmentally friendly packaging materials. The high price of polymeric AO will still remain a problem in the future. Moreover, it seems to be very difficult to synthesize one general purpose polymeric AO which would be compatible with several classes of conventional polymers. The fact the relative contents of active functional groups can be maintained approximatively constant in relation to the increasing molecular weight in stabilizers prepared by polyreactions should be exploited for practical applications. The improvement of the molecular architecture of oligomeric/polymeric stabilizers with the aim of obtaining better physical properties in the doped polymer is a continuing problem. It is however an iterdisciplinary problem which cannot be solved just on a chemical basis. Physical and physicochemical research must be encouraged to a much greater extent. Data published up to now are not sufficient to describe physical processes taking place in blends of a particular conventional polymer with a polymeric stabilizer. This is perhaps due to the low concentration level of the latter and experimental difficulties connected with this fact. However, the incompatibility problems observed in blends containing polymeric stabilizers can be solved only on a physical experimetal basis, using chemically well-defined couples of stabilizer/polymer.

6 References

1. Hoffmann W (1989) Rubber technology handbook. Hanser, Munich
2. Domininghaus H (1976) Die Kunststoffe und ihre Eigenschaften. VDI Verlag, Duesseldorf
3. Pospíšil J, Klemchuk PP (1990) In: Pospíšil J, Klemchuk PP (eds) Oxidation inhibition in organic materials, vol 1. CRC Press, Boca Raton, p 1
4. Gugumus F (1990) In: Pospíšil J, Klemchuk PP (eds) Oxidation inhibition in organic materials, vol 1. CRC Press, Boca Raton, p 61
5. Pospíšil J (1983) In: Jellinek HHG (ed) Degradation and stabilization of polymers, vol 1. Elsevier, Amsterdam, p 193
6. Pospíšil J (1990) In: Pospíšil J, Klemchuk PP (eds) Oxidation inhibition in organic materials, vol 1. CRC Press, Boca Raton, p 33
7. Kuczkowski J (1990) In: Pospíšil J, Klemchuk PP (eds) Oxidation inhibition in organic materials, vol 1. CRC Press, Boca Raton, p 247
8. Chan MG (1990) In: Pospíšil J, Klemchuk PP (eds) Oxidation inhibition in organic materials, vol 1. CRC Press, Boca Raton, p 225
9. Gugumus F (1990) In: Pospíšil J, Klemchuk PP (eds) Oxidation inhibition in organic materials, vol 2. CRC Press, Boca Raton, p 29
10. Sedlář J (1990) In: Pospíšil J, Klemchuk PP (eds) Oxidation inhibition in organic materials, vol 2. CRC Press, Boca Raton, p 1
11. Clough RC, Gillen KT (1990) In: Pospíšil J, Klemchuk PP (eds) Oxidation inhibition in organic materials, vol 2. CRC Press, Boca Raton, p 191
12. Sutker BJ (1990) In: Pospíšil J, Klemchuk PP (eds) Oxidation inhibition in organic materials, vol 2. CRC Press, Boca Raton, p 163
13. Owen ED (ed) (1984) Degradation and stabilization of PVC. Elsevier, London
14. Lorenz J (1983) In: Gaechter R, Mueller H (eds) Taschenbuch der Kunststoff-Additive, 2nd edn, Hanser, Munich, p 617
15. Pospíšil J (1979) In: Scott G (ed) Developments in polymer stabilization, vol 1. Elsevier, Barking, p 1
16. Pospíšil J (1984) In: Scott G (ed), Developments in polymer stabilization, vol 7. Elsevier, London, p 1

17. Pospíšil J (1980) Advan Polym Sci 36: 69
18. Pospíšil J (1981) In: Allen NS (ed) Developments in polymer photochemistry, vol 2, Elsevier, Barking, p 53
19. Pospíšil J (1989) In: Patsis AV (ed) 11th Intern. Confer. on stabilization and controlled degradation of polymers, 24–26 May 1989. Luzern, Proc p 163
20. Pospíšil J (1988) Polym Degrad Stab 20: 181
21. Pospíšil J (1985) In: Patsis AV (ed) 7th Intern. Confer. on stabilization and controlled degradation of polymers, 22–24 May 1985, Luzern. Published (1989) Technomic, Lancaster, p 55
22. Pospíšil J (1990) In: Pospíšil J, Klemchuk PP (eds) Oxidation inhibition in organic materials, vol 1. CRC Press, Boca Raton, p 173
23. Vyprachtický D, Pospíšil J, Sedlář J (1990) Polym Degrad Stab 27: 227
24. Pospíšil J (1988) Angew Makromol Chem 158/159: 209
25. Pfahler G, Loetzsch K (1988) Kunststoffe-German Plast 78: 142
26. Gupta A (1982) In: Patsis AV (ed) 4th Intern. Confer. on stabilization and controlled degradation of polymers, 24–26 May 1982, Luzern
27. Billingham NC (1990) In: Pospíšil J, Klemchuk PP (eds) Oxidation inhibition in organic materials, vol 2. CRC Press, Boca Raton, p 249
28. Chemical additives for the plastics industry. Properties, applications, toxicologies (1987). Noyes data corporation, Park Ridge
29. Hawkins WL, Worthington MA, Matreyek W (1960) Ind Eng Chem, Prod Res Devel 3: 277
30. Moisan JY (1985) In: Comyn J (ed) Polymer permeability. Elsevier, London, p 119
31. Tocháček J, Sedlář J (1989) Polym Degrad Stab 24: 1
32. Latos EJ, Sparks AK (1969) Rubber J 151(6): 18
33. Gendek TP, Hatton TA, Reid RC (1989) Ind Eng Chem, Res 28: 1036
34. Tucker RJ, Susi PV (1984) Polym Prepr 25(1): 34
35. Rotschová J, Pospíšil J (1990) In: Pospíšil J, Klemchuk PP (eds) Oxidation inhibition in organic materials, vol 2. CRC Press, Boca Raton, p 347
36. Albarino RV, Shonhorn H (1973) J Appl Polym Sci 17: 3323; (1974) 18: 635
37. Kulich DM, Wolowitz MD (1987) Polym Mater Eng Sci 57: 669
38. Minagawa M (1989) Polym Degrad Stab 25: 121
39. Henman TJ (1982) World list of polyolefin stabilizers. Kogan Page, London
40. Vidal A, Feder M, Papirer E (1989) Inter. Conf. Rubber '89, 28 Aug.–1 Sept. 1989. Prague. Abstr E 31, Summaries of papers, vol II, p 54
41. Ruffing NR, Amos JL (1962) US Pat 3.017,426
42. Bartl H, Falbe J (eds) (1987) Makromolekulare Stoffe. Methoden der organischen Chemie (Houben-Weyl), Band E 20. G. Thieme, Stuttgart
43. Kuczkowski J, Gillick JG (1984) Rubber Chem Technol 57: 621
44. Keskey WH, Johnson MR (1987) US Pat 4.690,995
45. Beshimov BM, Yariev OM, Drhalilov AT (1986) Uzb Khim Zh (5): 38; (1987) Chem Abstr 106: 67712
46. Vogl O (1984) Makromol Chem, Suppl 7: 1
47. Tirrell DA (1981) In: Pappas SP, Winslow FH (eds) Photodegradation and photo-stabilization of coatings, ACS Symp Ser 151: 43
48. Grosso P, Vogl O (1986) J Macromol Sci, Chem A23: 1041
49. Cassidy HG, Kun KA (1965) Oxidation-reduction polymers. Interscience, New York
50. Horwath JW, Burdon JR, Mayer GE, Naples FJ (1974) Appl Polym Symp 25: 187
51. Horwath JW (1979) Elastomerics 111(8): 19, 62
52. Mayer GE, Kavchok RW, Naples FJ (1973) Rubber Chem Technol 46: 106
53. Kline RH, Miller JP (1974) Rubber Chem Technol 46: 96
54. Tocker S (1967) Makromol Chem 101: 23
55. Minagawa M, Akutsu M, Fujiwara H, Kaikishi M (1979) Ger Offen 2.835,937
56. Vogl O, Albertsson AC, Jovanovic Z (1985) In: Klemchuk PP (ed) Polymer stabilization and degradation, ACS Symp Ser 280: 197
57. Kleiner EK, Dexter M (1976) US Pat 3.957,920

58. Cottman KS (1979) US Pat 4.168,387
59. Dale JA, Ng SYW (1978) US Pat 4.078,091
60. Fertig J, Goldberg AL, Skolrechi M (1966) J Appl Polym Sci 10: 663
61. Osawa Z, Suzuki M, Ogiwara Y, Sugaya J, Hirama K, Kasuga T (1971) J Macromol Sci A5: 275
62. Nir Z, Vogl O (1982) J Polym Sci, Polym Chem Ed 20: 2735
63. Bailey D, Tirrell D, Vogl O (1976) J Polym Sci, Polym Chem Ed 14: 2725
64. Loffelman FF, Brady TE (1982) US Pat 4.356,287
65. Kubota N, Shibata T, Ryozo A (1982) Eur Pat Appl 65,655
66. Conetta TE, Malherbe RF, Winter RAE (1985) Eur Pat Appl 157,734
67. Beshimov BM, Yariev OM (1986) Uzb Khim Zh (4): 35; (1987) Chem Abstr 106: 51,098
68. Siegman A, Yanai S, Dagan A (1988) In: Price D, Iddon B, Wakefield (eds) 1st Int. confer. chem. appl. of bromine and its compounds. Elsevier, Amsterdam, p 339
69. Nguyen NA, Maréchal E (1988) J Macromol Sci, Rev Macromol Chem Phys C28: 187
70. Eichenauer H, Pischitzman A, Hott K (1987) Ger Offen 3.542,468
71. Mitsubishi Rayon Co., Ltd. (1983) Jpn Kokai Tokkyo Koho 83 215, 410
72. Callais PA, Kamath VR, Sargent JD (1988) proc., 15th water-borne higher-solids coat. symp. USA; (1988) Chem Abstr 109: 112122
73. Weinstein AH (1978) Ger Offen 2.735,178
74. Weinstein AH (1977) Rubber Chem Technol 50: 641, 650
75. Kline RH (1978) Ger Offen 2.747,444
76. Vogl O, Albertsson AC, Jovanovic Z (1984) Polymer 26: 1288
77. Parker DK, Schulz GO (1989) Rubber Chem Technol 62: 732
78. Barba NA, Shur AM, Korzha ID (1986) Sint Fiz Khim Issled Koord Polim Soedin −: 80; (1987) Chem Abstr 107: 78388
79. Beshimov BM, Maslamov BA, Kiryushkin SG (1988) Vysokomol soedin 30B: 706
80. Wilette GLW, Hanauer RA (1978) US Pat 4.086,319
81. Kurimura Y, Takao K (1988) J Chem Soc, Farad Trans I 84: 841
82. Neidlinger HH, Schissel P (1986) Solar Energ Mater 14: 327
83. Yagi M, Nakahara Y, Takatori K, Najima T (1988) Jpn Kokai Tokkyo Koho 63 139,958
84. Oo KM, Tahan M (1977) Eur Polym J 13: 915
85. Munro HS, Banks J, Bottino FA, Pollicino A, Recca A (1987) Polym Degr Stab 17: 185
86. Munro HS, Banks J, Recca A, Bottino FA, Pollicino A (1986) Polym Degr Stab 15: 161
87. Won KS (1988) Eur Pat Appl 253,011
88. Gupta A, Scott G, Vogl O (1982) Polym Prepr 23(1): 2191
89. Gomez PM, Fu SK, Gupta A, Vogl O (1985) Polym Prepr 26(1): 100
90. Bailey D, Vogl O (1976) J Macromol Sci-Rev Macromol Chem C14: 267
91. Lorenz DH, Gruber BA (1982) US Pat 4.321,396
92. Shibata T, Sudo T, Okura K (1988) Jpn Kokai Tokkyo Koho 63 260,962
93. Recca A, Libertini E, Pinocchiaro P, Munro HS, Clark DT (1988) Macromolecules 21: 2641
94. Yang CP, Hsiao SH (1988) J Appl Polym Sci 36: 1221
95. Li SJ, Albertsson AC, Gupta A, Bassett W, Vogl O (1984) Monatsh Chem 115: 853
96. Xi F, Bassett W, Vogl O (1984) Makromol Chem 185: 2497
97. Tocker S (1963) US Pat 3.113,907
98. Sumida Y, Vogl O (1981) Polym J (Japan) 13: 521
99. Schwedel C, Moszner N, Schweirer D (1986) Ger (East) 249,030
100. Miyazawa T, Endo T, Okawara M (1985) J Polym Sci, Polym Chem Ed 23: 1527
101. BF Goodrich Co. (1979) Jpn Kokai Tokkyo Koho 79 153,850
102. Myers TN (1989) PCT Int. Appl. 89 01,474; (1989) Chem Abstr 111: 154349
103. Fujita T, Kurumada T, Toda T, Yoshida T (1981) J Polym Sci, Polym Chem Lett 19: 609
104. Yang CP, Wang SS (1989) J Polym Sci, Polym Chem Ed 27: 3551
105. Potin C, Pleurdeau A, Bruneau CM (1982) Double Liaison-Chimie Peint. 29(322): 15,282, (324): 35

106. Schaaban AF, Mahmoud AA, Messiha NN (1988) J Appl Polym Sci 36: 1191
107. Skripnichenko LN, Prokopchuk NR (1989) Vestsi Akad. Navuk BSSR, Ser Khim Navuk (1): 113; (1989) Chem Abstr 110: 213873
108. Dickstein W, Vogl O (1985) J Macromol Sci, Chem A22: 387
109. Heublein G, Knoeppel G, Stadermann D (1988) Polym Bull 20: 109
110. Svjatkina LI, Gaintseva LL, Kurov GN, Skvortsova GG (1984) Vysokomol Soedin 26A: 1178
111. Dale JA, Leonard WJ (1977) US Pat 4.028,342
112. Tocker S (1962) Brit Pat 893,507
113. Floyd JC, Plank DA (1978) US Pat 4.032,510
114. Furia TE, Bellanca N (1976) J Amer Oil Chemist's Soc 53: 132
115. Grimatsu G, Tani K, Mitamura H, Saito M, Katsuo K (1986) Jpn Kokai Tokkyo Koho 86 296,024
116. Kikuchi N, Hawakami H, Saito T, Miyajima H (1986) Jpn Kokai Tokkyo Koho 61 145,222
117. Kikuchi N, Kawakani H, Saito T (1987) Jpn Kokai Tokkyo Koho 87 45,546
118. Brassat B, Buysch HJ, Matner M, Striegler H (1983) Ger Offen 3.138,180
119. Arimatsu V, Tani K, Mitamura H, Saito M, Kaji A, Katsue K (1989) US Pat 4.824,929
120. Sasaki M, Yamakawa M, Hirose T, Isayama K (1987) Jpn Kokai Tokkyo Koho 87 223,288
121. Buysch HJ, Kussy S, Boehmke G (1986) Ger Offen 3.444,884
122. Gloth RE, Tazuma JJ, Smith RA (1984) US Pat 4.463,170
123. Phillips LN, Russel EW, Thomas DK, Wright WW (1968) Brit Pat 1.100,111
124. Sabaa MW, Madkour TM, Assin AA (1988) Polym Degr Stab 22: 195, 205
125. Nakahama S, Hirao A (1990) Progr Polym Sci 15: 299
126. Andrews GD (1985) US Pat 4.522,990
127. Gomez PM, Neidlinger HH (1987) Polym Prepr 28(1): 209
128. Iwata T, Sasaki J (1971) Canad Pat 976,796
129. Fujisawa T, Kakutani M (1974) J Polym Sci, Polym Lett Ed 12: 557
130. Naumova SF, Isakovich VN, Duobinis N (1983) Vestsi Akad Navuk BSSR, Ser Khim Navuk (5): 85
131. Lustoň J, Gumis J, Maňásek Z (1973) J Macromol Sci-Chem A7: 587
132. DiBattista P (1982) Eur Pat Appl 60,559
133. Lustoň J, Smiešková E, Vašš F (1988) Czech Pat 254, 695
134. Soma T (1979) Jpn Kokai Tokkyo Koho 79 29,400
135. Leistner WE, Minagawa M, Kubota N, Shibata F, Arata R (1984) Eur Pat Appl 118, 578
136. Maňásek Z, Vašš F, Lustoň J (1984) Czech Pat 212,508
137. Cassidy HG, Wegner G, Nakabayashi (1971) US Pat 3.600,411; (1972) US Pat 3.707,488
138. Hanabatake H, Kondo K (1986) Jpn Kokai Tokkyo Koho 86 47,749
139. Mitra S, Mitra SB (1987) Polym Mater Sci Eng 56: 614
140. Mitsubashi K (1988) Jpn Kokai Tokkyo Koho 88 2.488,885
141. Cassandrini P, Tozzi A (1977) Ger Offen 2.363,143
142. Son PN (1982) Eur Pat Appl 47,967
143. Vašš F, Lustoň J, Maňásek Z (1981) Czech Pat 223,422
144. Subramanian RV, Garg BK (1978) Org Coatings Plastics Chem 39: 572
145. Schleier G, Jaehme J, Trapp H, Schmidt HV (1986) Ger Offen 3.426,367
146. Ritter H, Rosenkranz HJ, Preuss R, Oertel H (1979) Ger Offen 2.756,338
147. Kishida K, Sasaki I, Mori H (1983) Ger Offen 3.228,544
148. Boor J (1970) J Appl Polym Sci 14: 2558
149. Ivanov S, Shopova N, Jossifov C, Shopov I (1987) Acta Polym 38: 303
150. Kotlyarevskij IL, Markina VK, Terpugova MP (1985) USSR Pat 1.177,316-7
151. Yoshikawa T, Inaike T, Sano N (1974) Jpn Kokai Tokkyo Koho 7 445,060
152. Hirsh RH (1978) Ger Offen 2.756,051
153. Blinne G, Muench V (1984) Ger Offen 3.307,051

154. Ruehlmann K, Jansen I (1985) Ser Chem, Univ Im A Mickiewicza Poznanin —: 47; (1986) Chem Abstr 104: 131201
155. Kimura H, Mori I, Murai B (1985) Jpn Kokai Tokkyo Koho 60 166,327
156. Clark BC, Chafin TC, Hunter GLK (1979) J Amer Oil Chemist's Soc 56: 894
157. Constanzi S, Zavattini O, Bersellini J, Gusoni D (1989) Eur Pat Appl 301,705
158. Morawetz H (1949) Ind Eng Chem 41: 1442
159. Huglin MB, Knight GJ, Wright WW (1972) Makromol Chem 152: 67, 83, 105
160. Abe Y, Matsumura S, Osakura K, Kascura H (1986) Yukagaku 35: 751; (1986) Chem Abstr 105: 209791
161. Langsley GW (1976) US Pat 3.985,710; (1977) ibid 4.059,569
162. Tajima K, Imai K, Takahashi H (1988) Jpn Kokai Tokkyo Koho 88 238,164
163. Seiffarth K, Schulz M, Goermar G, Bachmann J (1989) Polym Degr Stab 24: 73
164. Pastor SD, Adorisio P (1985) US Pat 4.617,336
165. Baer V, Mercz J, Szvoboda J, Pollack ZB, Matyas J (1977) US Pat 4.025,631
166. Patel MR, Patel MM (1986) J Macromol Sci-Chem A23: 1505
167. Parmar JS, Patel MR, Patel MM (1983) J Macromol Sci-Chem A20: 79
168. Lad MJ, Patel HS, Patel SR (1986) Eur Polym J 22: 207
169. Patel HS, Daniel D (1983) J Macromol Sci-Chem A20: 453
170. Tanaka M, Tenjin YO, Hayashi E (1968) Jpn Kokai Tokkyo Koho 68 09,386
171. Koltsova TY, Volkov VA (1988) USSR Pat 1.428,734
172. Storn BH, Kuczkowski JA (1987) US Pat 4.707,300
173. Kong Soo Kim, Yong Woo Lee, Hee Doo Lee (1985) Takuman Hwahakhe Chi 29: 543; (1986) Chem Abstr 104: 19898
174. Nesterovich VM (1985) Vestsi Akad Navuk BSSR, Ser Khim Navuk (3): 111; (1985) Chem Abstr 103: 142793
175. Lewandowitz G, Dominiak A, Zuk A (1987) Polimery 32: 242
176. Losev YP, Nesterovich VM, Tozmina TS (1984) Vestsi Akad Navuk BSSR, Ser Fiz-Energ Navuk (4): 97
177. Karrer F (1983) French Pat 2.511,382
178. Ruebner J, Frommelt H, Krueger H (1987) Plaste Kaut 34: 123
179. Fuhr K, Mueller F, Eicher T (1986) Ger Offen 3.504,169
180. Papava GS, Borisov GB, Varbanov S (1988) Acta Polym 39: 719
181. Wolfe JR (1981) Rubber Chem Technol 54: 988
182. Taijin Ltd. (1983) Jpn Kokai Tokkyo Koho 83 142,910
183. Hammer H, Eichenauer H, Roos E, Juergens E (1987) Ger Offen 3.531,497
184. Dean PR, Kuczkowski JA (1986) Rubber Chem Technol 59: 842
185. Podkosotelny W, Rudr W (1987) Pol Pat 170,722
186. Showa DKK (1986) Jpn Kokai Tokkyo Koho 60 84,377
187. Weawer MA, Hilbert SD, Pruett WP, Coates CA (1988) US Pat 4.826,903
188. Vašš F, Maňásek Z, Lustoň J (1986) Czech Pat 220,996
189. Cantatore G (1983) Eur Pat Appl 124,486
190. Gugumus F (1988) Eur Pat Appl 290,386
191. Leistner WE, Minagawa M, Nakahara Y, Haruna T, Nishimura A (1984) Eur Pat Appl 109,993
192. Winter RAE, Malherbe RF, Fu FTY (1984) US Pat 4.439,565
193. Durišinová L, Maňásek Z, Belluš D (1967) Plaste Kaut 14: 387
194. Onishi K, Kuzuno T (1986) Jpn Kokai Tokkyo Koho 86 55,119
195. Imamura T, Matsumoto T, Ichihashi E, Nakatomari Y (1986) Jpn Kokai Tokkyo Koho 61 261,320
196. Cipriani G, Mariano A (1988) Eur Pat Appl 254,360
197. Sato M, Yokoyama M (1980) Eur Polym J 16: 79
198. O'Brien WL (1977) US Pat 4.062,829
199. Mueller F, Leitr E, Eichenauer H, Ott KH, Buysch J (1988) Ger Offen 3.717,451
200. Kuno S, Tamano Y, Ota K (1985) Jpn Kokai Tokkyo Koho 60 208,341
201. Valdiserri LL, Woodbury RP (1985) US Pat 4.546,180
202. Mayer N, Pfahler G, Wiezer H (1982) Ger Offen 3.029,176

203. Rueger C, Noack R, Habicher W, Schwetlick K (1981) Ger (East) 146,958
204. Kishore K, Annakutty KS, Malick M (1988) Polymer 29: 756, 762
205. Diafoil Co, Ltd. (1984) Jpn Kokai Tokkyo Koho 84 38,230
206. Leistner WE, Minagawa M, Nakahara Y, Haruna T (1978) US Pat 4.104,217
207. Devaux AFLG, Maniotte PG (1984) Eur Pat Appl 128,890
208. Gomez PM, Hu L, Vogl O (1986) Polym Bull 15: 135
209. Sakai K (1986) Jpn Kokai Tokkyo Koho 86 81,454
210. Cantatore G, Borzatta V (1988) Eur Pat Appl 255,990
211. Mutterer F, Berger K (1987) Eur Pat Appl 233,153
212. Lai JT, Son PN (1987) US Pat 4.639,479
213. Lustoň J, Vašš F, Maňásek Z, Vaššová G (1988) Czech Pat 252,983
214. Wang RHS, Irick G (1978) US Pat 4.087,405
215. Mellisaris AP, Mikroyannidis JA (1989) Eur Polym J 25: 275
216. Gerenkova IA, Sergeev VA, Vdovina LI (1987) Vysokomol Soedin 29B: 851
217. Ceresa RJ (1973) In: Ceresa RJ (ed) Block and graft copolymerisation, vol 1. J Wiley,
     New York, p 42
218. Munteanu D (1987) In: Scott G (ed) Developments in polymer stabilization, vol 8.
     Elsevier, London, p 209
219. Johnson MR (1986) US 4.704,470
220. Parks G (1979) US 4.158,955
221. Yachigo S, Sasaki M, Takabasho Y, Kojima F, Takada T, Okita T (1988) Polym
     Degr Stab 22: 63
222. Nakamura Y, Mori K (1988) Jpn Kokai Tokkyo Koho 63 81,145
223. Sakurai H (1986) Jpn Kokai Tokkyo Koho 61 130,356
224. Parks CR (1978) US Pat 4.087,619
225. Cottman KS (1986) Eur Pat Appl 185,606
226. Sharma YN, Naqvi MK, Gawande PS, Bhardwaj IS (1982) J Appl Polym Sci 27: 2605
227. Karrer F, Hoffmann P (1984) Eur Pat Appl 101,411
228. He M, Hu X (1987) Polym Degr Stab 18: 321
229. Shkolnik S, Rajbenbach LA (1982) J Appl Polym Sci 27: 4199
230. Pospíšil J (1990) Angew Makromol Chem 176/177: 347
231. Howarth H (1963) Rubber World 149(1): 54
232. Scott G (1977) Plast Rubber Process 2(2): 41
233. Cain ME, Gazeley KF, Gelling IR, Lewis PM (1972) Rubber Chem Technol 45: 204
234. Kogan LM, Krol VA, Kopylcova NB, Kuzmina LP, Sokolova OS (1988) Zh Prikl
     Khim 61: 1119
235. Chakraborty KB, Scott G, Yaghmour H (1985) J Appl Polym Sci 30: 189
236. Scott G (1981) In: Scott G (ed) Developments in polymer stabilization, vol 4, Elsevier,
     London, p 181
237. Otomo S, Kobayashi Y, Sako T, Yamamoto Y, Yamashita S (1983) J Appl Polym
     Sci 28: 3671; (1984) 29: 89
238. Aglietto M, Alterio R, Bertani R, Galleschi F, Ruggeri G (1989) Polymer 30: 1133
239. Kaplan ML, Kelleher PG, Bebbington BH, Hartless RL (1973) J Polym Sci, Polym
     Lett Ed 11: 357
240. de Jonge CRHI, Hope P (1980) In: Scott G (ed) Developments in polymer stabilization,
     vol 3, Elsevier, Barking, p 21
241. Malherbe R, Rasberger M (1977) Ger Offen 2.727,385
242. Cantor SE (1981) Brit Pat 1.583,652
243. Scott G (1987) In: Scott G (ed) Developments in polymer stabilization, vol 8, Elsevier,
     London, p 209
244. Mori K, Nakamura Y (1984) Rubber Chem Technol 57: 665
245. Scott G (1979) In: Scott G (ed) Developments in polymer stabilization, vol 1, Elsevier,
     Barking, p 309
246. Wozny JC (1984) Eur Pat Appl 108,396
247. Rosenberger S (1980) Eur Pat Appl 17,614
248. Nakamura Y, Mori K, Akaishi F (1977) Rubber Chem Technol 50: 660

249. Buysch HJ, Witte J, Zsentivanyi Z (1984) Ger Offen 3.430,510
250. Buysch HJ, Zsentivanyi Z, Witte J (1985) Ger Offen 3.324,194
251. Engels W, Hammer H, Brueck D, Redetzky W (1989) Rubber Chem Technol 62: 609
252. Al-Malaika S, Honggokusumo S, Scott G (1986) Polym Degr Stab 16: 25
253. Kolawole EG, Scott G (1981) J Appl Polym Sci 26: 2581
254. Sohma J (1989) Progr Polym Sci 14: 451
255. Ghaemy M, Scott G (1981) Polym Degr Stab 3: 253
256. Pác J, Petruj J, Sedlář J (1976) Czech Pat 182,585
257. Minoura Y, Jamamoto Y, Sako T, Oinoto S, Satoyoshi K (1980) Nippon Gomu Kyokaishi 53: 625, 631
258. Gregory JT, Morris RE (1964) Brit Pat 948,549
259. Russell E, Vail LGMC (1978) US Pat 4.107,144
260. Parker RC (1981) US Pat 4.247,664
261. Roggero A, Clerici M, Bartolini G (1987) Eur Pat Appl 2.410,055
262. Oda S, Sakaguchi S (1984) Jpn Kokai Tokkyo Koho 60 161,993
263. Chandra R (1983) Polymer 24: 229
264. Hepburn C, Amu A (1984) Rubber World 190(2): 49, 54
265. Bebikh GF, Saraeva VP (1976) Vysokomol Soedin A18: 461
266. Bartulin J, Zanza H, Parra ML, Rivas BL (1986) Polym Bull 16: 293
267. Oertel H, Uhrhan p, Lautzsch R, Roos E, Schweer H (1978) Ger Offen 2.642,461
268. Kobayashi A, Satake Y, Usui S (1987) Jpn Kokai Tokkyo Koho 87 177,029
269. Khardin AP, Tuseev AP, Valdman AI (1983) USSR Pat 1.016,289
270. Onari Y (1986) J Appl Polym Sci 31: 1663
271. Carraker CE, Gehrke TJ, Giron DJ, Cerutis DR, Mollay HM (1983) J Macromol Sci-Chem A19: 1121
272. Zaffaroni A (1976) US Pat 3.994,828
273. Frangatos G (1973) US Pat 3.714,045
274. Merdan J, Crican L, Luca M, Balin S (1983) Rev Roum Chim 28: 757
275. Schrall AL, McLaury MR (1984) J Appl Polym Sci 29: 3883
276. Ramey CE, Thompson RE, Rostek CJ (1988) Eur Pat Appl 293,253
277. Weinshenker NM, Dale JA (1977) US Pat 4.054,676
278. Parker DK, Burlett DJ (1984) US Pat 4.452,939
279. Itagaki K, Ito T, Watanabe J (1987) Jpn Kokai Tokkyo Koho 87 195,031
280. Berdahl DR, Nye SA, Yeager GW (1987) US Pat 4.668,739
281. Anaud RC, Chander K, Varma IK (1989) J Polym Mater 6: 73
282. Varma IK, Chander K, Anaud RC (1985) J Macromol Sci-Chem A22: 1075
283. Kratz MR, Hendricker DG (1986) Polymer 27: 1641
284. Kontinas AA, Kalfoglou NK (1983) J Appl Polym Sci 28: 123
285. Al-Mehdave MS, Stuckey JE (1989) Rubber Chem Technol 62: 13
286. Barba NA, Shur AM, Mamole SF (1978) Izv Akad Nauk MoldSSR, Ser Biol Khim Nauk (4): 76
287. Cantatore G (1978) Ger Offen 2.752,740
288. Kamorosa EA, Domnina NS, Shagov US (1988) Vestn Leningr Univ, Ser 4. Fiz Khim (2): 120; (1988) Chem Abstr 109: 74481
289. Jayawardena S, Reyx D, Durand D. Pinazzi CP (1984) Makromol Chem 185: 19, 2089
290. Olejnikova GA, Kirpichev VP (1978) Zh Prikl Khim 51: 471
291. Toyobo Co., Ltd. (1983) Jpn Kokai Tokkyo Koho 83 132,039
292. Mizutani Y, Kusumoto K (1975) J Appl Polym Sci 19: 713
293. Sustic A, Zhang C, Vogl O (1987) Polym Prepr 28(2): 226
294. Allmer K, Hult A, Ranby B (1989) J Polym Sci, Polym Chem Ed 27: 3419
295. Citovický P, Chrástová V, Mejzlík J, Sedlář J (1983) Plasty Kaučuk 20: 267
296. Citovický P, Balog M, Kosík M (1987) Chem Pap 41: 125
297. Beresnev VV, Stepanov EA, Grigoriev EI, Kirpichnikov PA (1988) USSR Pat 1.388,401
298. Farrar RC (1983) US Pat 4.377,666
299. Jayabalan M (1982) Angew Makromol Chem 104: 31
300. Kayama K, Kikuchi Y, Muraki T (1988) Jpn Kokai Tokkyo Koho 88 218,746
301. Frangatos G (1972) US Pat 3.663,439
302. Domnina N, Gorokhova L, Shagov V (1982) Zh Prikl Khim 55: 884

303. Beresnev VV, Stepanov EA, Kirpichnikov PA (1985) Izv Khim 18: 546
304. Moriya O, Arai S, Endo T (1988) J Polym Sci, Polym Chem Ed 26: 2573
305. Jansen I, Hahn M, Schilling H, Delling D, Ruehlmann K (1988) Acta Polym 39: 139; (1988) 40: 116, 121
306. Okawara M, Taui S, Imoto E (1965) Kogyo Kagaku Zasshi 68: 223
307. Tocker S (1963) US Pat 3.165,497
308. Lo J, Lee SN, Pearce EM (1984) J Appl Polym Sci 29: 35
309. O'Donnell TW, Olson DR (1985) US Pat 4.520,074
310. Holden DA, Jordan K, Safarzadeh-Amiri A (1986) Macromolecules 19: 895
311. Gupta A, Liang RH, Yavronian HA (1988) US Pat 4.749,726
312. Miranda PM, Factor A (1989) J Polym Sci, Polym Chem Ed 27: 4427
313. Dubin PL, Leonard WJ (1977) Plast Eng 33(10): 29
314. Horák D (1989) Chem Listy 83: 1256
315. Guyot A (1989) React Polymers 10: 113
316. Gugumus F (1981) Res Discl 209: 357
317. Klemchuk PP (1980) In: 2nd Intern. confer. on advances in stabilization and controlled degradation of polymers, 2–4 June 1980. Luzern
318. Lustoň J, Maňásek Z (1977) J Appl Polym Sci 21: 915
319. Goh SH, Lee SY (1986) Thermochim Acta 101: 27
320. Knobloch G (1990) Angew Makromol Chem 176/177: 333
321. Scott G, Tavakoli SM (1988) Plast Rubber Process Appl 9: 59
322. Scott G (1988) Food Addit Contam 5(Suppl. 1): 421
323. Ehrend H, Morche K (1984) Rubber World 191(1): 24
324. Jacquet B, Mahieu C, Papantonion C (1977) Rev Gen Caout Plast 54(575): 85
325. Yassin AA, Sabaa MW, Mohamed NA (1985) Polym Degr Stab 13: 167
326. Dainippon Ink Chemicals, Inc. (1983) Jpn Kokai Tokkyo Koho 83 215,410
327. Adam GA, Jaward KA, Matoog A (1984) Thermochim Acta 80: 317
328. Day M, Suprunchuk T, Cooney JD, Wiles DM (1987) J Appl Polym Sci 33: 2041
329. Akelah A (1986) J Mater Sci 21: 2977
330. Matsumura S, Kasama H, Asakura K, Yoshikawa S (1988) Yakagaku 37: 654
331. Adams A, Braun D (1980) J Polym Sci, Polym Lett Ed 18: 629
332. Ranogajec F, Mlinac M, Dvornik I (1981) Radiat Phys Chem 18: 511
333. Al Malaika S, Ibrahim AQ, Scott G (1988) Polym Degr Stab 22: 233
334. Hodgeman DKC (1979) Polym Degr Stab 1: 155
335. Ozawa A (1988) Jpn Kokai Tokkyo Koho 63 35,660
336. Sustic A, Albertsson AC, Vogl O (1987) Polym Mater Sci Eng 57: 231
337. Liu R, Wu S, Li S, Xi F, Vogl O (1988) Polym Bull 20: 59
338. Chmela S, Maňásek Z, Hrdlovič P (1985) Polym Degr Stab 11: 233, 339
339. Nakahara Y, Kubota M, Katanai M (1988) Jpn Kokai Tokkyo Koho 88 83,169
340. Gupta A, Scott GW, Klinger D (1981) In: Pappas SP, Winslow FH (eds) Photodegradation and photostabilization of coatings, ACS Symp. Ser. 151: 27
341. Gupta A, Yavronian A, di Stefano S, Merrit CD, Scott GW (1980) Macromolecules 13: 821
342. Gupta A, Sarbolonki MN, Huston AL, Scott GW, Pradellock WP, Vogl O (1986) J Macromol Sci-Chem 23: 1179
343. Neidlinger HH, Steffeck MR, Goggin R (1987) Polym Prepr 28(1): 205
344. Jiang Y, Wu S, Sustic A, Xi F, Vogl O (1988) Polym Bull 20: 161
345. Ghiggino KP, Scully AD, Bigger SW, Yandelt MD (1988) J Polym Sci, Polym Lett Ed 26: 505
346. Gupta A, Scott GW, Klinger D, Vogl O (1983) In: Gebelin CG, Williams DJ, Deanin RD (eds) Polymers in solar energy utilization, ACS Symp Ser 220: 191
347. Cen J, Wen Z, Zhou X, Zhu L, Hu X (1985) Gaofenzi Tongxun (5): 384; (1986) Chem Abstr 104: 89713
348. Yang Y, Li G, Feng S, Gao Z, Wu S (1989) Geofenzi Xuebao (1): 37; (1989) Chem Abstr 111: 79185

Editor K. Dusek
Received September 19, 1990

Degradation and Stabilization of Ethylene-Propylene Copolymers and Their Blends: A Critical Review*

S. Sivaram and R. P. Singh,
Division of Polymer Chemistry, National Chemical Laboratory,
Pune-411 008, INDIA

Ethylene propylene copolymers and their blends exhibit diverse degradation behavior under the influence of light, heat and radiation. In spite of many papers in this area, little, if any, mechanistic data on degradation and stabilization of this important class of materials is available in the literature. The present paper reviews the published literature in this area organised under five distinct class of materials, namely, thermoplastic, elastomeric, and heterophasic copolymers, thermoplastic elastomer and blends. Of this, elastomeric ethylene-propylene copolymers appears to have been most exhaustively studied. Very few studies have reported on thermoplastic copolymers, both random as well as heterophasic as well as thermoplastic elastomers and blends. Specific mechanisms of degradation and stabilization of each of these classes of materials are discussed.

In general, there is a paucity of information on the relationship between polymer structure and degradation kinetics. This becomes especially critical in multiphase polymers like heterophasic copolymers, thermoplastic elastomers and blends. This review should stimulate research in this important area, which could ultimately lead to polymers with better photo-thermal and radiation resistance as well as more effective stabilizers.

* NCL Communication No. 4904

1 Introduction

Copolymers of ethylene and propylene have come to stay as important materials with diverse practical applications. They span the full range of polymeric properties, from soft elastomers to hard thermoplastics depending on the relative composition of the two monomers and the manner of their enchainment. Ethylene-propylene copolymers are manufactured commercially using Ziegler-Natta catalysts [1]. For the purposes of this discussion, we will treat these copolymers in terms of three distinct classes of materials:
- Thermoplastic copolymers.
- Elastomeric copolymers.
- Thermoplastic elastomers.

Typical physical properties of these classes of materials are given in Table 1.

1.1 Thermoplastic Copolymers

The crystalline copolymers of propylene with ethylene are thermoplastic materials. They can be further classified as random and block or heterophase copolymers.

1.1.1 Random Copolymers [2]

Random copolymers of propylene and ethylene normally contain less than 6 mole% ethylene which occurs in a random sequence in the copolymer. Compared to homopolymers, random copolymers are softer materials with a lower flexural modulus and higher impact strength.

1.1.2 Block/Heterophase Copolymers [3, 4]

Block or heterophase copolymers comprise chemically dissimilar (e.g. A = ethylene, B = propylene) terminally connected segments. The segmental arrangement can very from an A-B structure containing only two segments, or an AB-A copolymer containing three segments to multi block such as $+A-B+_n$ containing many segments. Other variants of these structures are termed terminal or tapered block copolymers. The former, represented as $+A-(A-B+]_n$, is one which the block segments comprise homopolymer A units and random copolymers of A-B units. Tapered block copolymers are those in which terminal monomer B is added while the polymerization reactor still contains a small amount of monomer A. Since Ziegler-Natta catalysts have finite life times, pure blocks of the $+A-B+$ or $+A-BA+$ type cannot be produced. Most commercially significant "block" copolymers belong to either the terminal or tapered type. These are more appropriately called heterophase copolymers because they consist of a mixture of tapered or terminal block copolymers with crystalline homopolymers and amorphous ethylene-propylene copolymers. Commercial heterophase copolymers contain typically less than 16 mole% ethylene. Heterophase propylene-ethylene copolymers provide a good balance of properties, namely, rigidity and impact, especially at low temperature.

Table 1. Typical physical and mechanical properties of different types of EP copolymers

Type	C_2H_4 mole%	T_g/T_m °C	Crystallinity %	Tensile strength kg cm^{-2}	Izod impact strength kg cm cm^{-1} 23 °C	Elongation at break %	Flexural modulus kg cm^{-2}	Hardness, shore A	Brittle temperature °C
Random copolymers (thermoplastic)	<6	–	<70	200–240	2.5–8.0	–	6500–10000	–	–
Random copolymers (elastomeric)	65	–50 to –70	Absent	650	–	600	–	65	–
Heterophase copolymers	<16	165–170	55–60	200–230	8.0–12.0	–	13000–14000	–	–25 to –45
Thermoplastic olefinic elastomers	50	–	–	70–140	–	150–300	–	65–92	–

1.2 Elastomeric Copolymer

The elastomeric ethylene-propylene copolymers (EPR) [5, 6] are also random copolymers but have an amorphous structure with a typical rubber-like elasticity and high elongation upon deformation. Amorphous character is achieved if the structure of the polymer is essentially random with a minimum of molecular regularity and a moderately high ethylene content. Ethylene content in EPR's are typically about 65 mole%.

1.3 Thermoplastic Elastomers

Thermoplastic olefin elastomers (TPOs) are a recent class of olefin copolymers which have the processing advantage of thermoplastics combined with the physical properties of chemically vulcanized elastomers. These materials need both the elastomeric soft segments and thermoplastic hard segments. Proper combinations of the hard and soft segments are required to introduce suitable molecular tie points to produce a network structure similar to those established by chemical crosslinks. A variety of compositions have been explored as useful TPOs. Examples include propylene/α-olefin copolymers and their melt blends with isotactic polypropylene, stereoblock polypropylene, graft copolymers of crystallizable monomer segments onto elastomer molecules (e.g. pivalolactone-g-EPDM) and dynamically vulcanized elastomer-thermoplastic blends (e.g. EPDM(i-PP blends in the presence of EPDM curatives). A large variety of products with diverse property profiles have been reported in the literature.

1.4 Methods of Synthesis

Ethylene-propylene copolymers are synthesized by simultaneous, sequential or intermittent addition of one of the monomers into a slurry of liquid monomer or gas phase using Ziegler-Natta catalysts. Improvements in catalysts have substantially simplified the process of manufacture and improved the quality of the products [7, 8]. In commercial practice, copolymers are made in a continuous process either in hydrocarbon slurry (stirred tank), bulk propylene (loop reactor) or in gas phase (fluidized bed reaction) [9]. Random copolymers are manufactured by the simultaneous introduction of both monomers in the reactor. In a typical block/heterophase copolymer synthesis, monomer A (e.g. propylene) is fed into a reactor along with chain transfer agents (e.g. hydrogen) containing the catalyst and the reaction medium, if any (e.g. hexane), at the desired polymerization temperature. After a given residence time, the reaction medium is degassed to remove unreacted propylene and the slurry in hydrocarbon (or powder in the case of liquid monomer or gas phase process) is conveyed to another reactor where a mixture of monomer A and B in definite proportions are introduced. The reaction is reinitiated by the active catalyst residues bound to the homopolymers, coming from the previous reactor. The resulting final polymer has a heterogeneous character since chain transfer processes are prevalent during synthesis. Commercial processes involve a minimum of two reactors, the first for homopolymerization and the second for copolymerization. The two reactors can be of the slurry type, or a loop followed by fluid bed, or two fluid bed reactors, all in a series.

The properties of the copolymers are dependent both on the nature of catalysts employed and the type of reactor and process used for their manufacture. Consequently, commercial polymers obtained from different sources must be compared with caution.

Elastomeric copolymers are made by either solution or suspension process using a vanadium based catalyst along with alkyl aluminum compound as cocatalyst. In the suspension process propylene is used as a diluent, whereas in the solution process hexane is used as diluent. Superior catalysts based on supported titanium compounds have further improved the suspension process in recent years. In the conventional suspension process, ethylene, propylene and catalysts are fed continuously to a stirred reactor at 20 °C and $12 \, \text{kg cm}^{-2}$ total pressure. Diethylzinc is used to control molecular weight.

Polyolefin based thermoplastic elastomers are either made by reactor copolymerization or physical blending of crystalline polypropylene with EPDM in an extruder under conditions where the optimum level of chemical crosslinking of the EPDM phases could be promoted.

2 Degradation of Polymers

The mechanism of degradation and stabilization of ethylene-propylene copolymers broadly resembles that of polyolefins; yet there are major differences arising mainly from their distinct chemical composition and structure. Although voluminous literature has accumulated in this area [10–12], there is no concise compilation and organization of this literature.

There is great interest at present in the photooxidative, thermal and radiation induced degradation of polymeric materials. In broad terms, the degradation involves modification that is often (but not always) detrimental to the performance of the polymeric materials to varying extents depending on their chemical structure. All the commercial organic polymers degrade in air when exposed to solar radiation but there are conflicting views on the mechanism of degradation and stabilization which is a very important problem from both a scientific and an industrial point of view. Moreover, a better understanding of their mechanism is necessary to achieve better stabilization. The degradation of elastomers and thermoplastics may be induced by a number of factors which are described in brief in the following paragraphs.

2.1 Initiators of Degradation

The prime factors causing degradation of polymers are UV light and the heating effects of solar radiation. Additional relevant factors include ozone, atmospheric contaminants and induced radiation. It is the impurities present in the commercial polymers that initiate photochemical reactions because pure saturated polyolefins should be unaffected by their exposure to sunlight. The main light absorbing species may be carbonyls [13], metallic impurities [14], dienes [15], trienes [16], hydroperoxides and oxygen-polymer charge-transfer complexes [17, 18]. A wide range of derivatives of aldehyde, ketones, quinones, inorganic metal oxides and salts effectively sensitize the photooxidative thermal degradation by the process of hydrogen-atom abstraction [19] or electron transfer. The polymers, upon exposure to gamma or other high energy radiations are fragmented directly to free radicals, excited molecular fragments or may be crosslinked into a rigid three dimensional network. The radiation degradation of polyolefins has been extensively studied [20–22] both in air and vacuum.

2.2 General Mechanism of Degradation

Polyolefin degradation comprises of initiation, propagation and termination steps

2.2.1 Initiation

The formation of the polymer radical $(R^{\cdot})$

$$RH \longrightarrow R^{\cdot} + H^{\cdot} \tag{1}$$

is necessary for rapid polymer oxidation. The reaction may be initiated by physical factors, e.g. UV radiation, ionizing radiation, heat, ultrasonics, mechanical treatment etc., or by chemical factors such as catalysis, direct reaction with oxygen, singlet oxygen, atomic oxygen or ozone.

2.2.2 Propagation

In propagation, the degradative chain is prolonged by the following methods:

2.2.2.1 Formation of Polymer Hydroperoxide

The macroradicals $(R^{\cdot})$ formed during the initiation can easily react with oxygen molecule producing peroxy radicals [23]

$$R^{\cdot} + O_2 \longrightarrow ROO^{\cdot} \tag{2}$$

The peroxy radical can abstract hydrogen from another polymer molecule to form polymer hydroperoxide

$$ROO^{\cdot} + RH \longrightarrow ROOH + R^{\cdot} \tag{3}$$

The peroxy radicals are strongly resonance stabilized and are relatively electrophilic species abstracting tertiary bonded hydrogen in preference to secondary or primary bonded.

2.2.2.2 Decomposition of Polymer Hydroperoxide

The polymer hydroperoxides are decomposed by light irradiation [24] according to the following equations:

$$ROOH \longrightarrow \dot{R} + \dot{O}OH \tag{4}$$

$$ROOH \longrightarrow R\dot{O} + \dot{O}H \tag{5}$$

The decomposition of hydroperoxides can also be induced by raising the temperature and is promoted by metal catalysis. Free radicals formed during the reaction (4) and (5) can also take part in radical induced decomposition of hydroperoxides

$$ROOH + \dot{R} \longrightarrow R\dot{O} + ROH \tag{6}$$

$$ROOH + \dot{O}H \longrightarrow RO\dot{O} + H_2O \tag{7}$$

2.2.2.3 Formation of Hydroxyl Group

The hydroxyl groups [25, 26] are formed in the reaction between alkoxy polymer radicals (R\dot{O}) and other polymer molecules

$$R\dot{O} + RH \longrightarrow ROH + \dot{R} \tag{8}$$

The hydroxyl groups may be formed along the polymer chain or on its end groups. The carbonyl and aldehyde groups are formed from the scission of alkoxy radicals or by the decomposition of hydroperoxy radicals. The reaction between two polymer alkoxy radicals also produces a carbonyl and hydroxyl group by disproportionation:

$$\begin{array}{ccc} R & \dot{O} & R \\ | & | & | \\ -CH-CH-CH-CH_2- \end{array} + \begin{array}{ccc} R & \dot{O} & R \\ | & | & | \\ -CH-CH-CH-CH_2- \end{array} \longrightarrow \begin{array}{ccc} R & O & R \\ | & \| & | \\ -CH-C-CH-CH_2- \end{array} + \begin{array}{ccc} R & OH & R \\ | & | & | \\ -CH-CH-CH-CH_2- \end{array} \tag{9}$$

2.2.2.4 Decomposition of Carbonyl Group

The carbonyl groups thus formed are further decomposed by a Norrish type I or II reaction.

Norrish type I: In the primary process, the bond between the carbonyl group and adjacent α-carbon is unimolecularly cleaved to produce radicals

$$RCOR^1 \xrightarrow{h\nu, \Delta} \begin{cases} R\dot{C}O + \dot{R}^1 \longrightarrow \dot{R} + CO + \dot{R}^1 & \tag{10} \\ \dot{R} + R^1\dot{C}O \longrightarrow \dot{R} + CO + \dot{R}^{'1} & \tag{11} \end{cases}$$

Norrish type II: It is a non-radical intramolecular process. The abstraction of a H-atom from the γ-carbon results in its subsequent decomposition into an olefin and an alcohol or an aldehyde

$$R_2\overset{\gamma}{C}H\text{-}\overset{\beta}{C}R_2\text{-}\overset{\alpha}{C}R_2\text{-}\overset{O}{\overset{\|}{C}}\text{-}R^1 \xrightarrow{h\nu,\Delta} R_2C{=}CR_2 + R_2C{=}C\overset{OH}{\underset{R^1}{\diagdown}} \tag{12}$$

This reaction may also involve intramolecular β-hydrogen atom transfer

$$R\text{-}\overset{O}{\overset{\|}{C}}\text{-}\overset{CH_3}{\underset{R^1}{\overset{|}{C}H}} \xrightarrow{h\nu,\Delta} R\text{-}\overset{O}{\overset{\|}{C}}\text{-}H + H_2C{=}CHR^1 \tag{13}$$

2.2.3 Termination

The termination of the radical chain is due to reactions of free radicals with each other by combination, in which inactive products are formed

$$R O\overset{\cdot}{O} + R\overset{\cdot}{O} \ \rule[0.5ex]{0pt}{1ex} \tag{14}$$

$$R O\overset{\cdot}{O} + R^{\cdot} \ \rule[0.5ex]{0pt}{1ex} \longrightarrow \text{inactive products} \tag{15}$$

$$\overset{\cdot}{R} \ + \overset{\cdot}{R} \ \rule[0.5ex]{0pt}{1ex} \tag{16}$$

When the oxygen pressure is high, the termination reaction almost exclusively followed Eq. (14). In the solid state, when sufficient oxygen concentration cannot be maintained in the system, the termination reaction (Eq. 15) becomes significant. The polymer radicals may be coupled mutually (Eq. 16) and form crosslinks with polymer radicals. These processes are dependent on the chemical and physical structure of irradiated polymers.

3 Degradation of Ethylene-Propylene Copolymers

3.1 Thermoplastic E-P Copolymers

The sodium p-styrene sulfonate and methyl methacrylate have been grafted [27] onto an E-P copolymer (10 mol% C_2H_4) fiber by irradiation (1.92×10^5 rad/h) at 80 °C. Asaka et al. [28] studied the γ-radiative degradation of random E-P copolymers and found that degradation was greater at a lower rate over a longer period than at a higher rate for a shorter period. The γ-ray induced oxidative degradation [29] of E-P copolymers under pressurized oxygen was found to reduce the radiation resistance period for polymer testing. The depth of the oxidation region was proportional to the square root of (oxygen pressure/dose rate). The y-irradiation of EP copolymers [30] at 30–175 °C gave n-butane and isobutane fraction is addition to hydrogen and other volatile hydrocarbons. The isobutane formation is due to isomerization of n-butane or recombination of methyl and isopropyl radical in E-P copolymer radiolysis. The yield of G-values derived from the slopes of yields versus dose plots, of n-butane increased slightly from 30–100 °C but quite rapidly from 100–175 °C as shown in Fig. 1. The yield of n-butane

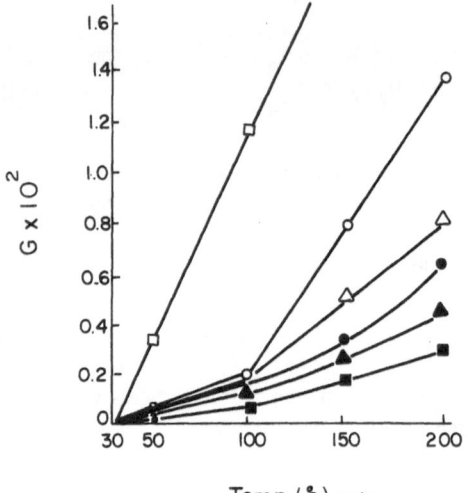

Fig. 1. Temperature dependence of G-values of *n*-butane and isobutane from γ-irradiation of E-P copolymer (○) *n*-butane; (●) isobutane; CH_3SH: (△) *n*-butane, (▲) isobutane; $(C_2H_5)_3N$: (□) *n*-butane, (■) isobutane

increased by a factor of ∼3 over this range whereas isobutane increased about 50% which is a confirmation of the recombination of methyl and isopropyl radicals. The marked temperature dependence of *n*-butane yields clearly indicates the radiation induced thermal depolymerization in the present system.

Busico et al. [31] investigated the thermal history of random E-P copolymers with low ethylene content (>20 mole%). The melting temp. (T_m), crystallinity percentages and melting enthalpies (ΔH_m) are reported in Table 2. As for the propylene homopolymer, the T_m of the annealed samples are higher with high annealing temperatures and reach finally a limiting value. The T_m and ΔH_m are lower for the higher ethylene content in the copolymer. It appears that only *i*-PP chains develop crystallinity with lower concentration of ethylene. On the other hand, the observed drop in the crystallinity beyond 10 mol% ethylene indicates the limit in the concentration of defects (e.g. ethylene units).

Table 2. Melting temperatures T_m, melting enthalpies ΔH_m and crystallinity of E-P copolymers, heated at 200 °C and annealed for 5 min at different temperatures

Sample	Ethylene	Heated at 200 °C			Annealed for 5 minutes			
		T_m	ΔH_m	X-ray Cryst.	T_a	T_m	ΔH_m	X-ray Cryst.
	mol%	°C	kal g^{-1}	%	°C	°C	kal g^{-1}	%
E-P Copolym.	4	151	19.83	63	156	154	27.71	77
E-P Copolym.	6	143	19.59	72	152	135	21.26	79
E-P Copolym.	10	139	17.92	65	152	120	20.78	68
E-P Copolym.	16	113	8.84	52	137	116	19.11	67
E-P Copolym.	20	110	4.06	43	147	120	12.19	54

The degradation [32] of cable made up of E-P copolymers, due to fast neutrons or γ-rays was virtually the same upto the absorbed dose of 1.5×10^6 Gy. Arnaud et al. [33] have studied the photooxidation of the physical mixture of low density polyethylene-polypropylene and found that photochemical behaviour of the mixture changed gradually from that of pure low density polyethylene to pure isotactic polypropylene. The rate of hydroperoxidation increased with increasing propylene contents. Recently, the present authors [34] have studied the photo-oxidative degradation of heterophase E-P copolymers with polychromatic irradiation at 30 °C. Upon irradiation of E-P copolymer (11.8–13.2 mol% C_2H_4) films in the solid state it was found that enthalpy of melting (ΔH_m) decreased with increasing ethylene content and irradiation times whereas crystallinity increased correspondingly (Table 3). The crystallinity increase with irradiation times is a clear indication that the copolymer behaves like i-PP.

Table 3. The changes of melt enthalpy (ΔH_m) and crystallinity in E-P copolymers in the solid state after polychromatic irradiation (>290 nm) at 30 °C

Sample	Irrdn.	Ethylene	T_m	ΔH_m	X-ray Cryst
	h	mol%	°C	kal g^{-1}	%
EPF 30R	0	11.82	169.8	24.32	45.7
	550	11.82	158.4	20.86	61.3
EPQ 30R	0	13.19	166.7	19.95	49.7
	550	13.19	155.6	17.49	58.9

Although it is reasonable to expect differences in degradation behaviour between random and heterophase copolymers, literature seldom defines clearly the nature of copolymers studied. Well characterized heterophase E-P copolymers, however, are of very recent origin.

3.2 Elastomeric E-P Copolymers (EPR)

EPR (68 and 32 mol% C_2H_4) was pyrolysed [35] up to 530 °C in an evacuated pyrolysis unit. Both the polymers degraded into ethylene and propylene monomers along with several aliphatic hydrocarbons. There is an intense degradation with decreasing ethylene contents. Sieron and Murry [36] carried out their investigations at 205 °C on EPR (65 mol% C_2H_4) in both nitrogen and air, and proved that oxidation rather than thermal degradation is the prime cause of degradation at elevated temperatures. The kinetics of benzoyl peroxide decomposition of EPR (equimolar ratio of C_2H_4 and C_3H_6) was studied by Manaress [37]. The number of crosslinks per gram were detemined by measuring the swelling ratios at equilibrium after 24 h immersion of the elastomer in CCl_4 at 30 °C and the variations of the peroxide concentration were also recorded for 75, 85 and 95 °C.

The crosslinking of EPR in the presence of radicals from peroxide decomposition (benzoylperoxide, dicumylperoxide etc.) by light or heat, is attributed to the attack on the secondary CH_2 moieties and the generation of polymer radicals which couple either with similar secondary polymer radicals or polymer radicals generated by attack on the tertiary CH moieties on the propylene units in the chain.

$$
\underset{\overset{|}{CH_3}}{\overset{\overset{CH_3}{|}}{C_6H_5-C}}-O-O-\underset{\overset{|}{CH_3}}{\overset{\overset{CH_3}{|}}{C-C_6H_5}} \xrightarrow{hv,\Delta} 2\ \underset{\overset{|}{CH_3}}{\overset{\overset{CH_3}{|}}{C_6H_5-C}}-O^\cdot \xrightarrow{hv,\Delta} C_6H_5-\overset{\overset{O}{\|}}{C}-CH_3 + {}^\cdot CH_3 \qquad (17)
$$

$$
-CH_2-CH_2-\underset{\overset{|}{CH_3}}{\overset{\overset{CH_3}{|}}{CH}}-CH_2- + \underset{\overset{|}{CH_3}}{\overset{\overset{CH_3}{|}}{C_6H_5-C}}-O^\cdot \longrightarrow \underset{\overset{|}{CH_3}}{\overset{\overset{CH_3}{|}}{C_6H_5-C}}-OH + -CH_2-\overset{\overset{\cdot}{}}{C}H-\underset{\overset{|}{CH_3}}{\overset{\overset{\cdot CH_3}{}}{C}H}-CH_2- \qquad (18)
$$

$$
\text{or } -CH_2-CH_2-\underset{}{\overset{\overset{CH_3}{|}}{\overset{\cdot}{C}}}-CH_2-
$$

$$
2\ -CH_2-\overset{\cdot}{C}H-\underset{\overset{|}{CH_3}}{\overset{}{C}}H-CH_2- \xrightarrow{crosslinking} \begin{array}{c} CH_3 \\ | \\ -CH_2-CH-CH-CH_2- \\ -CH_2-CH-CH-CH_2- \\ | \\ CH_3 \end{array} \qquad (19)
$$

$$
-CH_2-CH_2-\underset{\overset{|}{CH_3}}{\overset{\overset{CH_3}{|}}{\overset{\cdot}{C}}}-CH_2- \xrightarrow{disproportionation} H_3C-CH_2- + -\underset{\overset{|}{\cdot}}{\overset{\overset{CH_2}{\|}}{C}}-CH_2- \qquad (20)
$$

The generation of a tertiary radical by the later route is the predominant reaction. The tertiary radicals preferentially undergo disproportionation, resulting in degradation rather than crosslinking. The number of crosslinks increased at room temperature after exposure of EPR to γ-rays radiolysis [38] at $-196\ ^\circ C$. It is because the life time of free radicals is less at room temperature than at lower temperature and the free radicals couple instantly at room temperature to cause crosslinking.

Slobodin et al. [39] confirmed that thermal decomposition of EPR (equimolar ratio) began at $170\ ^\circ C$ and ceased at $360\ ^\circ C$. A total 93.66% condensate products, 5.2% gas and 1.14% carbonaceous residue were obtained, mainly at $\sim 235\ ^\circ C$. The composition of the gaseous portion, determined by GLC was ethane-ethylene 1.25%, propane 0.81%, propylene 0.98%, butane-butylene 0.99% and butadiene 0.99% by wt. of EPR. The liquid products were separated into five fractions with boiling ranges of $100\ ^\circ C$, $100-150\ ^\circ C$, $150-200\ ^\circ C$, $200-250\ ^\circ C$, and $> 250\ ^\circ C$. The fractionation yielded pentane, 1-pentene, 2-methylbutane, 2-methyl-1-butene, 2-methyl-2-butene, isoprene and piperylene of C_5 hydrocarbon; and hexane, 1-hexane, 2-methylpentane of C_6 hydrocarbons. Based on these data, the thermal degradation was proposed to proceed via a free-radical mechanism. A free radical

$$
\text{of the type } -CH_2-\underset{\overset{|}{CH_3}}{\overset{\overset{CH_3}{|}}{CH}}-CH_2(CH_2)_n-\overset{\cdot}{C}H_2 \text{ was assumed to be formed first}
$$

which then either disproportionated with another similar radical saturated, unsaturated or conjugated hydrocarbons with the same carbon skeleton, or the active centre of the backbone shifted to another position of the chain which subsequently, led to the formation of paraffins, olefins and diolefins with shorter chains than the initial radical

$$-CH_2-\overset{\overset{\displaystyle CH_3}{|}}{CH}-CH_2-(CH_2)_n-\overset{\bullet}{C}H_2 \longrightarrow -CH_2-\overset{\overset{\displaystyle CH_3}{|}}{CH}-CH_3 + H_2C=CH(CH_2)_{\overline{n-2}}\overset{\bullet}{C}H_2 \quad (21)$$

Tobolsky et al. [40] proposed the mechanism of autooxidation by measuring the rates of oxygen absorption and chain scission for the benzoyl peroxide initiated oxidation of EPR (35 mol% C_2H_4). Benzoyl peroxide was found to be a good initiator for random radical polymerization initiation and, therefore, a sensitizer in thermal oxidation of EPR. The scission was random, and at the crosslink site it was negligible compared to the scission along the chain. Yu [41] studied the γ-ray induced chain scission kinetics of crosslinked EPR in vacuum and air. Kozlov and Torasova [42] degraded saturated EPR (equimolecular ratio) by ^{60}Co-ionizing radiations and confirmed the radical accumulation with increasing radiation dose in the solid state at −196 °C (Fig. 2). Kuzminskii et al. [43, 44] suggested that tetramethylthiuram disulfide acted as a sensitizer in γ-rays degradation of EPR and caused crosslinking also due to sulfur in it. The small doses of ^{60}Co γ-rays start depolymerization whereas the higher ones start crosslinking.

Kozlov [45] studied the effect of temperature from −196 to 50 °C on the decay of radicals formed during radiolysis of EPR in the solid state by ESR spectroscopy. The radical decay process at low temperature begins before movement of the macromolecules relative to one another and is dependent on the migration mechanism with a very low energy of activation. This process consists of migration of the 'secondary' ions, formed by the capture of charges by free-radicals. That the radical decay process involves a combination of reactions between the radicals

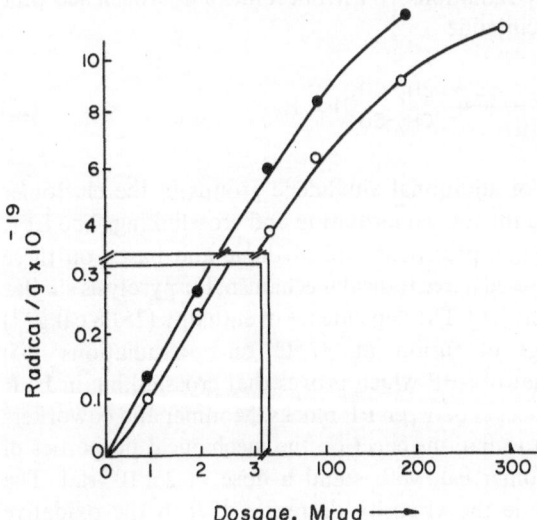

Fig. 2. Accumulation of free radicals during radiolysis of EPR films at −196 °C, (○) EPR, (●) EPR + 5 wt.% N-PMI

and 'secondary' ions of radical origin, was suggested. With increasing temperature $(-196$ to $100\,°C)$ in γ-ray radiolysis [46] of EPR, the rate of crosslinking increases without any appreciable increase in the degradation rate and this agrees with the literature findings. The EPR method was employed to study the role or sulfur in the formation of free radicals during E-P elastomer radiolysis and their annihilation on heating and photolysis at $-196\,°C$. It was confirmed that on heating the irradiated rubber, the free radical concentration increased and reached a maximum at T_g. Apparently these radicals are formed as polymer-sulfide radicals by interaction of macromolecular radicals with sulfur. Upon illumination of these irradiated samples with $200-700$ nm UV/visible light, the free radical decay increased especially with short wavelengths, at the same time the process of annihilation proceeded much more effectively in the presence of sulfur.

It was confirmed by viscosity measurements [48] that thermooxidative degradation of EPR always preceeds crosslinking. Thermogravimetric results [49] suggest that chain scission occurs preferentially to crosslinking and during pyrolysis, and the propane units are degraded to ethylene units. The ionizing irradiation [50] of EPR with ^{60}Co or ^{90}Sr at $-196\,°C$, rsulted in thermoluminescence and the area below the thermoluminescence curve was proportional to the absorbed dose. Copper conductors were coated with E-P rubbers by heating at $250\,°C$ with γ-irradiation [51] in an inert atmosphere. The crosslinking, degradation, gas evolution, formation of free-radicals, and unsaturation occuring in EPR during the irradiation [52] with γ-rays suggested that β/α was inversely proportional to the molecular weight of the elastomer (β and α are the degradation and crosslinking rates, respectively). The addition of sulfur intensified the radiation crosslinking.

Smetania and coworkers [53] proved that N-phenylmaleimide (N-PMI) acted as a sensitizer in radiation crosslinking of EPR (63 mole% C_2H_4) the rate of gel formation being directly proportional to the quantity of additive up to 20 Mrad dose. A comparison of radical ion concentration up to 3 Mrad dose is shown in Fig. 2. The sensitizing effect of the N-PMI is not of a free-radical or ionic nature but is due to the fact that under y-radiations, N-PMI becomes a hydrogen acceptor and is reduced to N-phenylsuccinimide

$$\begin{matrix} CH-CO \\ CH-CO \end{matrix}\!\!\!\diagdown\!\!\!N-C_6H_5 \xrightarrow{\ \ \gamma\text{-irradiation}\ \ } \begin{matrix} CH_2-CO \\ CH_2-CO \end{matrix}\!\!\!\diagdown\!\!\!N-C_6H_5 \tag{22}$$

thus promoting the formation of additional vinylidene groups in the elastomer which caused the acceleration of the radical formation and crosslinking. The EPR pyrolyzed at $800\,°C$ gives ethylene, propylene and methane and based on these results Mamedov et al. [54] proposed a free radical mechanism for pyrolysis similar to that of Slobodin and coworkers [39]. The dependence of enthalpy (25.08 cal g^{-1}) and entropy (0.0668 cal/deg/g) of fusion at $27\,°C$ on γ-irradiations [55] ($0-620$ Mrad) was similar to that of i-PP which proves that crosslinking in EPR (73 mol% C_2H_4) preferentially occurs between PP blocks. Monnier and coworkers [56] studied y-rays and neutron radiations effect on the mechanical properties of EPR and found that the elastomer can with stand a dose of 2×10^9 rad. The hydroxyl number increased while the viscosity decreased [57] in the oxidative degradation of EPR in toluene in the presence of AIBN with increasing concentration of AIBN and reaction time where AIBN acts as a radical initiator. The yield of both the crosslinks as well as chain scissions in EPR during γ-irradiation [58] in the temperature range of -196 to $100\,^\circ$C increased with increasing temperature but the chain-scission efficiency-increased faster than crosslinking.

Locke [59] reduced the molecular weight and intrinsic viscosity [η] of EPR ([η] $= 1.3$ at $25\,^\circ$C of 0.1 wt.% solution in toluene) by passing nitrogen for 16 h through two litres hexane containing 100 g EPR. In hexane solution, 8 g AlCl$_3$ in 40 ml cyclohexane was also added and the whole mixture was heated for 3.5 h at $60\,^\circ$C to give a pourable liquid ([η] $= 0.25$). The rates of oxygen absorption and formation of oxidation products were determined by Decker et al. [60] in γ-initiated oxidative degradation of EPR (37 mol%, 73 mol% and 86 mol% C$_2$H$_4$) thin films. The radiation yield for oxygen absorption and formation of hydroperoxides depend on dose rates and decreases sharply with increasing ethylene contents or crystallinity. The minute concentration ($> 9 \times 10^{-3}$ wt.%) of vanadium residues in the E-P rubber lowered [61] its resistance to thermal oxidative degradation. These metallic impurities are the primary centres to initiate thermal, photo or radiation degradation. The rate of γ-ray crosslinking [62] of EPR (equimolecular of C$_2$H$_4$) is increased with increasing pressure because the rate of propagation reaction was accelerated by radical addition to double bonds by reduction of the distance between polymer chains at higher pressure at the same time the probability of termination reaction was lessened by the decrease of molecular segmental mobility.

Makhlis [63] reported the radiation degradation of strained and unstrained EPR in air and vacuum from -196 to $200\,^\circ$C. The degradation rate of the elastomer increased rapidly with the applied mechanical stress whereas the activation energy decreased. The degradation rate increased when irradiation was carried out in an oxygen atmosphere as it is the carrier for propagating the chain. The thermo-oxidative degradation [64] of EPR in the temperature range of $170-500\,^\circ$C, revealed two stages of degradation i.e. $250-360\,^\circ$C and $340-480\,^\circ$C, respectively. The activation energy for the formation of individual products (based on MS) varies within the limit of $17-20$ kcal mol^{-1} in the first stage and $41-47$ kcal mol^{-1} for the second stage of degradation. Thus, it appears from the results of mass spectral analysis that the first stage involved primarily the degradation of oxidized structures which were in fact weak bonds. The EPR thermal degradation [65] in vacuum at $150-450\,^\circ$C proceeded in three states: $\leq 250\,^\circ$C the degraded rubber had a lower molecular weight but retained its original structure; at $300-350\,^\circ$C the degradation was attended by a rapid decrease in molecular weight and formation of various unsaturated structures whereas at $> 350\,^\circ$C the degradation gave low molecular weight products, having a wide range of molecular weights.

Makhlis et al. [66] proved that more than one degradative processes are operative in thermal-oxidative dagradation of EPR. The low activation energy also indicates the existence of weak bonds. The rate of thermal degradation of γ-irradiation cured vulcanizates can be effectively reduced by vacuum heat treatment at $200-250\,^\circ$C. During thermal degradation [67] at 65° and $> 350\,^\circ$C in vacuum, EPR exhibited an increasingly aggressive degradation with an increase the propene content because its increase caused reduction in molecular weight and unsaturation.

Degradation and increased wettability was observed by treatment of EPR in a glow discharge [68] in air. The weight loss and surface wettability due to glow discharge is proportional to the exposure time and proceeds at the rate of $(4-20) \times 10^3$ mg cm^{-2} min, and is independent of the sample thickness. The contents of CO, CO_2 and water vapours increased with the glow discharge exposure time [70]. With increasing exposure time to the UV light [69], the surface hardness increased and the impact strength decreased. Ito [71] carried out investigations on EPR under the combined environment of heat and radiation at temperature 120–170 °C and dose rate from 2.2×10^4 to 1.05×10^5 rad h^{-1}. The stress relaxation decay curves were Maxwellian over a period of time. Gillen and Clough [72] have aged EPR (65 mol% C_2H_4) in air and nitrogen at radiation dose 10^3–10^6 rad h^{-1} and established that in air, mechanical damage was stronger when the dose rate was lowered because at lower dose oxidative scission became more important. This result was also supported by the carbonyl content and swelling experiments both of which increased at lower dose rates.

Gueskens and Kabamba [73] photooxidised EPR (65 mol% C_2H_4) films in air and vacuum using 310 nm and 365 nm monochromatic light and indicated that Norrish type II reaction is mainly responsible for photooxidation but chain-scission (Norrish type I) is also operative. The γ-irradiation [74] of the rubber (73 mol% C_2H_4) in oxygen gave mainly chain scission. The oxygen consumption and yields of oxidation products during EPR (73 mol% C_2H_4) γ-ray irradiation [75] were studied by GC-MS at room temperature under pressure. The oxygen consumption (above the threshold) and oxidation products of oxidative irradiation are independent of oxygen pressure. The oxidation was found more progressive in the crystalline region which is contrary to the results of Decker et al. [60] probably because of different experimental conditions and techniques. The steam aging [76] caused much quicker property deterioration than hot water aging.

Bausquet et al. [77] investigated photocrosslinking of EPR grafted with benzophenone on its backbone. The grafted benzophenone acts as a photosensitizer for vulcanization and oxygen plays a substantial role in the vulcanization by enhancing crosslink formation and the alkoxy and hydroperoxy radicals act as crosslink sites. The benzophenone gives a ketyl radical after hydrogen abstraction at the tertiary position from the EPR chain.

$$C_6H_5\text{-}\underset{O}{\overset{\|}{C}}\text{-}C_6H_5 \xrightarrow{h\nu} {}^1(C_6H_5\text{-}\underset{O}{\overset{\|}{C}}\text{-}C_6H_5)^* \longrightarrow {}^3(C_6H_5\text{-}\underset{O}{\overset{\|}{C}}\text{-}C_6H_5)^*$$

$$\xrightarrow{EPR} -CH_2\text{-}CH_2\text{-}\underset{\underset{}{}}{\overset{\overset{CH_3}{|}}{C}}\text{-}CH_2- \ + \ C_6H_5\text{-}\underset{OH}{\overset{*}{C}}\text{-}C_6H_5$$

$$(23)$$

The crosslinking efficiency of benzophenone could be due to the formation of a more active radical site and/or to a more efficient sensitization of hydroperoxide decomposition. The insulating electrical wire and cable made of EPR were irradiated with $\leqq 200$ Mrad γ-ray [78] for vulcanisation purposes. The changes in mechanical properties of EPR after irradiation [79] in air, under oxygen and in vacuum were investigated. The tensile strength decreased sharply with dose in any environment and the decrease in elongation at break was stronger in vacuum than in oxygen while the properties were intermediate when irradiated in air. Gueskens and Kabamba [80] have confirmed that in photochemical oxidation of EPR (65 mol% C_2H_4), the complex between the neighboring hydroperoxide and ketonic group

$$
\begin{array}{c}
\overset{\delta^+}{H}---\overset{\delta^-}{O} \\
\overset{\delta^-}{O}---\overset{\delta^+}{C} \\
\underset{R}{\overset{|}{O}}\quad R^1 \quad R^1
\end{array}
\xrightarrow{h\nu}
\quad
R\overset{O}{\diagdown}R^1
\quad + \quad
\overset{O}{\underset{HO}{\overset{\|}{C}}}R^1
\tag{24}
$$

is photolysed, and only the hydroperoxides were photolysed by 365 nm monochromatic light and the ketones not at all. The dipole-dipole interactions probably weaken the $O-O$ bond. These authors suggested that alkoxy and alkyl fragments recombine to produce carboxylic acids. gillen and Clough [81] studied the mechanism of radiochemical degradation of EPR (65 mol% C_2H_4) by density gradient column which proceeded by the inhomogeneous mechanism depending on the dose rate. At a high dose rate (1.2 Mrad/h), the diffusion-limited degradation by oxygen was found but as the dose rate decreased, the oxidation region spread inward until the encompassed the entire sample. The second degradation mechanism involved copper-catalyzed oxidation which is due to interaction of the materials with the rubber.

Li and Guillet [82] thermally oxidised EPR (78 mol% C_2H_4) to produce varying amounts of keto and hydroperoxy group. They estimated the quantum yields ($\varphi_s = 0.2$–1.3) higher for the samples containing both carbonyl and hydroperoxide groups in comparison to those containing only ketones ($\varphi_s = 0.036$). Their results offered direct evidence that excitation energy is transferred from the keto group to the hydroperoxide, possibly by the exciplex mechanism. Aliguliev et al. [83] showed that the thermal degradation of EPR at 250–470 °C was a first-order reaction and estimated the degradation activation energy to be 60.68 kcal mol^{-1}. The oxygen consumption and yield of oxidation products during γ-irradiation [84] at room temperature were studied on EPR films of varying thickness. The G-values of oxygen consumption decreased with film thickness and it was confirmed that oxidation was controlled by diffusion of oxygen.

Clough et al. [85] reported diffusion limited heterogeneous degradation in EPR in an oxygen atmosphere. In the solid film, the oxidised and non-oxidised regions were distinguished by differences in the surface reflectivity and differences are very pronounced at a higher dose rate (10^6 rad h^{-1}) where the elastomer undergoes crosslinking; in contrast, at a lower dose rate, where oxygen permeation is complete, predominantly scission is observed. Buckalew [86] compared the electron and photon degradation of EPR and observed that the damage to the elastomer was proportional to the absorbed energy irrespective of the particle type. Arnaud et al. [33] studied the photooxidation of various grades of E-P elastomers

(50–83 mol% C_2H_4). In the elastomer, the isolated hydroperoxides appeared on the PP units surrounded by PE sequences and the associated hydroperoxides are formed on the PE segments. The rate of hydroperoxidation was an increasing function of the propylene contents but even at high propylene contents, the carbonyl evolution resembled photooxidised polyethylene.

Bousquet and Fouassier [87–89] have investigated photosensitized oxidative degradation of EPR (equimolar ratio) by grafting low molecular weight benzophenone and its p-substituted derivatives onto an EPR backbone and confirmed that the polymer photoproducts played a key role in photooxidative degradation. Their results also revealed that grafting of benzophenone groups preferentially enhanced the extent of oxidation products formation and chain scission with respect to crosslinking. Recently, ZnO, TiO_2, and CdS photocatalyzed oxidation of an E-P elastomer (82 mol% C_2H_4) in the solid state has been studied by Lacoste et al. [90, 91]. It was observed that under polychromatic ($\lambda > 300$ nm) or monochromatic ($\lambda = 365$ nm) UV light, the introduction of the pigments modified the course of photooxidation. Under UV irradiation, these adsorbing pigments generate electron positive hole pairs (excitons). It is well known that electrons are trapped by chemisorbed molecular oxygen

$$O_{2(ads)} + e^- \xrightarrow{\quad\quad} O^-_{2(ads)} \tag{25}$$
$$\text{or}$$

$$ZnO + O_{2(ads)} \xrightarrow{h\nu} (ZnO)^{+} + O^-_{2(ads)} \tag{26}$$

The positively charged holes which are not annihilated through recombination react on the pigment surface either with a surface hydroxyl group or with hydroxyl anion ($\bar{O}H$) which supplies hydroxyl ($\dot{O}H$) radicals. On the surface, annihilation of $\bar{O}_2$ (ads) by the positive hole leads to different excited forms of oxygen [92] i.e. O_2^* excited molecular oxygen, O atomic oxygen or O* excited atomic oxygen. If water is present, neutralization of $\bar{O}H$ by a positive hole ($\oplus$) supplies $\dot{O}H$ radicals [93]. The perhydroxyl radical ($H\dot{O}_2$) can be formed through protonation of O_2^- or through annihilation of $\dot{O}H$ with O_2^- (ads)

$$\oplus + {}^-OH_{(ads)} \xrightarrow{\quad\quad} \dot{O}H_{(ads)} \tag{27}$$

$$O_2^- + H^+ \xrightarrow{\quad\quad} H\dot{O}_2 \tag{28}$$

$$\dot{O}H + O^-_{2(ads)} \xrightarrow{\quad\quad} H\dot{O}_2 + O^-_{(ads)} \tag{29}$$

The reaction as a whole is as follows:

$$H_2O + O_2 \xrightarrow[h\nu]{ZnO/TiO_2} \dot{O}H + H\dot{O}_2 \tag{30}$$

in which water and oxygen are constantly being consumed to destroy the polymer. Thus ZnO, TiO_2, and CdS accelerate the process of EPR photodegradation which is accelerated due to reactive species which can attack macromolecules chemically.

3.3 Thermoplastic Elastomers

Practically no studies of degradation have been carried out on thermoplastic elastomers.

4 Mechanisms of Ethylene-Propylene Copolymer Stabilization

The ethylene-propylene copolymer (thermoplastics and rubbers) stabilization may be achieved by the following mechanisms:

4.1 Light Screeners

The screeners are interposed as a shield between the radiation and the polymer. They function either (a) by absorbing the radiation before it reaches the photoactive species in the polymer or (b) by limiting the damaging radiation penetration into the polymer matrix. Reflection of radiation can be achieved by selection of suitable paints, coatings, pigments or by metallizing [94, 95] the surface.

Heat resistant [96] EPR is made by mixing the elastomer with carbon black, MgO and curing at 160 °C in the presence of sulphur and tetrachloro-*tert*-butyl peroxide. In this composition, carbon black acts as a light screener as well as a filler, tetrachloro-*tert*-butyl peroxide is a free-radical initiator for crosslinking, and the stabilizing effect of MgO is largely due to a non-chain reaction. The metal particles may form a new type of polymer containing metal atoms in its chain. The non-chain inhibition includes removal or deactivation of initially agents. An EPR (equimolar ratio) of highly increased modulus, hardness, abrasion resistance and of stretchability was obtained [97] by curing with 3 wt.% *tert*-butylperoxide at 163 °C where the furnace carbon was used as a screener and filler. Amberg and Robinson [98] have considerably increased the tensile strength of EPR (73 mol% C_2H_4) by vulcanizing with dicumyl peroxide (a free-radical initiator) and re-inforcing with HAF or SRF black. To achieve high strength at elevated temperatures (200 °C), Sieron and Murray [36] compounded EPR (73 mol% C_2H_4) with carbon-black. The stability of a sample containing HAF black, EPC black, and SAF black in the presence and absence of antioxidants was compared (Table 4) and it was estimated that the particle HAF-black gives the best strength without any antioxidant. The tensile strength and crosslinking density of HAF black mixed EPR (66 mol% C_2H_4) was increased [99] to a greater extent upon addition of sulfur into its matrix during radiation curing with γ-irradiation or with an electron beam, furthermore, the maximum tensile-strength was obtained at a lower radiation

Table 4. Effect of carbon black, antioxidants and curing on tensile strength of EPR

No.	Ingredients	Curing temp. °C	Curing time min	Tensile strength kg cm^{-2}
1.	50 phr HAF + 4 phr Di-Cu-P	160	30	190
2.	50 phr HAF + 4 phr Di-Cu-P + 1 phr S	155	20 40 60	197 233 201
3.	50 phr HAF + 3 phr D-Cu-P + 1 phr stearic acid	160	60	115
4.	50 phr HAF + 3 phr Di-Cu-P + 1 phr stearic acid + 0.3 phr sulfur	165	20	179
5.	50 phr EPC-black + 2.5 phr ZnO + 5 phr Di-Cu-P + 0.4 phr S	177	15	169
6.	50 phr EPC-black + 2.5 phr ZnO + 5 phr Di-Cu-P + 2 phr antioxidant Agedite white Antioxidant-2246 Mercapto-benzothiazole	177	15	194 247 257
7.	50 phr HAF	γ-irrdiation doses	20 Mrad 40 Mrad 60 Mrad	55 83 101
8.	50 phr HAF + 0.5 phr S	γ-irrdiation doses	20 Mrad 40 Mrad 60 Mrad	84 127 124

dose than the corresponding control experiments, the reason may be that at a higher dose rate the chain-scission may be dominant over crosslinking.

An EPR (equimolar ratio) was stabilized [100] against photo- and thermo-oxidative degradation by milling stearic acid, carbon black, ZnO, dicumyl peroxide, and tri(isopropylbenzene)polycarbodiimide at 151 °C for 35 min. The EPR composition [101] was made by addition of an organic peroxide as a crosslinking agent and MgO which acts both as a screener and a non-chain inhibitor. The powdered Vulcafixe, Lutetia and granulated Vulcafix organic pigments were added to EPR to give coloured products which did not fade during vulcanization or exposure to light [102]. The functions of these pigments are both screening and absorption.

Recently, Lacoste et al. [90, 91] have investigated the photoaging phenomena of pigmented EPR for fundamental as well as practical reasons. It has been emphasized that with a low pigment content, the inner filter effect prevails over the photocatalytic influence with any pigment. It was suggested that the carbonyl groups formed are adsorbed onto the pigment. The photo-irradiation of EPR in the presence of pigments showed that the filter effect alone was observed for low pigment concentrations and the photocatalytic influence was superimposed only for higher concentrations. In the unpigmented polymer, the deterimental effect on the mechanical properties was observed from the beginning of the irradiation but in the presence of pigment a slight improvement was observed just after short exposure. Such an increase of physical properties can only be understood if a reinforcing effect exists between oxidized groups and the pigments (Fig. 3). Thus, these absorbing pigments afford an efficient control of the initiation rate and ensure the photochemical protection of the intermediate products.

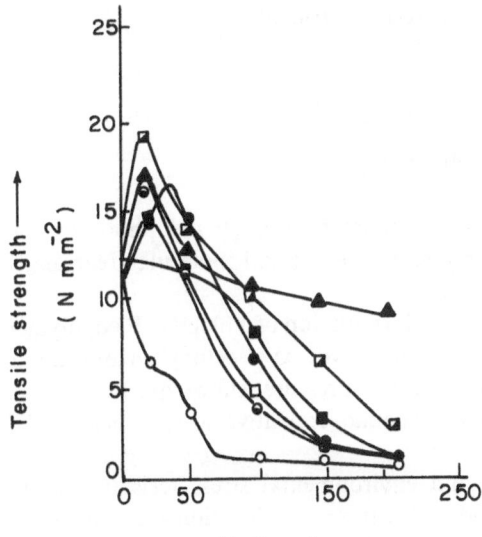

Fig. 3. Variations in mechanical properties during polychromatic irradiation of EPR films (o) EPR, (◐) EPR + 0.5 wt.% ZnO, (●) EPR + 2 wt.% ZnO, (■) EPR + 5 wt.% ZnO; (□) EPR + 0.5 wt.% TiO$_2$, (◪) EPR + 2 wt.% TiO$_2$, (▲) EPR + 5 wt.% TiO$_2$

4.2 Ultra-Violet Absorbers

The hydroxybenzophenones and hydroxybenzotriazoles are important groups of UV stabilizers/absorbers in use since both the groups adsorb strongly in the UV region. These compounds also appear to have the ability to quench [103] excited polymers. Furthermore, it has been suggested that hydroxybenzophenones might operate in part as radical scavengers [104]. The photostabilization mechanism [105, 106] of o-hydroxybenzophenones and o-hydroxybenzotriazoles is believed to be a rapid tautomerism of the excited state

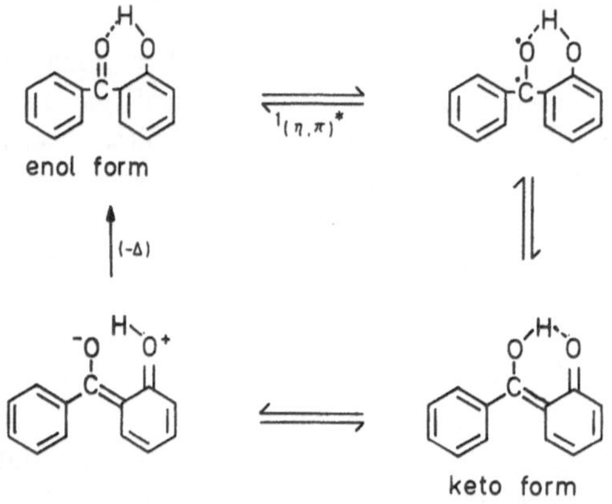

enol form

keto form

(31)

The more basic the hetero-atom (O) in the ground state, the more light stable is the compound. It is assumed that in the ground state the 'enol' form is energetically preferred, whereas the reverse is true for the first excited singlet. An argument for this viewpoint is the fact that in the excited state phenol becomes much more acidic whereas the hetero-atom 'O' becomes more basic than in the ground state. The same mechanism is applicable for o-hydroxybenzotriozole

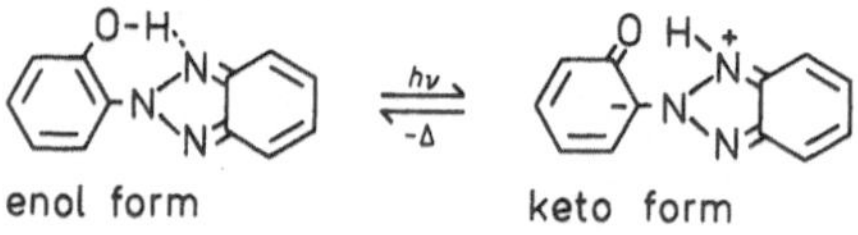

enol form   keto form

(32)

In addition to this effect, UV absorbers function through the photophysical mechanism, internal conversion, intersystem crossing and molecular rearrangements which are discussed in Sect. 4.7.

Kimura et al. [107] made a light resistant formulation of a block E-P copolymer by extruding it at 220 °C with $Mg(OH)_2$, and 2-hydroxy-4-octoxybenzophenone. The injection molded product (at 800 kg cm^{-2} pressure) shows fire retardation also due to MgO filler and the benzophenone derivative is an efficient UV absorber.

Kusunomi [108] made EPR heat and environmental stress resistant by introducing into the matix carboxylic acid derivatives which function via photo-Fries rearrangement. Kato [109, 110] added hydroxybenzamide or its derivatives with or without metal. He furthermore improved the thermal degradability of EPR with tetrazine derivatives also of the type I:

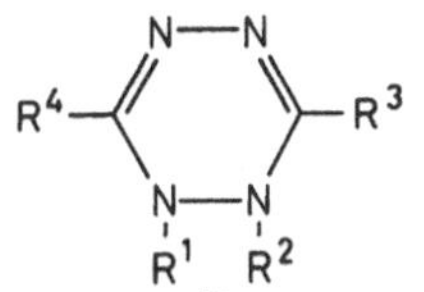

I

where R^1, R^2 = H or aryl with or without alkyl, aryl, aralkyl, alkoxy or halogen substituents; R^3, R^4 = alkyl, aryl or aralkyl. He found that the rubber retained 81% of its initial tensile-strength and 73% elongation after 4 days aging of the vulcanizate at 140 °C by compounding it with 80% whitetex, ZnO, stearic acid, sulfur and 1,2-dihydroxy-3,6-diphenyl-1,2,4,5-tetrazine. The tetrazine derivatives stabilize the polymer by absorbtion and resonance while the whitetex acts as a filler. Treshchalov and coworkers [111] used s-triazine and s-heptazine as stabilizers against thermal degradation of EPR. The use of 4,6-diallylloxy-2-(ethylamino)-s-triazine, 4,6-diallyloxy-2-(octadecylamine)-s-triazine and N,N-bis(4,6-di-alloyloxy-2-s-triazinyl)piperazine in the peroxide catalyzed crosslinking of EPR increased its thermal stability [112]. The stabilization mechanism of these additives is similar to those of benzophenone and benzotriazole.

Derbisher et al. [113] described the results of an experimental investigation concerning the effects of dicarboxylic acid and their dihydrazines. The enhanced oxidative thermal aging was exhibited when EPR was vulcanized at 165 °C in the presence of dicarboxylic acid and its dihydrazide. The increased aging resistance was attributed to polycondensation of dicarboxylic acid with dihydrazide to form a product readily cyclizable at > 200 °C into a stable 1,3,4-oxidiazole structure (II)

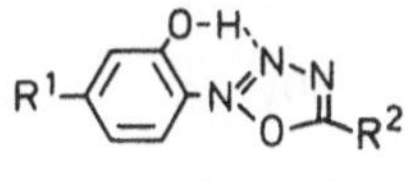

II

but open modification of EPR either with dicarboxylic acid or its dihydrazide, no increase in oxidative thermal degradation was observed.

4.3 Antioxidants

The antioxidant (AH) may inhibit oxidation processes by the following proposed mechanism

$$RH \xrightarrow[\text{UV radiation}]{\text{sunlight/heat}} \dot{R} + \dot{H} \xrightarrow{O_2} R\dot{O}\dot{O} + \dot{O}OH \qquad (33)$$

$$\dot{R} + AH \longrightarrow RH + \dot{A} \qquad (34)$$

$$ROO + AH \longrightarrow ROOH + \dot{A} \tag{35}$$

$$R\dot{O}O + \dot{A} \longrightarrow \text{stable products} \tag{36}$$

$$2\dot{A} \longrightarrow \text{stable products} \tag{37}$$

The antioxidant acts as a chain terminating agent. The antioxidants react faster with peroxy radicals (RO$\dot{O}$) than with macroradicals ($\dot{R}$) and the activity depends upon their structure [114–116].

The copper catalyzed degradation of E-P copolymer (3–40 mol% C_2H_4) is prevented [117] by azimidobenzene or phenothiazine or their derivatives and 0.1–5 wt.% [4,4'-thiobis(3-methyl-6-*tert*-butylphenol)] and N-phenyl-β-naphthylamine. The later two compounds are powerful antioxidants and functions as follows:

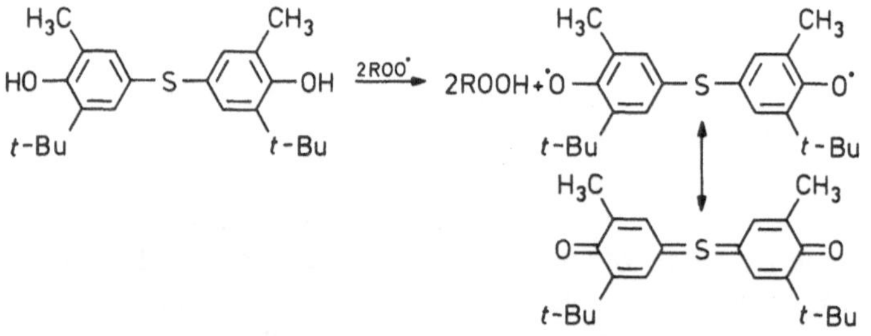

$$\tag{38}$$

Thus the diradical is resonance stabilized with a diquinoid structure. During the reaction sulfur is liberated as SO_2 which reacts with hydroperoxide

$$\tag{39}$$

The N-phenyl-β-naphthalamine also donates its H-atom to stabilize the macro-radical

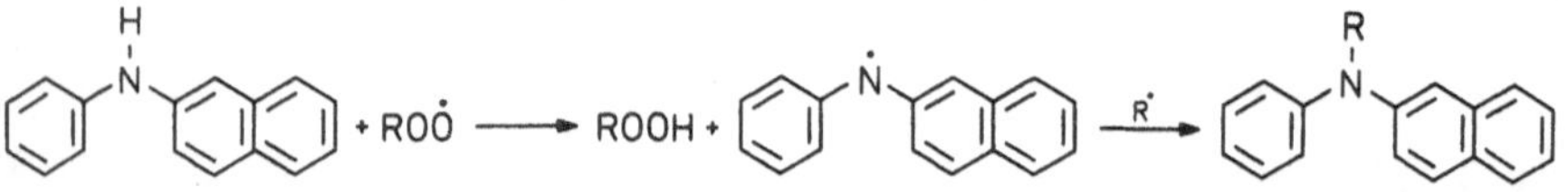

$$\tag{40}$$

the free-radical thus formed is resonance stabilized and also acts as a radical trap. The heat stability and processability of block E-P (7.32–20.93 mol% C_2H_4) is improved [118] by treating the copolymer with butylated hydroxytoluene and 2,5-dimethyl-2,5-bis(*tert*-butylperoxy) hexane. The first compound acts as an antioxidant while the second one is a chain terminator.

The analogs of 3,5-di-*tert*-butyl-4-hydroxytoluene [119] have been used as thermal-oxidative inhibitor for EPR. When sulfur is used as a coagent in peroxide-cured EPR, crosslinking proceeds in a manner which results in a high crosslink density with low elongation at break and reduced tensile strength. On the other hand, phenol and aromatic amine type antioxidants [120] in the presence of EPC-black tend to give a lower crosslink density with high tensile strength and elongation at break (Tab. 4). Field [121] made (88 mol% C_2H_4) oxidation resistant and electrical insulating EPR the addition of 2,5-bis(*tert*-butylperoxy)-2,5-di-methyl hexane or benzyl-α-methylbenzyl peroxide as a crosslinking agent and 4,4'-thiobis(6-*tert*-butyl-*m*-cresol) as an antioxidant. The thermal and ozone resistant [122] EPR (72 mol% C_2H_4) has been prepared by incorporating a mixture of poly(ethylene episulfide), poly(propylene episulfide) and ethylene-episulfide-propylene-episulfide copolymer. The organic compounds [123] contain sterically hindered phenols and the amine groups are effective stabilizers against heat and UV radiation, and function as antioxidants in the rubber matrix.

The effectiveness of *p*-cresol/formaldehyde condensate [124] of the type III

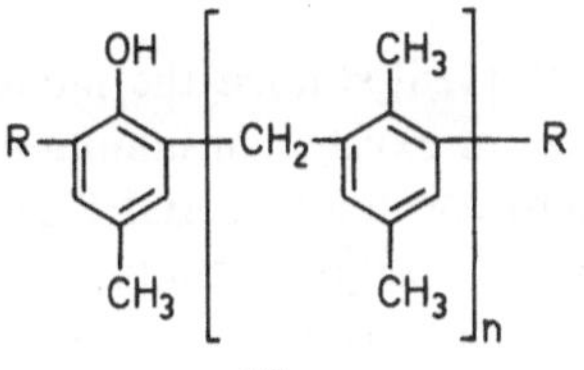

III

(where R = H, Me, CMe_3 and n = 0, 1, 2, 4) as an antioxidant in EPR consisting of approximately equimolar proportion of ethylene and propylene increases with increasing molecular weight of the additive but the interference with the cure becomes weaker. The thermal oxidative degradation of EPR was prevented [125] by a mixture of Neozone D and vanadyl triacetylacetonate. During thermal exposure, the acetylacetonate compound is decomposed and the products may prevent oxidation via the non-chain inhibition reaction. In a similar way, the metal may convert oxygen into the ionic form and thereby transfer it to the inactive state.

$$M^{2+} + O_2 \longrightarrow M^{3+} + O_2^{\bar{}}$$ (41)

The heat and ozone resistant [126] EPR was made by incorporating acrylic rubber, dicumyl peroxide, triallyl cyanurate, ZnO and carbon-black into the matrix. Triallyl cyanurate increases the crosslinking efficiency probably due to an addition reaction between polymeric and allyl radicals and leads to stable chemical crosslinks. Thus ozone because there is no unsaturation cannot initiate a degradation reaction. Digteva et al. [127] prepared sealants for use at high temperature by adding aromatic diaminodisulfide, MgO, ZnO and carbon black in EPR. The aromatic diaminodisulfide is an antiozonant and functions both as an antioxidant and a

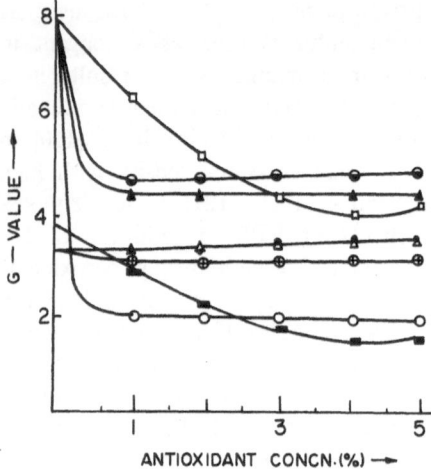

Fig. 4. G-values of O_2 consumption and evolved H_2 against wt.% of antioxidants at dose rate 0.2 Mrad/hr up to 20 Mrad DPPD; (○) O_2, (⊕) H_2; Irganox 1010: (◓) O_2, (●) H_2; NBC: (▲) O_2, (△) H_2; PFR: (□) O_2, (■) H_2

hydroperoxide decomposer. Arakawa et al. [128] studied the radiation induced oxidative degradation of EPR. The G-values of oxygen consumption and evolved gases in presence of N,N'-diphenyl-p-phenylene diamine (DPPD), Ni-di-butyl-di-thiocarbamate (NBC), Irganox 1010 and propyl fluoranthene (PFA) are shown in Fig. 4. All the antioxidants decreased oxygen consumption and also the amount of gaseous products. DPPD is the most effective in the reduction of EPR oxidation. The effects of these antioxidants has also been confirmed in swelling experiments by the same authors [74] which demonstrated that chain scission correlates with oxygen consumption results. The G-values become nearly constant beyond 0.5 wt.% of antioxidants while they are independent of the percentage up to 2.5 wt.% of PFA. These results indicate that 0.5 wt.% of antioxidant is effective up to 60 Mrad and then loses its activity, which means that the antioxidant is consumed during γ-irradiation. Probably PFA absorbs high energy from the polymer chains and converts it into thermal energy thus reducing radical formation which induces oxidative degradation.

The addition of aromatic diamino disulfides [129] e.g. 2,2′-dithiobis]N-(2-naphthyl)aniline], 2,2′-dithiobis]]2-anilinophenyl)-2-propyl]phenol], etc. to peroxide-vulcanized EPR at 170 °C increased the heat and sealing capacity of the rubber. These additives also increase the low temperature resistance of the rubber seals (from −60 to 150 °C). These compounds are antiozonants, the mechanism of which is described by the so called scavenger theory. According to this theory, antiozonants are effective because they diffuse to the surface and scavenge ozone due to their greater reactivity. The second view of the antiozonant mechanism is believed to be due to diffusion to the surface and then a spreading out over the surface to form a protective film and only the ozonized anti-ozonant is believed to make an effective film. the 2,4,6-tri(antioxidant group)-1,3,5-triazine [130, 131]

of the general formula IV

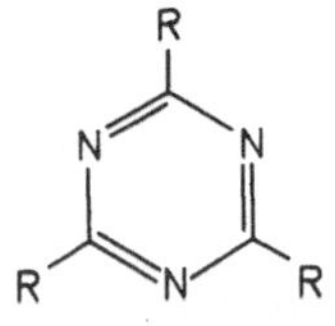

IV

(where R = p-PhNHC$_6$H$_4$NH$_2$ or p-PhNHC$_6$H$_4$O) have been found as effective antioxidants at higher temperature without any volatility in the EPR systems. These polyfunctional additives have no adverse effects on the crosslinking activity of the peroxide. The heat resistant [132] compound is obtained by kneading phenol derivatives of type V

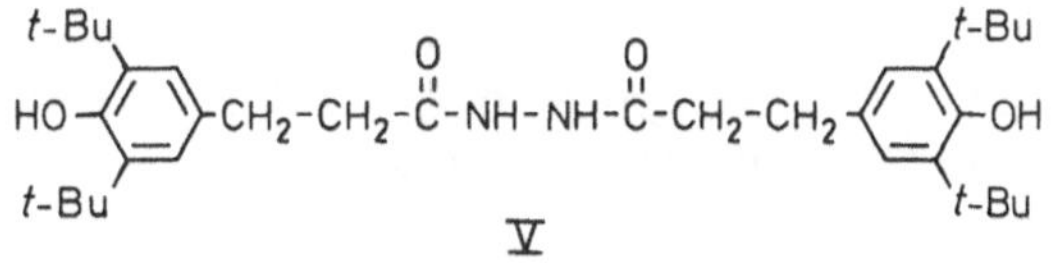

V

with the rubber. The phenol derivatives are well known to be effective antioxidants. The effectiveness of p-hydroxyneozone and diafen-N,N as antioxidants in oxidative and radiochemical thermal aging [133] of EPR was increased on modification with S$_2$Cl$_2$ where the sulfur atoms functioned as hydroperoxide decomposers.

4.4 Peroxide Decomposers

The salts of alkyl xanthates, N,N'-di-substituted dithiocarbamates and dialkyl dithiophosphates [43] are effective peroxide decomposers. Since no active hydrogen is present in these compounds, an electron-transfer mechanism was suggested. The peroxide radical is capable of abstracting an electron from the electron-rich sulfur atom and is converted into a peroxy anion as illustrated below for zinc dialkyl dithiocarbamate [44]

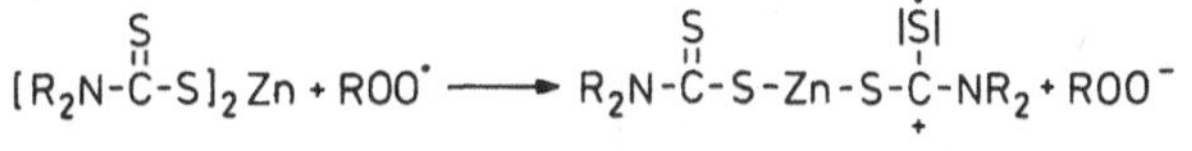

(42)

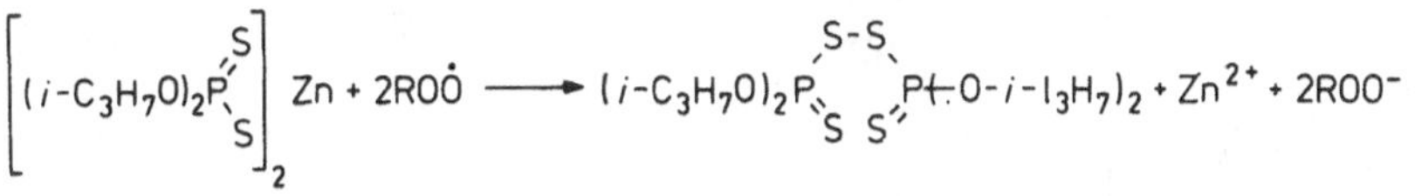

(43)

The E-P copolymer (2.98 mol% C$_2$H$_4$) was stabilized [134] against heat and light degradation by incorporation in the copolymer matrix dialkyl thiodipropionate or 2-(2'-hydroxyphenyl)-2,4,4-trimethyldialkyl-5,6-dinonyl or alkylidene-bis[2,2'-isopropylidene bis(4-nonylphenol)]. The dialkyl thiodipropionate decomposes the hydroperoxide in the following way

$$(RO-\overset{O}{\overset{\|}{C}}-CH_2-CH_2\overset{}{)_2}S + 3ROOH \longrightarrow 2RO-\overset{O}{\overset{\|}{C}}-CH=CH_2 + SO_2 + H_2O + 3ROH \quad (44)$$

and forms stable products while the phenol derivatives inhibit oxidation process by an antioxidant mechanism.

Schroeder and Leonard [135] obtained high impact EPR (73 mol% C_2H_4) by vulcanizing and blending it with metal oxide. The phosphonic group in the formulation acts primarily as hydroperoxide decomposer. The mechanical properties of EPR (42 mol% C_2H_4) were improved [136] by incorporating PbO, mercaptobenzthiazole, staybelite (a hydrogenated natural resin) and benzoyl peroxide in the polymer matrix. Mercaptobenzthiazole and its metal complexes are powerful catalysts for the decomposition of hydroperoxides

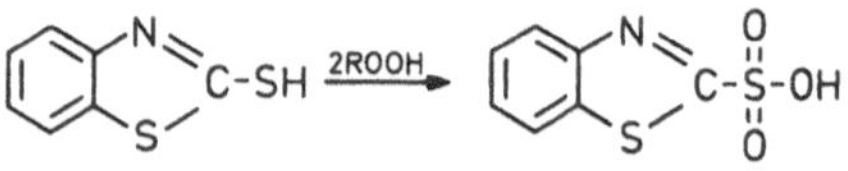

$$(45)$$

The benzthiazole sulfonic acid is the major endproduct; it is relatively stable and may form a complex with PbO. The weatherability [137] is enhanced by the addition of ZnO and 2-mercaptobenzimidazole with small amounts of conventional antioxidants where mercaptobenzimidazole acts as a peroxide decomposer as well as antioxidant by donating hydrogen

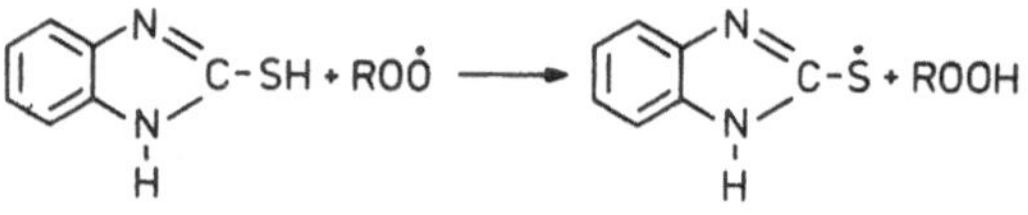

$$(46)$$

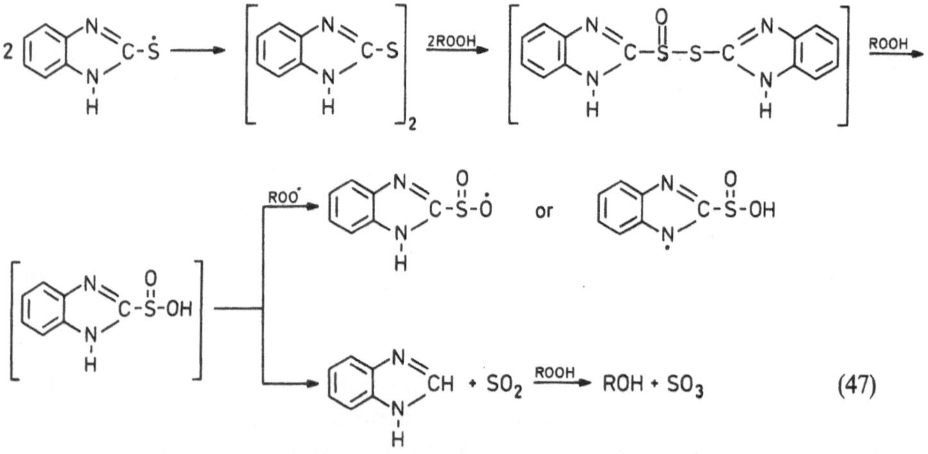

$$(47)$$

ZnO acts as a vulcanization accelerator and may also form a complex with mercaptobenzimidazole

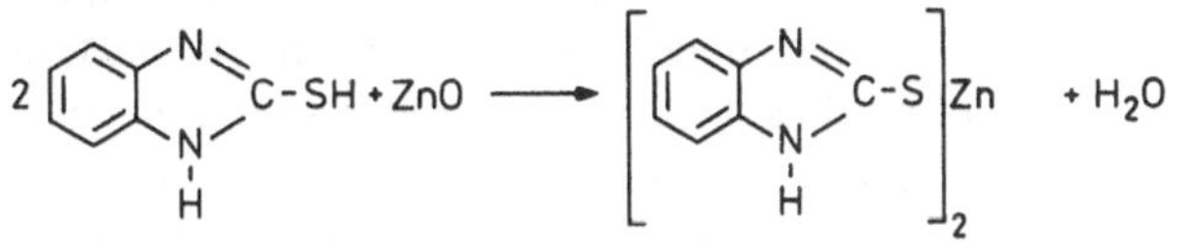

$$2 \left[\underset{\underset{H}{N}}{\overset{N}{\bigcirc}} C\text{-SH} \right] + ZnO \longrightarrow \left[\underset{\underset{H}{N}}{\overset{N}{\bigcirc}} C\text{-S} \right]_2 Zn \quad + H_2O \tag{48}$$

which acts as an antioxidant and peroxide decomposer.

Brams et al. [138] prepared a typical adhesive for metal and cloth by mixing in EPR a mixture of ZnO, stearic acid, HAF black, petroleum ether, sulfur, tetramethylthiuram monosulfide and mercaptobenzothiazole. The tetramethylthiuram monosulfide is a new important curing agent but, in the presence of ZnO, it gives vulcanizates with high level resistance to ozone deterioration, UV degradation, and thermal oxidation and forms the corresponding zinc dialkylthiocarbamate

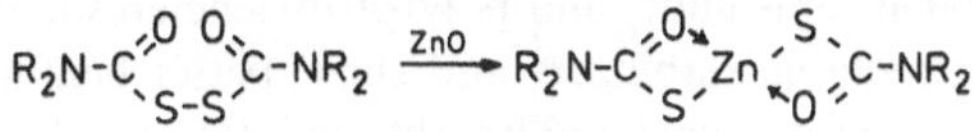

$$R_2N\text{-}C\overset{O}{\underset{S\text{-}S}{\overset{O}{\diagdown}}}C\text{-}NR_2 \xrightarrow{ZnO} R_2N\text{-}C\overset{O}{\underset{S}{\diagdown}}Zn\overset{S}{\underset{O}{\diagdown}}C\text{-}NR_2 \tag{49}$$

a hydroperoxide decomposer. Mercaptobenzothiazole is also a powerful hydroperoxide decomposer. The thermoplastic bitumen-rubber compound for road surfacing with improved heat, UV radiation and ozone resistance [139] was prepared by kneading EPR, ZnO, chalk, paraffin oil and tetramethyl thiuram disulfide. In the presence of ZnO, it gives a highly thermooxidative resistant vulcanizate with the formation of the corresponding zinc dialkyl dithiocarbamate (a powerful hydroperoxide decomposer)

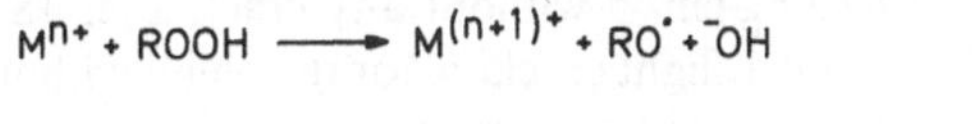

$$R_2N\text{-}C\overset{S}{\underset{S\text{-}S}{\overset{S}{\diagdown}}}C\text{-}NR_2 \xrightarrow{ZnO} R_2N\text{-}C\overset{S}{\underset{S}{\diagdown}}Zn\overset{S}{\underset{S}{\diagdown}}C\text{-}NR_2 \tag{50}$$

Hasebe et al. [140] prepared an antitracking and weather resistant EPR by mixing into its matrix carbon and titanium phosphate. The metal salt gets ionized and then decomposes the hydroperoxide and peroxy radicals:

$$M^{n+} + ROOH \longrightarrow M^{(n+1)+} + RO^{\bullet} + {}^{-}OH \tag{51}$$

$$M^{n+} + RO\overset{\bullet}{O} \longrightarrow M^{(n+1)+} + ROO^{-} \tag{52}$$

The heat stability [141] of EPR was increased by the addition of a bifunctional thiophosphate and the product had no offensive odour. Thiophosphate derivatives stabilize EPR by the mechanism of peroxide-hydroperoxide decomposition.

4.5 Nucleating Agents/Fillers

Remarkable stabilization may also be achieved by the use of chemically inactive fillers or nucleating agents. The nature of the interaction of fillers and polymer molecules is not completely understood. However, it is known that the filler restricts chain mobility and reduces the diffusivity of attacking agents in polymers which naturally prolongs their lifetime at elevated temperatures. There is not much difference between fillers and nucleating agents. The nucleating agents may crystallize the polymer and thus increase the stability of the polymer. They should possess a crystalline structure and should melt above the polymer melting point. The nucleating agents should have the capacity to absorb the polymer on its surface and make the polymer matrix insoluble at and below the polymer melting point.

Moteki et al. [142] have increased the dimensional stability of block E-P copolymer (11.5 mol% C_2H_4) by mixing it with polyethylene, finely ground waste paper and butyl stearate-glycerine ester resin (residue from terpentine distillation). The composition after mixing had a tensile strength of 335 kg cm^{-2}, Izod impact strength 3.1 kg cm/cm and a decomposition temperature of 138 °C.

EPR was stabilized [143] against thermal and photooxidation degradation by the addition of poly(ethylene episulfide) and poly(propylene episulfide) which are non-volatile and non-extractable stabilizers and show better heat stability than the furnace black. The transparent, heat and weather resistant flexible rubber [144] was prepared by incorporating SiO_2 (reinforcing agent) and dicumyl peroxide (curing agent) in EPR at 160 °C. The calcium sulfite is a suitable filler [145] for EPR (37.5 mol% C_2H_4) which gives high heat and sunlight mechanical strength and adhesion to plastics. Ohara and coworkers [146] made heat, weather, water and solvent resistant elastic foam by extruding a mixture of EPR, asphalt, sulfur, $CaCO_3$ and processing oil at 150 °C. Asphalt acts as a reinforcing agent. Decomposition of carbonate involves the formation of oxide,

$$CaCO_3 \longrightarrow CaO + CO_2 \uparrow \tag{53}$$

that can play a role of an active filler and CO_2 functions as a blowing agent. The mixture [147] of EPR, EPDM, chlorosulfonated polyethylene, vinyl silane, carbonblack, condensate of acetone dibenzylamine, 2-(mercaptotalyl imidazole) and polybutadiene after vulcanization at 350 °C, shows a high temperature resistance and has tensile strength of 82.3 kg cm^{-2} without any crack after 18 h at 200 °C, compared with 55.3 kg cm^{-2} and a slight cracking for the blend without imidazole. The phenol cumylphenol oligomer (Aluroflen) shows two for four times higher heat stability [148] than that of conventional heat stabilizer (Neozone D) in EPR. The reason is a nucleating and plasticizing effect of Alurofen in the rubber.

4.6 Radical Scavengers/Free-Radical Traps

Since the radicals are the most responsible species for the degradation reactions, scavenging of radicals is effective for protecting polymers against photo-, thermal and radiation degradation. The mercaptans [149] and aromatic amines [150] are the most common class of radical scavengers. The scavenging by such compounds proceed through the donation of hydrogen atom to radical sites in the polymer

$$R^\cdot(\text{or } RO\dot{O}) + SH \longrightarrow RH(\text{or } ROOH) + \dot{S} \tag{54}$$

Mercaptans first act as electron acceptors and then transfer the negative charge to the H atom.

Any substance capable of reacting with free radicals to form products that do not reinitiate the oxidation reaction could be considered to function as free-radical traps. The quinones are known to scavenge alkyl free radicals. Many polynuclear hydrocarbons show activity as inhibitors of oxidation and are thought to function by trapping free radicals [47]. Addition of R˙ to quinone or to a polynuclear compound on either the oxygen or nitrogen atoms produces adduct radicals that can undergo subsequent dimerization, disproportionation or reaction with a second R˙ to form stable products.

The peroxide-quinone dioxime or peroxide-dibenzoyl-quinone dioxime cured EPR is much more heat-stable [151] than those of ovenaging samples because both crosslinking as well as radical trapping is done simultaneously. The aliphatic primary diamine [152] or aromatic primary diamine [153] were subjected to polycondensation separately with a quinone to give a Schiff base polymer of type VI or VII i.e.

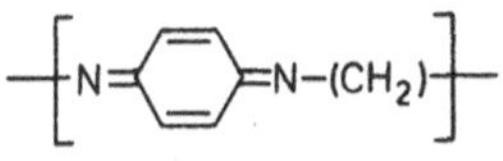

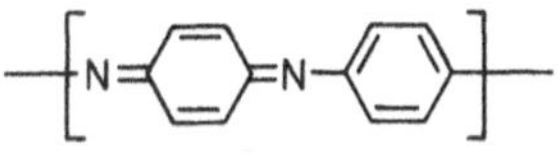

which was mixed with p-phenylenediamine derivatives to give an antioxidant and free-radical trap for EPR

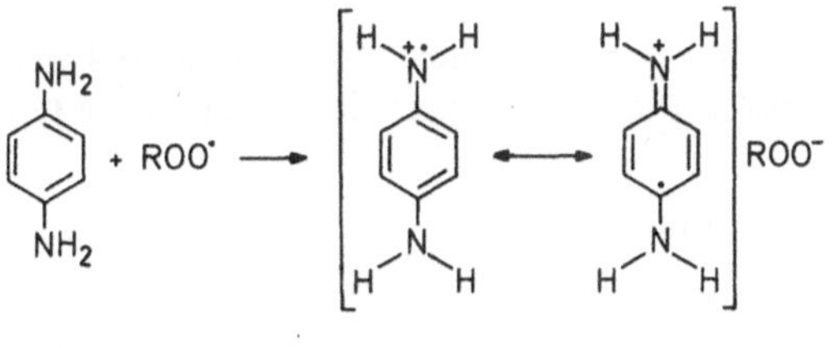

$$\tag{55}$$

The m-phenylenedimaleimide [66] has been found to have a stabilizing effect on the thermooxidative degradation of EPR vulcanizates. This compound prevents formation of unsaturation by reacting randomly with any double bond in the polyene. Thus, it terminates the active site of polymer degradation and removes the destabilizing influence of conjugated double bonds in further decomposition of the adjacent units. The condensation product [154] of 2,4-dihydroxybenzylidene-2-naphthylamine with S_2Cl_2 is the effective antiradiant and antidegradant for EPR. The naphthylamine derivative was chemically attached to the macroradical which makes the polymer undegradable. The mechanism is as follows

$$R^{\cdot} + HN\overset{R^1}{\underset{R^2}{\diagdown}} \longrightarrow RH + {}^{\cdot}N\overset{R^1}{\underset{R^2}{\diagdown}} \tag{56}$$

$$R^{\cdot} + {}^{\cdot}N\overset{R^1}{\underset{R^2}{\diagdown}} \longrightarrow R-N\overset{R^1}{\underset{R^2}{\diagdown}} \tag{57}$$

where $R^{\cdot}$ is a macroradical. Thus, this compound acts as an antioxidant and a radical trap.

The methyl/ethyl mercaptan (radical scavenger) caused a large decrease ($\simeq 50\%$) in yield of isobutane with a comparable increase in the yield of n-butane but the methylamine/triethylamine (cation scavenger) caused greater decrease in the isobutane yield than the mercaptans [30] (Fig. 1). The correlation between n-butane and isobutane yield, suggests that a free-radical or cationic intermediate is formed during the γ-rays radiolysis of E-P copolymer and EPR. Tanimoto et al. [155] obtained excellent weatherability by mixing hindered amine light stabilizers (HALS) in EPR and TPO matrices. The hot pressed sheet shows no discolouration and fold cracks even after 1000 h exposure to a sunlight weatherometer. The stabilizing effectiveness of HALS is believed to depend upon their ability to form a stable nitroxyl radical which then scavenges macroradicals produced during oxidative degradation. The mechanism [156] of stabilization is proposed to be as follows

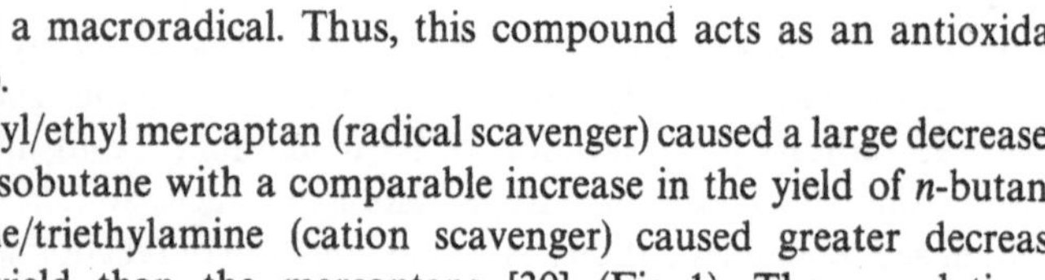

$$\overset{\diagup}{\underset{\diagdown}{\bigcirc}}N-H \xrightarrow[\Delta/h\nu/air]{ROOH} \overset{\diagup}{\underset{\diagdown}{\bigcirc}}N-\dot{O} + R\dot{O} + H_2O \xrightarrow{RH/O_2} \overset{\diagup}{\underset{\diagdown}{\bigcirc}}N-OOH + R^{\cdot}$$

$$\overset{\diagup}{\underset{\diagdown}{\bigcirc}}N-O-R \xleftarrow{R^{\cdot}} \overset{\diagup}{\underset{\diagdown}{\bigcirc}}N-\dot{O} + \dot{O}H \tag{58}$$

Some workers suggested that nitroxyl radical stabilizes the polymer by forming a hydrogen-bonded complex with a hydroperoxide group of the type VIII

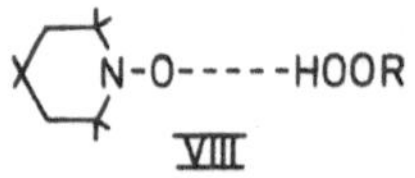

$$\overset{\diagup}{\underset{\diagdown}{\bigcirc}}N-O-----HOOR$$
$$\underline{VIII}$$

and by charge-transfer complex [157] with a macroperoxy radical

$$R\dot{O}O + HALS \longrightarrow ROO^- ----(HALS)^+ \longrightarrow [ROO----HALS]----OOR + H^+$$
$$\longrightarrow HALS + products \tag{59}$$

The stability of the charge-transfer complex is supposedly enhanced in solid polymers due to reduced mobility of the macroperoxy radicals. During the thermal processing, nitroxyl radicals are regenerated [158]:

$$\langle \overline{N}-O-R + RO\dot{O} \longrightarrow \langle \overline{N}-\dot{O} + ROOR \tag{60}$$

By this mechanism, each nitroxyl radical is believed to terminate many free radical chains.

4.7 Excited-State Quenchers

Although for many years chain termination, UV absorption and peroxide decomposition were the traditional methods for photo, thermal and radiation stabilization, metal chelates and quenchers also are in current use. It has been reported that metal chelates with a variety of ligands are excellent quenchers for excited states [105, 159]. In the solid state, transfer of energy occurs by resonance or dipole-dipole interactions. The photo-oxidative degradation is promoted by electronically excited oxygen molecule (commonly called singlet oxygen $^1O_2$) formed in polymers. The $^1O_2$ can also be generated in polymers by energy transfer from electronically excited carbonyl group to dissolved (chemisorbed) molecular oxygen [94]. The quenching of $^1O_2$ is also necessary for effective stabilization and nickel-chelates [160–162] have proved to be effective quenchers for excited states of $^1O_2$. The quenching may occur in one of the two ways:

4.7.1 Energy Transfer

The energy transfer may result in a reactive or non-reactive quencher molecule-

$$^1O_2^* + S \longrightarrow O_2 + S^* \tag{61}$$

$$S^* \longrightarrow S + h\nu \tag{62}$$

$$^1O_2^* + S \longrightarrow SO_2 + h\nu \tag{63}$$

$$RH^* + S \longrightarrow [RH ---- S]^* \longrightarrow \text{photophysical processes} \tag{64}$$

where S^* is an excited singlet or triplet state of the stabilizer (quencher). The quencher must be capable of dissipating most of the accepted energy harmlessly, i.e. without destroying the polymer bonds or its own structure.

The various light-emission processes are best described with reference to the Jablonski diagram. The diagram for a simple carbonyl compound is given in Fig. 5. The absorption process leads to the formation of excited singlet states (S_1, S_2, etc.). A very rapid depopulation of the upper excited states occurs by internal conversion through vibrational relaxation processes. After deactivation to the first excited singlet state (S_1), several processes are possible for the molecule to reduce

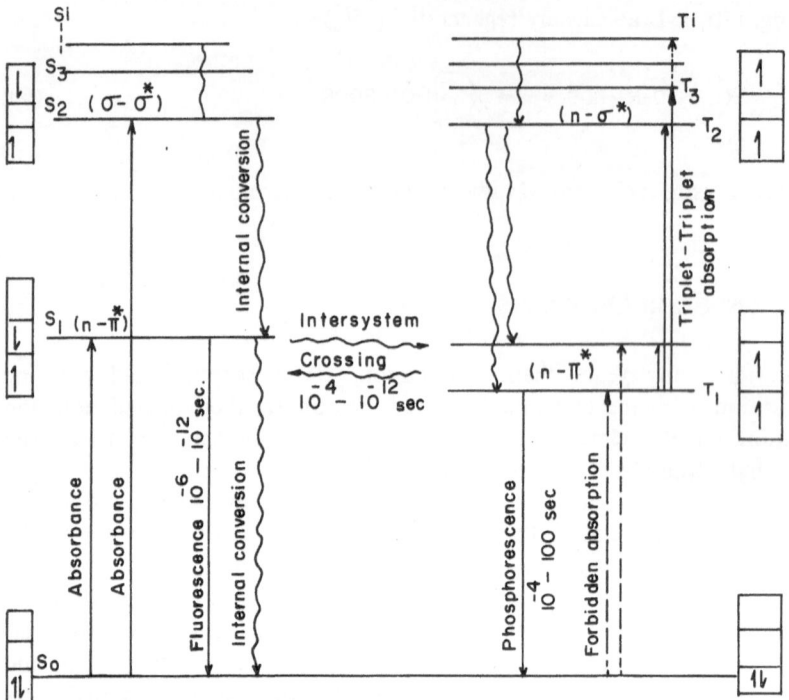

Fig. 5. Energy level diagram of a carbonyl compound

its excess electronic energy. It may react chemically, resulting in the formation of a new chemical species or a free radical. The first excited singlet state may lose its excess energy by emitting a photon of light and this emission is known as fluorescence. It is also possible for the first excited singlet state to lose its energy by some non-radiative process. Thus, internal conversion to the ground state may occur or, alternatively, since overlap of higher vibrational levels is possible, a non-radiative transition from S_1 to T_1 may occur by electron-spin reorientation. This transition from a singlet to a triplet state of a molecule is termed intersystem crossing.

The triplet molecule loses any excess vibrational energy to its surroundings and goes to the lowest vibrational level of the T_1 state. This molecule also loses its excess energy through chemical reaction or by a non-radiative process of intersystem crossing or by internal conversion. On the other hand, if the excited molecule in T_1 state has not returned to the ground state, it will emit a photon and return to some vibrational level of the ground state. This emission process $(T_1 \rightarrow S_0)$ is known as phosphorescence and occurs at longer wavelengths than those at which fluorescence occurs.

4.7.2 Formation of the Excited-State Complex

The most reasonable explanation for the quenching mechanism is the exciplex formation between excited chromophore (carbonyl or ketone) and the ground state of the quencher. The mechanism involves energy transfer from both the singlet and triplet excited states to the ground state of the quencher. Coupling of the excited carbonyl electronic energy and the quencher vibrational energy results in the 'exciplex' intermediate [163] and leads to subsequent deactivation of the excited state species to yield the observed products. These emissions occur from excimer and exciplex species [164, 165]. The mechanism is illustrated in the following scheme.

The excimer emission occurs from an excited associated complex (D^*) formed between a species in the excited singlet state (S^*) and a similar ground-state (S_0) species. The excimer is also called a dimer and is short-lived

$$S^* + S_0 \longrightarrow (D)^* \longrightarrow S + S + h\nu \tag{65}$$

This emission occurs at longer wavelengths than the normal fluorescence. The exciplex emission, on the other hand, occurs from an excited associated complex formed between an excited species and a different ground-state species.

$$S^* + B \longrightarrow (SB)^* \longrightarrow S + B + h\nu \tag{66}$$

The nature of the light emissions is influenced by the way in which the absorbed energy is transferred through the polymer matrix. In crystalline polymers, 'exciton migration' is possible as all molecules lose their energetic individuality and all electronic and oscillation levels are coupled [166]. Thus, new 'exciton absorption and emission bands' are formed and the excitation energy can move along the chain

$$\sim A^*\text{-}A\text{-}A\text{-}A\sim \longrightarrow \sim A\text{-}A\text{-}A\text{-}A^*\sim \tag{67}$$

The impurities may capture this migrating exciton and base its excess energy. The mutual annihilation of two or more 'triplet excitons' occurs in the same polymer chain and delayed fluorescence is observed.

In addition to the above mechanism, the metal chelates can influence the process of quenching through electron-transfer

$$ROO^{\cdot} + (\text{metal chelate}) \longrightarrow ROO^- + (\text{metal chelate})^{+*} \tag{68}$$

and formation of charge-transfer exciplex intermediate [167]

$$R\overset{\cdot}{O}O + (\text{metal chelate}) \longrightarrow [ROO^- \text{-}\text{-}\text{-}\text{-} (\text{metal chelate})^{+*}] \tag{69}$$
$$\downarrow$$
$$\text{non-reactive products}$$

Thus, chelates are able to dissipate the radiation harmlessly as infrared radiations or heat through resonating structures. Carlsson and Wiles [104] have confirmed in their studies that the quencher slowly migrate through the solid polymer destroying the hydroperoxide group.

EPR was stabilized by azomethine dyes [168] by rolling them at 160 °C for 5 h under pressure. Smith et al. [169] have found that these dyes function as quenchers of $^1O_2$. The azomethine derivatives may be considered as α-substituted anilines

$$Et_2N - \bigcirc - \underset{\underset{X}{\overset{\|}{C}}}{N \cdot} Y$$

and thus $^1O_2$ quenchers. Kato [170] further improved thermal stability of EPR by incorporating bis(1,3-propanediamine) copper chloride or bis(salicylaldoximal) nickel [171] or 8-hydroxyquinaldine copper complex [172]. These metal chelates deactivate the triplet state and function as quenchers for singlet oxygen. The thermal oxidation resistant EPR sheets were prepared by kneading metal complexes of 8-hydroxyquinoline [173] or dimethylglyoxime [174] or hydroxyben-zophenone [175] in the presence of a crosslinking agent. These chelates function as photo-physical processes. The radiation-resistant [176] rubber composition was made by vulcanization of EPR and incorporating clay as a filler and 5,6-dimethylacenaphthylene as an energy quencher and a radical trap.

4.8 Combined Effects/Synergism

The stabilization does not occur through a single process but a combination of all the above proposed processes and their combined effect is known as synergism when the cooperative action is greater than their individual effects [177] taken independently. The mechanism of synergism is unknown. The synergism [178] between UV absorbers (hydroxybenzotriazoles) and antioxidant [tris(3,5-di-*tert*-butyl-4-hydroxyphenyl)phosphite] is presumably due to the activity of the phosphite to react with free radicals or oxidation products (hydroperoxide or peroxide) which may be formed even in the presence of ultra-violet stabilizers.

A synergistic effect was observed [179] when a mixture of 2,5-dimethyl-2,5-di-(*tert*-butyl peroxy) hexane, triallyl cyanurate, Irganox 1010 and dibenzylidene sorbitol, was incorporated in the random E-P copolymer matrix. Triallyl cyanurate increases the radiation crosslinking efficiency. Katagawa and Kawada [180] made radiation, impact-resistant compositions by extruding block E-P copolymers (4.5 mol% C_2H_4) with tris(2,4-do-*tert*-butylphenyl)phosphate, bis(p-methylben-zylidene)sorbital, triallyl isocyanurate and condensate of succinic acid-N-(2-hydroxyethyl)-2,2,6,6-tetramethyl-4-hydroxypiperidine. This synergistic mixture acts as a peroxide decomposer, crosslinking agent and free-radical trap. High impact strength TPO has been prepared [181] by incorporating in an E-P block copolymer with 5 wt.% EPR 10 wt.% aminosilane treated glass strands, 10 wt.% talc and the necessary antioxidants and lubricant. A sheet of TPO had an impact strength of 6 kg cm cm^{-2} as compared to 4 kg cm cm^{-2} for a reference EPR.

The carbon black and amines can be used to provide synergistic stabilization [182] for EPR against radical induced oxidation. This is because the carbon black acts as a free radical trap in addition to the screening effect and amine functions as an antioxidant. The peroxide, oxime and isocyanate provided synergistic effects [183] and improved the heat stability of the EPR. In this case, the oxime terminates the propagation step by scavenging free radicals, oxy radicals and peroxy radicals. Sugas and Sato [184] made heat resistant EPR by compounding it with dicumyl peroxide, sulfur, talc, ZnO, stearic acid, carbon black, 2,2,4-tri-methyl-1,2-dihy-droquinoline and dithiobis(benzimidazole) in a roll mill at 90 °C. In this composition, dithiobis(benzimidazole) functions primarily as peroxide decomposer, dihy-droquinoline as an antioxidant and other gradients as screener, absorber and filler etc. The compounding of HAF-black filled EPR [185] with 2-mercaptoben-zimidazole (antioxidant and peroxide decomposer), N,N'-diethyl-2-benzothiazole-sulfenamide(peroxidedecomposer), Santocure (2-mercaptobenzothiazole sulfen-amide), Santacure MOR, Captax or Altax in the presence of p-quinone dioxime (free radical trap) followed by vulcanization with dicumyl peroxide led to improved heat resistance and physico-mechanical properties. Mizura and Matsuura [186] improved the heat resistance and workability of EPR at higher temperature by mixing with it with p-benzoquinone, dicumyl peroxide and 2-mercaptoben-zimidazole. Iwao et al. [187] made a light-resistant polymer composition from a mixture of 2-(2'-hydroxy-5'-methylphenyl)-benzotriazole-5-carboxylic acid chlori-de (UV absorber) and poly(ethylene oxide). This blend after mixing it with 3,5-di-*tert*-butyl-4-hydroxytoluene (a powerful antioxidant) gives a sheet with improved UV light resistance.

The EPR with a good high temperature and oil resistance [188] was pre-pared by compounding silica, trimethylpropane, trimethyl acrylate, MgO, ZnO, Chlorowax LV70, Irganox 1010 and dicumyl peroxide at 88 °C. In this composition, silica is a reinforcing agent, MgO a filler, Irganox 1010 a chain-breaking antioxidant and acrylate an absorber. The diol bis(phenylamino-phenyl diethylamidothio-phosphate) [189] of the general formula IX

$$\underset{C_6H_5}{\overset{H}{\underset{\displaystyle}{}}}N-\!\!\!\underset{}{\bigcirc}\!\!\!-O-\underset{\underset{Et_2N}{|}}{\overset{\overset{S}{\|}}{P}}-O-Z-O-\underset{\underset{Et_2N}{|}}{\overset{\overset{S}{\|}}{P}}-\!\!\!\underset{}{\bigcirc}\!\!\!-N\underset{C_6H_5}{\overset{H}{}}$$

$$\mathbf{IX}$$

has been used as stabilizer in EPR (where Z = hydroquinone or triethylene glycol). This compound functions as an antioxidant, peroxide decomposer and vulcanizing agent. Kato [190] made weather- and water-resistant EPR by mixing ferrocene, siloxane and/or quinoline into polymer matrix. Ferocene reduces absorption of light by photo-rearrangement and an energy transfer mechanism, siloxane a reinforcing agent and quinoline as an antioxidant. A mixture of EPR, ZnO, sulfur, stearic acid, Sb_2O_5, talc and pentabromophenyl acrylate, was rolled at 120 °C and after cooling, was mixed with diisopropylphenyl-hydroperoxide and dicumyl peroxide at 160 °C to give a sheet having good resistance to γ-irradiation [191]. The whole mixture is synergistic to each other. The photostabilization mechanism of acrylate derivatives is largely unknown but these derivatives are known as absorbers and act by photo-Fries rearrangements; peroxidehydroperoxide deri-vatives of acrylates are initiators for vulcanization. In Japan, a specimen having good resistance to radiochemical degradation [192] has been prepared by rolling a mixture of 2,2,4-tetramethyl-1,2-di-hydroquinoline (antioxidant), dicumyl per-oxide, traces of sulfur, calcinated clay(filler) and 5,6-di-methyl acenaphthylene at 100 °C with EPR matrix. Acenaphthylene absorbs and immediately re-emits it at longer wavelengths in the form of fluorescent light or heat. It also acts as a radical scavenger.

Antipova and coworkers [193] increased the hydrolytic stability and adhesive strength of EPR by adding an epoxy resin, dicarboxylic acid anhydride and dialkyl tin dicarboxylate. The dialkyl tin dicarboxylate does not react with the double bonds in keto allyl groups but at radical sites, thus the replacement of labile hydrogen atoms by the two electronegative groups of organo tin ompounds results in an increase in EPR stability.

5 Degradation and Stabilization of E-P Blends

Growing interest in polymer blends is due to the feasibility of varying properties of the materials within wide limits which may modify tensile strength, impact strength, etc. Numerous papers have appeared in connection with the preparation and mechanical properties of polyolefin blends. However, there appears to be a lack of information regarding oxidative stability of this type of blend.

Musso and Guzzetta [194] made thermoplastic blends by calendering a 15:85 composition of polyethylene and polypropylene at 240 °C. The blend was made more thermally stable by incorporating TiO_2 and phthalocyanine green which function as screen, screen and radical trap, respectively. The antioxidant-2246 [bis(2-hydroxy-3-*tert*-butyl-5-methylphenyl)methane] was mixed into the blend of PE and saturated amorphous copolymer of polyolefin at 140 °C to make it thermally stable [195]. The low temperature properties of *i*-PP were greatly improved by the addition of linear PE. The solutions of both the homopolymers were mixed at elevated temperature. The better properties were the result of blending and crosslinking [196] between *i*-PP and PE. The ozone-resistant vulcanizates of PE-PP blends [197] were made by blending then with 45 wt.% of an unsaturated elastomer. This vulcanizate is a substitute for natural rubber and cracks appeared at the surface after 200 h exposure in weatherometer. Lehare [198] improved the impact strength of a PE-*i*PP blend by the addition of 2 wt.% of EPR and 2 wt.% of TiO_2 or $CaCO_3$. In the absence of the rubber or when it is not dispersed in the composition prior to the addition of the additive, no improvement is obtained. Stanko et al. [199] have compared structural changes by electron microscopy in hd/ld PE blends before and after γ-irradiation. At a higher dose rate, the degradation occurs resulting from changes in crystallinity. The impact strength of the PE-PP blend was improved by incorporating 13 wt.% EPR which plays the role of a toughening agent [200].

Tsukamoto and Suzuki [201] extruded hdPE, PP, EPR at 240 °C to give a non-blocking and good dimensional heat stable blend. Okita et al. [202] prepared heat sealable blends by extruding 97 wt.% *i*-PP, 3 wt.% ldPE and 0.02 wt.% dibenzylidenesorbitol which acts as a heat stabilizer and a radical trap. Murata and Akimoto [203] studied thermal degradation of PE-PP blends and found close agreement between observed volatilization rates of the blends and of their constituent polymers. The effect of blending affects the decomposition rate but not the intramolecular reaction. The effect of mixing on the decomposition rate is interpreted in terms of an intermolecular radical transfer between different polymers. The PE-PP blends were made electrically insulating by incorporating 12 wt.% EPDM and 1 wt.% N,N-di-β-naphthyl-1,4-phenylene diamine [204]. The blend was radiochemically vulcanized and the additive acted as an antioxidant and radical trap. The light resistant blend [205] was obtained by extrading a mixture of PE (100 parts), PP (90), TiO_2 (0.8), bis(2,2,6,6-tetramethyl-4-piper-idyl)sebacate (0.06) BHT (0.1), Irganox 1076 and oleic acid (0.1). These additives make the blend light-resistant by their synergistic action. In this blend, no degradation was observed after irradiation in a weatherometer for > 2000 h but in the absence of TiO_2 and sanol, degradation was observed after 100 h.

Sadrmohashegh et al. [206] have showed by carbonyl formation and loss of toughness that ldPE blends are photo-oxidised more readily than PE alone. During their investigations, they confirmed that crosslinking also takes place between the two phases of the polyblend. The mixing of EPDM improves the mechanical performance but increases the rate of photodegradation. Radiation-resistant sheets [207] of a PP-EPR blend were made by hot pressing 30 parts PP and 100 parts EPR. The molding at higher temperatures causes crosslinking because of which the tensile strength of the blend is 83 kg cm^{-2} after γ-irradiation for 50 h, compared with 32 kg cm^{-2} and 40 kg cm^{-2} for the neat blend and EPR, respectively. The blends of PP with statistical EPR flow more readily than do the blends with E-P block copolymers [208]; presumably the structural irregularity limits internal interactions. The unfavourable effect of UV irradiation on mechanical properties of EPR is due to its unorganised structure with respect to the oxygen attach which promotes photo oxidation. The heat-resistant and weather-proof [209] roofing sheets are composed of EPR-butyl rubber (70:23) blends which are stabilized by mixing thiuram, Captax, plasticizer and bis(3,6-di-tert-butyl-catechol) sulfide or bis(3,6-di-tert-butyl catechol) phosphite. Phosphite derivatives attack at the β-carbon atom of a conjugated system of unsaturated ketone with subsequent formation of a stable keto-phosphonate

$$\sim \underset{\underset{O}{\|}}{C}-CH=CH\sim + P(OR)_3 \; \rightleftharpoons \; \sim \underset{\underset{-O}{|}}{C}=CH-\underset{\underset{+P(OR)_3}{|}}{CH}\sim \; \xrightarrow[-RCl]{+HCl} \; \sim \underset{\underset{O}{\|}}{C}-CH_2-\underset{\underset{O=P(OR)_2}{|}}{CH}\sim \qquad (70)$$

The formation of keto-phosphonate structure within macromolecule leads to the removal of internal unsaturation. Triallyl cyanurate and ionizing irradiations [210] made a E-P block copolymer-PE blend thermally stable. Triallyl cyanurate increases the crosslinking density probably due to addition reactions between polymeric and allyl radicals produced by ionizing radiation. The addition of 2,2,4-trimethyl-1,2-di-hydroquinoline and bis[4(1-methyl-1-phenylethyl)phenyl]-amine stabilized a PE-EPDM blend against heat [211]. Popov et al. [212] studied the ozone effect on PE-iPP blend. The oxidation rate was detected in relation to morphology, segmental mobility and influence of the compound on the blend. The limiting step is the abstraction of hydrogen from a main chain carbon-atom which then undergoes transhybridization from the sp^3 to sp^2 state

$$-\overset{|}{\underset{|}{C}}-H + \cdot O_3 \longrightarrow [-\overset{|}{\underset{|}{C}}\cdot + \; \cdot OH + O_2] \begin{cases} \text{liberation from a cage} \\ \\ \text{reaction products} \end{cases} \tag{71}$$

The degree of oxidation, degree of crystallinity and amount of each component change with segmental mobility of the macromolecules. The ozone-oxidation increases along with the degree of orientation, leading to reduced segmental mobility of the macrochains. The increase in the stability is probably due to reduced solubility of gases owing to the increasing density of the irregular regions as well as direct influence of segmental mobility affecting the kinetics of chemical reactions with the participation of the macromolecules. Figure 6 shows the degree of oxidation in PE-PP blend films in relation to the composition of the blend as a function of time. It can be seen from the plots that the relative oxidation stability of higher hdPE contents decreases as the exposure time increases. The plots are similar for low, medium and advanced degree of oxidation.

The peeling strength of PE-PP blends is increased by welding its sheet ultrasonically [213]. Rizzo et al. [214] have irradiated ldPE-*i*-PP blends with 125 Mrad and have studied structural modifications. For blends containing small amounts of one component, the decrease of crystallinity on increasing γ-irradiation

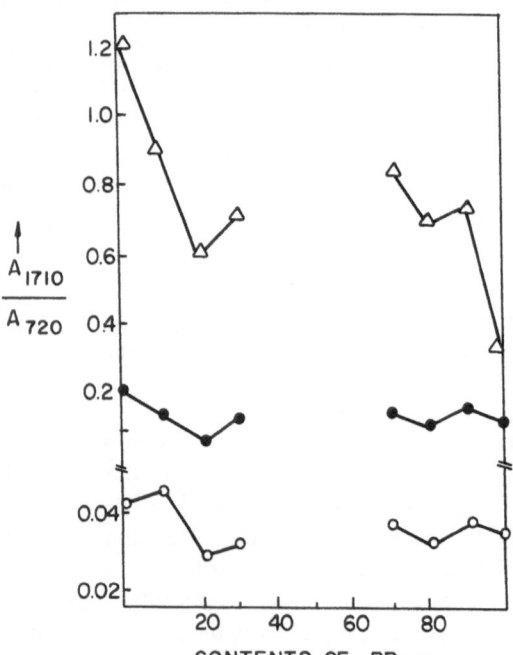

Fig. 6. Degree of degradation versus PP contents in a PE-PP blend (○) 9.3 h, (●) 16.5 h, (△) 120 h

dose and difficulty of the crystallization from the melt were enhanced for this component. Meizzel and coworkers [215] made thermally stable blends by calendering a mixture of 8 wt.% PE, 86 wt.% PP, 6 wt.% EPR and 0.02 wt.% sodium benzoate wher ethe later acts as a nucleating agent. An appealing method for stable blend preparation was given by Gueskens et al. [216]. In their procedure, they made the blend from 54% PP, 36% ldPE and 10% hydroperoxidized PP (0.03 mol OOH/kg) at 150 °C. The triallyl cyanurate was used as a stabilizer and the blend was crosslinked by heating at 200 °C for 6 min to give the dimensional stability. Silanes [217] have been used as reinforcing agents in polymer blends.

Heat-resistant [218] soft foams were prepared from the blends of hdPE with E-P random copolymers. The azodicarbanamide acts as a thermal antioxidant and the crosslinking of the blend was increased by electron beam radiations and foamed at 225 °C with 2320% expansion. A blend of 35 wt.% PE-PP (8:92), 15 wt.% E-P block copolymers, and 50 wt.% EPDM showed accelerated weathering resitance [219] $\geq$ 1000 h probably due to crosslinking between constituents of the block copolymer, polyblend and EPDM. The effect of filler and thermodynamic compatibility on kaolin-filled PE-PP blend was studied by Lipatov and coworkers [220]. The thermodynamic interaction parameter (χ) decreased and thermodynamic stability increased by filler addition. the degree of crystallinity decreased with increasing thermodynamic compatibility of the components due to sharp decrease in the phase separation rate during cooling.

The PE-PP blends are susceptible to attacks by chemically aggressive ingredients in the electrolytes. The zinc/bromide battery contains small amounts of free bromine in the electrolyte which can brominate and oxidise the polyblend. Arnold and Clough [221] have showed that oxidation-degradation of the polyblend was related to the direct action of bromine in the aqueous electrolytes. Their results illustrates that PE-PP blends have a strong affinity for bromine which is consistant with the fact that bromine is highly soluble in most organic solvents. They showed that bromine is not only absorbed by the polyolefins but also reacts with them to initiate their degradation. The IR spectrum of the blend which was aged for six months at 55 °C in synthetic cathalyte ($ZnBr_2$, Br_2 and quaternary ammonium bromide) indicates the prersence of a C$-$Br bond at 500–600 cm^{-1} and a carbonyl group at 1720 cm^{-1}. These results indicated that bromination occurred and was accompanied by oxidation. Some improvement in the stability was realized by incorporation of Irganox 1010 in the blend matrix. The combination of Agerite and Irganox 1010 synergised the present blend system. The γ-irradiation [222] led to an increase in the crosslink formation at the interface of the PE-PP blend. The crosslinking was accompanied by the formation of defects in the crystalline lattice of the PE which led to a decrease in the orthorhombic pseudohexagonal transition temperature. In the pseudohexagonal modification, defects increased the crystalline parameter and decreased the packing density, whereas irradiation decreased the crystalline parameter and increased the packing density. Wang et al. [223] compared the slow and fast cooling rates of thermograms. The fast cooling yielded only one thermogram while slow one yielded exotherms of both PE and PP and this is because of restricted block movement due to covalent bonds between blocks and crystallization process in the block copolymer. The crosslinking of PE-PP blend was also carried out by electron beam irradiation with subsequent elongation [224]. With increasing PP content, the blend becomes much weaker which is attributed to the considerable scission of main chains of PP because crosslinking occurs preferentially in the amorphous region of PE. Triallyl isocyanurate or diallyl phthalate play an important role for improvement in high temperature resistance due to enhanced crosslinking.

6 Future Developments

Ethylene-propylene copolymers and their blends exhibit diverse degradation behaviour under the influence of light, heat and radiation. In spite of the considerable literature, little is still understood about the fundamental processes that occur during degradation in these polymers. Studies aimed towards understanding the relationship between polymer structure and degradation behaviour are very scanty. Although much work has been reported on elastomeric copolymers, no comprehensive studies on heterophase thermoplastic E-P copolymers have appeared in the literature. Heterophase thermoplastic copolymers have technologically useful properties such as high impact even at low temperatures. Recent developments in Ziegler-Natta catalysis have enabled synthesis of well defined heterophase polymers with superior two-phase morphologies. Study of thermal and photooxidative degradation of such heterophase copolymers is a fertile area of study. There is an urgent need to understand the relationship between morphology and the aging reaction rate and the distribution of additives between the two phases. New stabilizer molecules may have to be designed which provide effective protection to such multiphase polymers. No information exists on the degradation and stabilization aspects of thermoplastic olefin elastomers in the literature.

With better understanding of the mechanism of degradation and stabilization, there undoubtedly will be an increased effort to produce polymers with greater photothermal/radiation resistance and also more effective stabilizers to achieve this end. Thus, the study of degradation and stabilization aspects of ethylene-propylene copolymers and their blends (and generally of multiphase polymer blends) appears to be both intellectually stimulating and of practical importance.

7 References

1. Boor J Jr (1979) Ziegler-Natta catalysts and polymerization, Academic Press, New York, p 587
2. Young LJ (1961) J Polym Sci 54: 411
3. a) Allport DC, Janos WH (1973) Block copolymers, Applied Science Publishers, London
   b) Noshay A, McGrath JE (1977) Block copolymers, An overview and critical survey, Academic Press, New York
4. Bier G (1961) Angew Makromol Chem 73: 186

5. Kennedy JP, Tornqvist EGM (1969) In: Natta G, Dall'Asta G (eds) Polymer chemistry of Synthetic Elastomers, Ethylene — Propylene Rubbers, Interscience Publishers, New York
6. Schroeder HE (1987) Ed. Holden G, Legge NR, Thermoplastic Elastomer: Comprehensive Review, Publisher Carl Hanser, Munich
7. Chandrasekhar V, Srinivasan PR, Sivaram S (1988) Ind J Technol 26: 53
8. Galli P, Simonazzi TS, Duce DD (1988) Acta Polymerica 39: 81
9. SRI Report Polyolefins through the 80's, Stanford Research Institute, Stanford, California (1983) Vol I–IV
10. Fox RB (1972) Prog Polym Sci 1: 47
11. Pinkerton DM (1972) Proc R Aust Chem Inst 33
12. Rabek JF, Ranby B (1975) Photodegradation, Photooxidation and Photostabilization of Polymers, Interscience, New York
13. Charlesby A, Partridge RH (1965) Proc Royal Soc A 283: 312
14. Uri N (1970) Israel J Chem 18: 125
15. Amin MU, Scott G, Tillakeratne LMK (1975) Eur Polym J 11: 85
16. Heacock JF, Mallory FB, Gay FP (1968) J Polym Sci A-1: Polym Chem Ed 6: 2921
17. Wood DGM, Kollman TM (1972) Chem Ind, 423
18. Milinschuk VK (1965) Vysokomol soed 7: 1293
19. Horper DJ, McKeller JF (1973) J Appl Polym Sci 17: 3503
20. Dole M (1974) Adv Rad Chem 4: 307
21. Tsuji K (1973) Adv Polym Sci 12: 131
22. Decker C, Mayo FR (1973) J Polym Sci Polym Chem Ed 11: 2847
23. Handy DG, Russel GA (1964) J Am Chem Soc 86: 2371
24. Carlsson DJ, Wiles DM (1969) Macromolecules 2: 587
25. Adams JH, Goodrich JE (1970) J Polym Sci: A-1, 8: 1269
26. Balaban L, Majer J, Vesely K (1969) J Polym Sci C(22): 1059
27. Toyo Spinning Co Ltd (1960), Japan, 8830
28. Asaka M, Yamanoto S, Yatsuhashi M, Hara G, Sakai M (1973) Fuzikura Tech Rev 5: 66
29. Geguchi T, Hashimoto S, Arakawa K, Hayakawa N, Rawakami W, Kuriyama I (1981) Radiat Phys Chem 17: 195
30. Ho Sy, O'Donnell JH (1984) Eur Polym J 20: 421
31. Busico V, Corradini P, Rosa CDe, Benedetto D (1985) Eur Polym J 21: 239
32. Kohno I, Yanokara M, Motonaga S, Kamitsabo H, Yatsuhashi M, Suematsu T, Kubayashi H (1986) IEEE Trans Nucl Sci, NS-33 (No 1): 707
33. Arnand R, Lemaire J, Jevanoff A (1986) Polym Degdn Stab 15: 205
34. Singh RP, Sivaram S (to be published)
35. Bua E, Manaresi P (1959) Anal Chem 31: 2022
36. Sieron JK, Murray K (1962) Rubber World 146: 61
37. Manaresi P (1963) Chim Ind (Milan) 45: 1488
38. Kozlov VT, Simpoziuma T (1963) Moscow, 220
39. Slobodin YM, Maiorova VY, Smirnova AM (1964) Vysokomol soed 6: 541
40. Tobolsky AV, Norling PM, Frick NH, Yu H (1964) J Am Chem Soc 86: 3925
41. Yu H (1964) Am Chem Soc, Div Polym Chem, Preprints 5: 545
42. Kozlov VT, Tarasova ZN (1966) Vysokomol soed 8: 943
43. Kuzminskii AS, Zakirova MA (1966) Radiats Khim Polim, Mater Simp, Moscow, 388
44. Kuzminskii AS, Lyubchanskaya LI, Yurtseva ES, Yudina GG (1966) Radiats Khim Polim, Meter Simp, Moscow, 301
45. Kozlov VT (1967) Vysokomol soed A9: 515
46. Kozlov VT, Yevseyev AG, Zubov PI (1969) Vyrokomol soed A11: 2230
47. Kozlov VT, Tarasova ZN, Klinshpont ER, Milinchuk VK, Dogadkin BA (1967) Vysokomol soed A9: 1541
48. Vinogradov GV, Ivanova LI (1968) Kauch Rezina 27: 15
49. Kayoko Negu (1968) Nippon Gomu Kyokaishi 41: 96
50. Max M (1968) Jad Energy 14: 198
51. Sumitomo Elec Ind, Ltd (1968) Fr 1,526,488

52. Kim IP, Egorov EV, Barkalov IM, Dogadkin BA, Tarasova ZN (1969) Khim Vys Energy 3: 427
53. Smetania LB, Leshchenko SS, Yegorova ZS, Starodubtsea DS, Klinshpont ER, Kaplunov MYa, Karpov LV (1970) Vysokomol soed A 12: 2401
54. Mamedov FV, Bilalov YA, Tagieva FM, Anashkin VI, Kolesnikov GS (1971) Uch Zap Azerb Inst Nefti Khim 9: 87
55. Melia TP, Salpadoru N, Wilkinson J (1972) Makromol Chem 154: 241
56. Monnier B, Voorde M Van de (1972) Ind At Spatiales 16: 39
57. Aliev VM, Bilalov YaM, Ibragimova SM (1972) Uch Zap Azerb Inst Nefti Khim 9: 75
58. Kozlov VT, Evseev AG, Gulanov GG, Zaeva MV, Kryukova AB, Smagin EN (1972) Radiats Khim 2: 358
59. Locke JM (1973) Ger Offen 2, 302, 936
60. Decker C, Mayo FR, Richardson H (1973) J Polym Sci, Polym Chem Ed 11: 2879
61. Piotrovsku KV, Romna MP, Stepanova VI (1974) Kauch Rezina, (4): 21
62. Saegusa T, Takehisa (1975) J Macromol Sci Phys B11: 389
63. Makhlis FA (1975) Khim Vys Energ 9: 271
64. Sokolousku AA, Borisova NN, Angert LG (1975) Vysokomol soed A17: 1107
65. Seidov NM, Delin MA, Kuliev RSh, Abasov AI, Mustafaev AM, Agakishieva MYa (1975) Azerb Khim Zh, (4): 99
66. Makhlis FA, Burenok VA, Sedov VV, Kuzminiskii AS, Svetlova GK (1977) Vysokomol soed A 19: 571
67. Seidov NM, Kuliev RSh, Basov AI, Mustafaev AM (1977) Azerb Khim Zh, (4): 99
68. Gilman AB, Khan AA, Shifrina RR, Kolotyrkin VM, Kozlov VT, Orlov VA (1979) Vysokomol soed B21: 220
69. Khan AA, Gilman AB, Sedov VV, Shifrina RR, Kolotyrkin VM, Kozlov VT (1980) Vysokomol soed B 22: 824
70. Racke HH (1980) Kunststoffe 70: 76
71. Ito K (1981) Radiat Phys Chem 17: 203
72. Gillen KT, Clough RL (1981) Radiat. Phys Chem 18: 679
73. Gueskens G, Kabamba MS (1982) Polym Degrdn Stab 4: 69
74. Seguchi T, Arakawa K, Hayakawa N, Watanabe Y, Kuriyama I (1982) Radiat Phys Chem 19: 321
75. Arakawa K, Seguchi T, Watanabe Y, Hayakawa N (1982) J Polym Sci Polym Chem Ed 20: 2681
76. Parray L (1982) Muanyag Gumi 19: 330
77. Bousquet JA, Haider B, Fouassier JP, Vidal A (1983) Eur Polym J 19: 135
78. Okamoto S, Hayakawe T, Takeya C (1983) Annu Rep Radiat Cent Osaka Perfect 24: 49
79. Seguchi T, Arakawa K, Ito M, Hayakawa N, Machi S (1983) Radiat Phys Chem 21: 495
80. Gueskens G, Kabamba MS (1983) Polym Degrdn Stab 5: 399
81. Gillen KT, Clough RL (1983) Radiat Phys Chem 22: 537
82. Li SKL, Guillet JE (1984) Macromol 17: 41
83. Aliguliev RM, Ibragimov KhD, Ovanesova GS, Aliev FA, Seidov NM (1984) Azerb Khim Zh, (1): 91
84. Arakawa K, Seguchi T (1984) Kobanshi Ronfunshu 41: 733
85. Clough RL, Gillen KT, Quintana CA (1985) J Polym Sci, Polym Chem Ed 23: 359
86. Buckalew WH (1986) Report, NUREG/CR-4543, SAND-86-0462, Order No TI 86010638, p 42
87. Bousquet JA, Fouassier JP (1987) Die Angew Makromol Chem 149: 1
88. Bousquet JA, Fouassier JP (1987) ibid 149: 19
89. Bousquet JA, Fouassier JP (1987) ibid 149: 45
90. Lacoste J, Singh RP, Boussand J, Arnaud R (1987) J Polym Sci PtA-Polym 25: 2799
91. Singh RP, Lacoste J, Arnaud R, Lemaire J (1988) Polym Degrdn Stab 20: 49
92. Herrmann JM, Disdier J, Pichat P (1977) Proc 3rd Intern Conf Solid Surface, Vienna, p 15
93. Bickley RI, Jayanti RKM (1974) Faraday Disc Chem Soc 58: 194
94. Heskins M, Guillet JE (1968) Macromol 1: 97

95. Pappillo PJ (1967) Mod Plastics 4: 31
96. Montecatini Societa Generale per l'Industria Mineria e Chimica (1959) Ital 606, 492
97. Cabot GL (1960) Brit 852, 035
98. Amberg LO, Robinson AE (1961) Ind Engg Chem 53: 368
99. Lal J, McGrath JE (1969) Rubber Chem Technol 36: 248
100. Farbenfabriken Bayer A-G (1965) Neth Appl 295, 731
101. Kanaya A, Konishi S, Imakitay (1975) Japan Kokai 75: 10, 342
102. Ozaneaux A (1975) Rev Gen Caoutch Plast 52: 885
103. Chaudet JH, Tamblyn JW (1961) SPE Trans 1: 57
104. Carlson DJ, Wiles DM (1974) J Polym Sci, Polym Chem Ed 12: 2217
105. Hiller HJ (1969) Eur Polym J, Suppl, p 105
106. O'Connell EJ (1968) J Am Chem Soc 90: 6550
107. Kimura H, Nakamura S, Ebinuma N (1977) Japan Kokai, 77130,846
108. Kusunomi I (1970) Japan, 70: 27, 904
109. Kato H (1974) Japan Kokai 74: 36, 745
110. Kato H (1974) Japan Kokai 74: 47, 439
111. Treshchalov VI, Makhalis FA, Gertsova NB, Zagranichnyi VI, Karlik VM, Golperin
     VA (1979) USSR, 648, 581
112. Kochnov IM (1979) Prom Sint Kauch, (2): 8
113. Derbisher VE, Kablov VF, Koroteeva AM, Ogrel AM (1983) Kauch Rezina, (1): 24
114. Bailey HC (1962) Ind Chem 38: 215
115. Brawn B, Perum AL (1962) SPE J 18: 250
116. Ingold KU (1961) Chem Rev 61: 563
117. Western Electric Co, Inc (1965) Neth Appl 6: 500, 129
118. Chisso Corp (1980) Jpn Kokai Tokkyo Koho 80: 139, 447
119. Rachinsku FYu, Slavachevskaya NM, Potapenko TG, Kremen MZ, Matveena EN
     (1963) Plast Massy, (7): 48
120. Howarth JT, Weinberg HW (1963) Rubber World, 149: 54
121. Field GB (1963) US 3: 110, 623
122. Solvay and Cie (1967) Fr 1: 489, 772
123. Mikhailov VV, Levin PI, Metveena EN, Otmakhova VM, Khinkis SS, Kuchenkova
     VA (1969) Sin Issld Eff Khimikatov Polim Mater, (3): 108
124. Huglin MB, Kay E, Knight GT, Wright WW (1972) Makromol Chem 152: 105
125. Aliev AF, Oganyan VA, Burdzhaliev DA, Seidov NM, Sukhguseinov TF, Abashin V,
     Gavyan MA (1973) Azerb Khim Zh, (4): 91
126. Kawada T, Nagasawa M, Yaeda Y (1979) Jpn Kokai Tokyo Koho 79: 157, 152
127. Digteva TG, Dontrov AA, Kondrateva VF, Granovskaya IM, Naumova SF, Mako-
     vetskii MI, Akulich (1980) 21 USSR, 757, 564
128. Arakawa K, Seguchi T, Hayakawa N, Machi S (1983) J Polym Sci, Polym Chem Ed
     21: 1173
129. Degteva TG, Dontsov AA, Kondrateva VF, Granovskaya IM (1984) Kauch Rezina,
     (8): 17
130. Nakamura N, Mori K, Tamura K, Satish Y (1984) Nippon Gomu Kyokaishi 57: 610
131. Kawaguchi Chemical Industry Co Ltd (1984) Jpn Kokai Tokkyo Koho JP 59: 168, 045
132. Tanimoto Y, Ikeda K (1985) Jpn Kokai Tokkyo Koho JP 60: 255, 839
133. Isakovich VN, Grachek VI, Smolyakov AV, Bukanova NN (1986) Kauch Rezina, (7): 4
134. Hercules Powder Co, (1960) Brit 851, 670
135. Schroeder JP, Leonard EC (1960) Brit 849, 058
136. Montecatini Societa Generale per l'Industria Mineraria e Chimica (1961) Brit 883, 763
137. Giulio E, Guglielmino A (1966) Proc. Inst. Rubber Ind. 12: 190
138. Brams SL, Frederick W, Riege RL (1969) US 3: 445, 318
139. Binder G, Krol H (1971) Ger. Offen. 1: 939, 926
140. Hasebe, M, Matsumoto Y, Takai T (1974) Japan 74: 41, 097
141. Tuseev AP, Bukalov VP, Potashova GN, Valdman AI (1979) Kauch Rezina, (6): 13
142. Moteki T, Yamaguchi K, Nakajima Y (1981) Ger Offen 3: 021, 776
143. Solvay and Cie (1968) Brit. 1: 116, 018

144. Usamoto T, Okika T, Hayashi A (1969) Ger Offen 1: 929, 584
145. Tomiyama S, Susuki R, Hosin H, Saito J, Goto H, Umehara K, Murakami K (1971) Ger Offen 2: 060, 384
146. Ohara N, Ishikawa M, Kusaka T (1975) Japan Kokai 75: 37, 863
147. Vestovich JE (1981) Fr Demande 2: 459, 266
148. Dishorski N, Dishovski M, Todorov S, Stankov S (1985) Khim Ind (Sofia) 57: 24
149. Heine HG, Rosenkranz HJ, Rudolph H (1975) Appl Polym Symp 26: 157
150. Davis A, Sims D (1983) Weathering of Polymers, Applied Sci Publishers New York, p 242
151. Minoru I (1968) Nippon Gomu Kyokaishi, 41: 583
152. Kato H (1975) Japan Kokai 75: 66, 542
153. Kato H (1975) Japan Kokai 75: 66, 541
154. Grachek VI, Naumov SF, Motolko GR, Kozlov NS, Smolyakov AV, Bukanova NN, Rudenko GA, Shingel IA (1982) Vestsi Akad Navuk USSR, Ser Khim Navuk, (5): 74
155. Tanimoto Y, Ikeda K, Hisada H, Kawakami K (1986) Jpn Kokai Tokyo Koho JP 61: 36, 345
156. Chakraborty KB, Scott G (1978) Chem & Ind (London) p 237
157. Grattan DW, Reddoch AH, Carlsson DJ, Wiles DM (1978) J Polym Sci Polym Chem Ed 16: 143
158. Gugumus F (1979) Dev Polym Stab (eds) Scott G, Appl Sci Publishers, London, p 261
159. Fors RP, Cowan DO, Hammond GS (1964) J Phys Chem 68: 3747
160. Schmit RG, Hirt RC (1963) J Appl Polym Sci 7: 1565
161. Guillopy JP, Cook CF (1973) J Polym Sci, Polym Chem Ed 11: 1927
162. Briggs PJ, McKeller JF (1968) J Appl Polym Sci 12: 1825
163. Ng HC, Guillet JE (1978) Macromol 11: 937
164. North AM (1975) Brit Polym J 7: 119
165. Somersall AC, Guillet JE (1975) J Macromol Sci Revs Macromol Chem C13: 135
166. Calvert JG, Pitts JN (1966) Photochemistry, John Wiley, New York
167. Singh RP (1985) Polym Degrdn Stab 13: 313
168. Matveena NM, Rachinski FYu, Kremen MZ, Potapenko TG (1961) Plast Massy, (2): 12
169. Smith Jr. WF, Herkstroeter WG, Eddy KL (1975) J Am Chem Soc 97: 2764
170. Kato H (1975) Japan Kokai 75: 94, 047
171. Kato H (1975) Japan Kokai 75: 94, 045
172. Kato H (1975) Japan Kokai 75: 94, 049
173. Kato H (1975) Japan Kokai 75: 94, 048
174. Kato H (1975) Japan Kokai 75: 94, 046
175. Kato H (1975) Japan Kokai 75: 100, 135
176. Furukawa Elec Co Ltd (1981) Jpn Kokai Tokkyo Koho 81: 84, 748
177. Olcott MS, Matill HA (1941) Chem Rev 29: 257
178. Ohnishi A (1971) Japan Pat 71: 42, 981
179. Hamada Y, Watanabe K (1986) Jpn Kokai Tokkyo Koho JP 61: 73, 711
180. Kitagawa S, Kawada H (1986) Jpn Kokai Tokkyo Koho JP 61: 130, 358
181. Hamada H, Kitamura K, Kodama K, Sasaki T (1976) Jpn Kokai 76: 136, 735
182. Yamashita Y, Funikawa J, Yamashiat S, Wada Y (1966) Nippon Gomu Kyokaishi 39: 446
183. Usamoto T, Kondo T (1973) US 3: 711, 454
184. Sugas K, Sato S (1973) Japan 73: 01, 427
185. Mironyak VP, Reikh VN, Nikandrova NE (1973) Issled Obl Fiz Khim Kauch Rezin 3: 77
186. Mizura H, Matsura T (1975) Japan Kokai 75: 104, 247
187. Iwao T, Sasaki E, Ito A (1977) Japan Kokai 77: 23, 144
188. Eldred RJ, Iobst SA, Ibu-Isa IA (1978) US 4: 066, 590
189. Tuseev AP, Valdman DI, Potashova GN, Valdman AI, Sizov SYu, Soboleuskaya VI, Loronina LV (1978) USSR 541, 480
190. Kato H (1978) Jpn Kokai Tokyo Koho 78: 146, 750
191. Furukawa Electrical Co. Ltd (1980) Jpn Kokai Tokkyo Koho 80: 82, 110

192. Japan Atomic Energy Research Institute (1980) Jpn Kokai Tokyo Koho 80: 106, 228
193. Antipova VF, Salnis K, Fainshtein RS, Katurkina NF, Gromova LN, Afanasev ID, Stepanova VI (1981) USSR 798, 136
194. Musso P, Guzzetta G (1957) Ital 570, 438
195. Montecatini Soceita General per l'Industria Mineraria e chimica (1962) Brit 886, 794
196. Esso Research Engg Co (1967) Brit 893, 540
197. Polym Corp Ltd (1963) Brit 939, 359
198. Lehare Jr JP (1964) US 3: 137, 672
199. Stanko D, Slavchew I, Kashtieva E (1974) Angew Makromol Chem 35: 1
200. Stamicarbon BV (1974) Neth Appl 72: 17, 487
201. Tsukamoto Y, Suzuki T (1978), Japan Kokai 78: 16, 748
202. Okita T, Togawa Y, Kimura J (1979) Jpn Kokai Tokkyo Koho 79: 04, 948
203. Musata K, Akimoto M (1979) Nippon Kagaku Kaishi (6): 774
204. Ishitani H, Saito E (1980) Jpn Kokai Tokyo Koho 80: 27, 353
205. Oji-yuka Synth Paper Co Ltd (1981) Jpn Kokai Tokyo Koho JP 81: 126, 155
206. Sadrmohashegh C, Scott G, Setoudeh E (1981) Polym Degrdn Stab 3: 469
207. Nishinippon Elec Wire and Cable Co Ltd (1981) Jpn Kokai Tokyo Koho JP 81: 127, 646
208. Thomas J (1981) Rev Gen Caoutch Plast 609: 53
209. Bars YuM, Dyatlova VP, Zhukova LF, Shtern VD, Eremeikina ZD, Stebleva NS, Ershov VV, Belostotskaya B, Voleva VB (1982) USSR Su 939, 487
210. Furakawa Elec Co Ltd (1982) Jpn Kokai Tokyo Koho JP 82: 09, 005
211. Tatsuta Elec Wire and Cable Co Ltd (1982) Jpn Kokai Tokyo Koho JP 82: 67, 640
212. Popov AA, Zaikov GE, Pracella M, Martuscelli (1982) Vysokomal soed A24: 2396
213. Mitsui Toatsu Chem Inc (1983) Jpn Kokai Tokyo Koho JP 58: 45, 244
214. Rizzo G, Spadaro G, Acierro D, Calderaro E (1983) Radiat Phys Chem 21: 349
215. Meissel L, Druzsbaczky G, Karger JK, Kollar L, Balajthy Z, Cser F, Erdei J (1984) Hung Teljes Hu 32, 143
216. Gueskens G, Bastin P, Debie B, Gromen M, Polart J, Delounois G (1984) Belg Be 899, 507
217. Trojna M, Reznik J, Rubes J, Franta J (1985) Czech CS 226, 268
218. Nishioka T, Kamijutsukaku S, Kawano H (1986) Jpn Kokai Tokyo Koho JP 61: 188, 431
219. Takimoto M, Takeuchi J, Yamazaki Y (1986) Ger Offen DE 3: 614, 464
220. Lipatov, YuS, Shifrin VV, Vasilenko OI, Krivko MS (1986) Vysokomol soed B28: 869
221. Arnold Jr C, Clough RL (1987) Polymer Prepr 28: 227
222. Antipov EM, Kuptrov SA, Popov VP (1987) Vysokomol soed B29: 466
223. Wang L, Qi Y, Chen D, Huang B (1987) Gaofenzi Xuebao (2): 81
224. Sawatari C, Matsuo M (1987) Polym J 119: 365

Editor H.-J. Cantow
Received August 8, 1990

Rheokinetics of Curing

A. Ya. Malkin and S. G. Kulichikhin
Research Institute of Plastics, Perovskii pr. 35, Moscow E-112, USSR 111112

The mechanisms and variations of rheological properties of network-forming polymers are considered. The variation in the relaxation state during curing determines which rheological method is to be employed at different stages of the process. In the flow region before the gelpoint, the viscosity at shear flow is the most informative and easily determined parameter, while after the gelpoint in the region of rubbery or glassy state, the components of dynamic modulus are more informative. Viscometric investigations show that sometimes microgels are formed in the reactive system before the gelpoint. Microgels represent branched and partly crosslinked molecules of the size of colloid particles.

The rheokinetic phenomenology of curing is associated with the structure and covers a wide range of polymers which form cured materials. The universality of the mathematical description of rheokinetics of curing of different products indicates that the physico-chemical phenomena associated with the formation of crosslinked structures are general.

Advances in Polymer Sciences 101
© Springer-Verlag Berlin Heidelberg 1991

1 Introduction

The formation of a network structure as a result of interaction of multifunctional compounds yields polymer materials in infusible and non-flowing states. In the general case, the process of formation of network polymers is called curing and the reacting mass passes through different physical and relaxation states typical for polymer systems – from viscous to rubbery or even glassy states. Curing is the principal process in the production of materials and items from epoxy and silicone oligomers, urea formaldehyde resins, polysters, and many others. Rheological characteristics of such systems and composite materials on their basis are one of the most important factors determining the conditions for production and applications.

Application of rheological methods to the analysis of the curing of reactive compounds made it possible to develop a well-founded pattern of variation of physical (relaxation) states of a curing material from a viscous liquid to a glassy solid [1–4]. The generalized „T-T-T diagrams" constructed by Gillham, which reflect different relaxation states of the curing material, are one of the fundamental results [5, 6]. However, this approach does not touch the kinetic aspect of the problem, although in studying the kinetics of these processes there is a limited number of experimental methods which are characterized by a sufficiently high response at all stages of structure formation of the cured material [7]. In addition, just the fact that curing materials change their rheological properties has the fundamental significance by allowing to separate them into a peculiar group of "rheokinetic media". The point is that the variation of rheological properties, immense in scale and range, reflects the kinetics of chemical transformations. The variation of molecular parameters of a substance is primary. The kinetics of this process can be studied by different methods, but as a result, this process leads to rheological transformations, so that the kinetics of chemical reaction should be estimated from rheological properties of the material.

Two principal cases can be pointed out: rheokinetics of production of linear or network polymers. In the first case, the viscosity variation of the reactive mass serves as a unified measure of conversion from the beginning to end of the process. In the second case, the reactive system becomes incapable of reversible deformations at a gelpoint, long before the curing is completed, i.e., the viscosity becomes indefinitely high but the process is continued.

In the general case, the variation of rheological properties during the process of curing can be represented by a diagram given in Fig. 1. The peculiarity of curing is clearly reflected in the presence of a special state – a gel point at the moment of time t*, which divides the whole process into two stages. The first stage is accomplished (although sometimes the particles of gel, microgel, can be formed before t*) by the formation of a network structure, covering the whole volume of the curing material or, in other words, by the formation of a giant molecule, whose dimensions are limited by the walls of the reaction vessel or dimensions of the product being formed. The first stage of the process can be called gelation and the second one curing, begond the gel point.

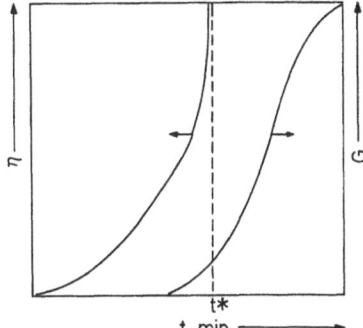

Fig. 1. Changes of rheological characteristics during the process of curing

The variation of the relaxation state of the reactive system of two curing stages determines the preferential use of different experimental rheological methods to characterize the curing material at different stages of the process. In the viscous-flow part before the gel point (t < t*), the viscosity at shear flow η is the most informative and easily measurable parameter, whereas after the gel point, in the field of rubbery or glassy state of the reactive system the main rheological properties are the components of the dynamical modulus, namely, the storage modulus G', the loss modulus G" and the loss tangent tan δ. A combination of these quantities, characterizing the variation of rheological properties of reactive systems, presents a complete rheokinetic pattern of the curing process.

2 Gel Point Determination

The gel point t* is one of the most important kinetic characteristics of curing, since it describes the attainment of a certain critical conversion responsible for the transition from the first to the second stage of the process. In the classical statistical theory of gelation developed by Flory [8], the gel point is characterized by the appearance in a reactive system of a macromolecule with an infinitely large molecular weight, $\bar{M}_w \rightarrow \infty$. Viscosity becomes infinite which corresponds to the above condition.

From these general statements, it follows that the gel point can be determined experimentally as a moment when the reactive system loses the possibility to flow. If the rotational viscometer is used, the curing composition tears of from the working surfaces of the device. It is evident that the moment of separation depends on the intensity of deformation of the material; moreover, the higher the shear rate the earlier the cessation of flow of the curing material is recorded. Therefore, the error of the gel point determination is determined by the conditions of viscometric measurements. As a rule, a total viscosity profile during gelation can be obtained with the shear rate variation within a few orders of magnitude and the moment of achieving the gel point is determined for the minimum shear rate. Strictly speaking, for the most exact determination of t* by a viscometry method it is necessary to carry out experiments at several shear rates with the subsequent

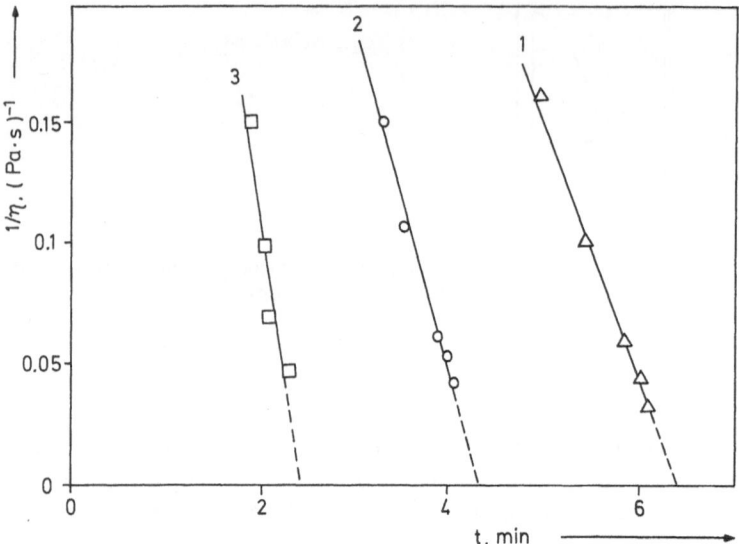

Fig. 2. Determination of the gel point by extrapolating the reciprocal of the viscosity to zero. Curing of silicone oligomer at T = 190 °C (1); 200 °C (2); and 220 °C (3)

extrapolation to zero rate. It should be emphasized that the effect of the shear rate on the moment of tearing off of the reactive mass from the working surfaces of the viscometer must not be identified with the effect of a mechanical field on the rate of the viscosity increase in the curing processes. This problem is important for the rheokinetics of the synthetic processes as well as for the transformation of polymers, and will be covered in Sect. 3.1.

In certain cases, the reactive system conserves the ability to flow up to the upper limit of the working range of the viscometer. Then, the gel point is often determined by extrapolating the dependence of the reciprocal viscosity on time in to zero time the final stage of gelation: $1/\eta \to 0$ [9–11]. Such an approach implies rather an exact determination of the gel point as a time of reaching an infinite viscosity, but there exists an uncertainty in extrapolation. An example of such a method is given in Fig. 2. According to some estimates, the error of the gel point determination by extrapolation does not exceed a few percent [12].

Another possibility of determining the gel point with the help of rheological methods is dynamical mechanical spectroscopy. Analysis of change of dynamic mechanical properties of reactive systems shows that the gel point time may be reached when tan δ or loss modulus G'' pass a maximum [3, 4, 13]. Some authors proposed to correlate the gel point with the intersection point of the curves of storage and loss moduli, i.e., with the moment at which tan $\delta = 1$ [14–16]. However, theoretical calculations have shown that the intersection point of storage modulus and loss modulus meets the gelation conditions only for a certain law of relaxation behavior of the material and the coincidence of the moment of equality $G' = G''$ with the gel point is a particular case [17]. The variation of the viscosity

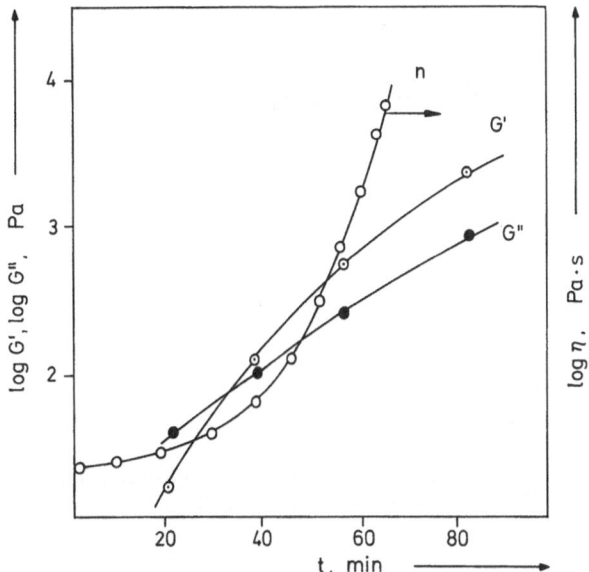

Fig. 3. Changes of viscosity η, elastic modulus G' and loss modulus G" in the process of thermal treatment of polyesterimide [18]. T = 350 °C

and components of dynamic modulus in the process of thermal treatment of fusible polyesterimide is shown in Fig. 3 [18]. G' becomes equal to G" if polyesterimide conserves its viscous state with a sufficiently low level of viscosity, whereas the gel point identified with the actual termination of the flow corresponds to a much longer reaction time.

According to our opinion, the most general gel point determination by means of rheological methods consists in comparison of viscosimetry with the attainment of a maximum of a loss tangent. Figure 4 gives an example of a gel point determination from tan δ and viscosity in curing of epoxy silicone oligomer [18]. The time of reaching a maximum of tan δ practically coincides with the moment of loss of the possibility to flow and corresponds to gelation. Similar results have been obtained on curing systems of different nature [6, 13, 19, 20].

Variation of tan δ determines the position of relaxation transitions irrespective of whether a polymer is linear or cured [21]. In linear polymers, such relaxation transition depends on temperature and molecular mass [21].

At the same time, in a number of cases a discrepancy is observed between the position of the maximum of G" and the point where viscosity goes to infinity, for example, during network-formation of silicone rubbers [22]. In this example, the difference between these two points reaches 7% in conversion which suggests that the gel point cannot be identified with the help of dynamical mechanical measurements [12].

The results of rheological investigation of the kinetics of gelation of physical nature, taking place as a result of phase separation of the corresponding polymer

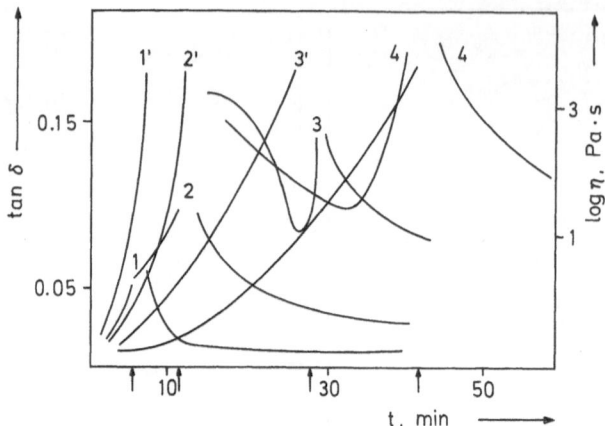

Fig. 4. Changes of viscosity and loss angle tangent upon curing of epoxy silicone oligomer [19]. T = 180 °C (1, 1′); 160 °C (2, 2′); 140 °C (3, 3′); and 120 °C (4, 4′)

solutions, point to a high information content of dynamical mechanical characteristics for gel point determination [24]. A change of macroscopic properties of the material does not differ from gelation caused by chemical reactions of multifunctional compounds. In the gelation process of the gelatin-water system, the viscoeleastic properties change from characteristics of a viscous fluid (G′ = 0) to a completely elastic body (G″ = 0), and here the extreme variation of the loss modulus shows that there exists a singular point which describes the transition from a viscoelastic liquid to a viscoelastic solid [24].

Thus, the comparison of the results of viscometric and dynamical mechanical measurements is the physically substantiated method for characterizing the gelpoint in curing compositions.

3 Viscometry of Curing

3.1 The Increase of Viscosity Near the Gel Point

One of the main problems in discussing the increase of viscosity during curing is the effect of deformation on the viscosity of reactive systems. This problem is of fundamental importance since any production operation is related with a certain mechanical action on the curing composition. The literature gives data on the effect of deformation conditions on variation of rheological properties moreover, it is shown that acceleration [25, 26], inhibition [27], or extreme change of the rate of the process [28] are possible. These contradictory results are explained in many cases by the effect of dissipative heating which is not taken into account in some experiments, but which, leads to acceleration but not to a change of the kinetic law of the curing reaction [29]. Quantitative estimates, based on the effect of deformation on temperature characteristics of reactive systems, show that an increase in the rate of curing for deformation with high shear rates is almost completely determined by thermal effects of dissipative heating and the resulting nonisothermality [30]. The validity of this conclusion is confirmed by an agreement between calculated and experimental dependences of the characteristic time of curing on shear rate [31].

If the isothermal regime of the curing process is not distorted, the deformation conditions in a general case do not have a pronounced effect on the viscosity variation. The absence of the effect of shear rate was confirmed by specially set up experiments with completely different chemical substances forming cured materials. This is illustrated by Fig. 5, which presents typical time dependences of viscosity in curing of epoxy silicone oligomer [32]. The values of viscosity obtained as a result of shear rate variation by a few orders of magnitude fall on a single dependence. Since similar results are obtained for different objects irrespective of whether polymerization [9, 10] or polycondensation [11, 13] is the curing process, we can conclude that this universality is of a sufficiently general character.

Viscometry is one of the most widespread methods of investigating initial stages of curing processes yielding information significant for applications in engineering. This is the reason why a rich experimental material has been accumulated concerning the mechanisms of viscosity variation in the initial stages of curing. Empirical exponential formulas are often suggested to describe the viscosity increase in the gelation process [34–36]:

$$\eta = \eta_0 \exp{(\theta t)}, \tag{1}$$

where η_0 is the initial value of the viscosity of the reactive system and θ is the constant.

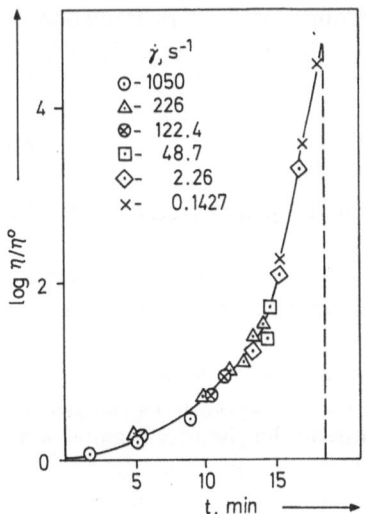

Fig. 5. Changes of viscosity in curing of epoxysilicone oligomer at different shear rates [32]. $T = 150\,°C$

Formulas of this type are used especially for practical purposes since they make it possible to predict, with a sufficient accuracy, the viscosity of reactive systems [35–39]. A limitation of such formulas is that formally they do not make allowance for the existence of the gel point for which $\eta \to \infty$; according to Eq. (1) at any finite time t, the viscosity is also finite. This has formal character since we can take for a gelpoint a certain level of value of the viscosity, for example, $\eta = 10^3$ or 10^4 Pa · and then t* is the time for which this level is achieved.

Attempts are also known to relate the type of time dependence of viscosity in the curing process to the kinetics of the reaction. Thus, upon curing of diglycidyl ester of Bisphenol A by triethanolamine, the viscosity curve η (t) was approximated by two linear segments [40]. The appearance of an inflection point is explained by the authors on the basis of the formation of a meshing network an the linearity of the η (t) dependence in the first party by the fact that a curing reaction is of a zeroth order.

Formally more strict is the approach to the calculation of viscosity increase based on the growth of molecular mass. For linear polymers viscosity is connected with molecular mass by the following relation [21]:

$$\eta = k\bar{M}^a , \tag{2}$$

where k is the constant depending on temperature and nature of the polymer, a ist the exponent having a "universal" value $a = 1$ for $\bar{M} < \bar{M}_c$ and $a = 3.5$ for $\bar{M} \geq \bar{M}_c$, and $\bar{M}_c$ is the critical value of the molecular mass corresponding to a change of the character of intermolecular interactions in polymers.

It is very difficult to establish an interconnection between rheological and molecular characteristics of curing materials due to the formation of complex branched molecular structures. The most complete and systematic investigations in this direction has been carried out for network formation of polyurethanes [41–43] and siloxane oligomers [44–46]. In these processes the molecular mass of oligomers was calculated from the concentration of functional groups using a branching theory, for instance, of isocyanate groups for network formation of polyurethanes. The following relation was obtained for polyurethane on the basis of ε-caprolactone and hexamethylene diisocyanate [41]:

$$\eta = Ae^{D/RT}(\bar{M}_w/\bar{M}_{w0})^{(C/RT+S)} , \tag{3}$$

where A, D, C and S are constants, $\bar{M}_{w0}$ is the initial weight-average molecular mass and R is the universal gas constant.

In a general case, Eq. (3) follows from Eq. (2) but it contains a greater number of constants describing the variation of viscosity and molecular mass of oligomer with temperature.

Generally speaking, the temperature dependence of the viscosity of reactive systems is determined by the activation energies of chemical reactions and viscous flow [47]. For this reason, the effect of temperature on the viscosity is ambiguous and the method of separating different contributions depends on the mechanism of reaction [47]. For polycondensation processes to which network formation of polyurethanes belongs, "apparent" values of the activation energy have been expressed in the following way [48]:

$$E_t = E - aU \qquad (4)$$

$$E_\eta = U - E/a, \qquad (5)$$

where E and U are the activation energies of viscous flow and chemical reaction, respectively; E_t and E_η are the apparent activation energies, corresponding to temperature dependences of the viscosity for a costant reaction time (E_t) or the time necessary for achieving a certain level of viscosity (E_η); a is the exponent in Eq. (2).

To simplify the situation and exclude the temperature dependence of viscosity, a relative variation of the parameter η/η_0 is considered, where η_0 is the initial value of the viscosity of the reactive system. This assumption is valid, if the activation energy of a viscous flow is constant during gelation, since in a general case E must vary with conversion [49].

It should be stressed that the viscosity changes during formation of polyurethanes even from bifunctional compounds can be correlated with gelation; most likely they are connected with the formation of a physical network then crosslinks arise from sufficiently strong specific interactions like hydrogen bonds [50]. An example of such a process is the reaction of macro (diisocyanate) with 3,3'-dichloro-4,4'-diaminodiphenylmethane [43].

The kinetics of equimolar polycondensation of two bifunctional monomers is usually described by a second-order rate equation. Integrating this equation, we obtain the following expression for a number-average degree of polymerization:

$$\bar{N} = 1 + x_0kt, \qquad (6)$$

where $\bar{N}$ is the number-average degree of polymerization, x_0 is the initial concentration of functional groups, k is the rate constant of the reaction. Figure 6 illustrates the variation of the degree of polymerization in the reaction of macro-(diisocyanate) and diamine. The $\bar{N}$ vs. t dependence cannot be approximated by a straight line but by two linear portions with different slope, which shows that the rate constant of the reaction varies during the process. Isothermal calorimetry [43] yields the same conclusions which indicate that even initial stages of curing cannot be described by a simple rate equation.

The dependence of viscosity on the degree of polymerization is also characterized by two linear portions (in log-log coordinates), and here the slope varies from 1 to 4.6 (Fig. 7). If the value of $a = 1$ is characteristic for a polymer with a low molecular mass, the value 4.6 exceeds the "universal" value of this exponent equal to 3.5. Considering the temperature dependence of viscosity of the reactive system, we see that Eqs. (4) and (5) are satisfied only if the exponent is equal to 4.6. Such a value of the exponent is not typical for linear polymers but usual for branched macromolecules.

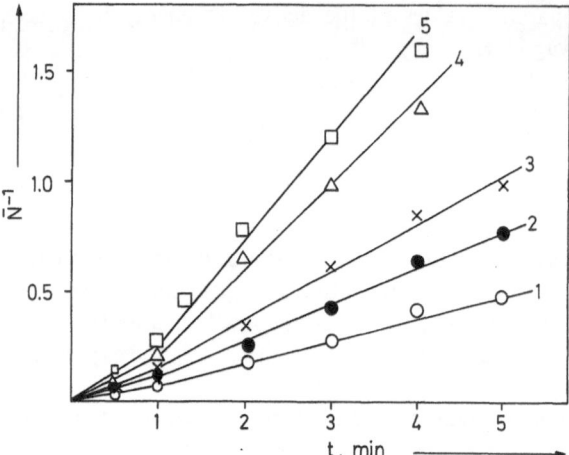

Fig. 6. Time dependence of the number-average degree of polymerization in polyurethane formation [43] at 60 °C (1); 70 °C (2); 80 °C (3); 90 °C (4); and 100 °C (5)

The peculiarities of the process described above are explained by the fact that branchings may create nuclei of insoluble fractions [43]. This assumption is confirmed by the data given in Fig. 8 where the viscosity increase is compared for network formation of polyurethanes from bi- and multi-functional diamine. The fact that the shape of the η (t) dependences is similar and the exponents $a = 4.6$ indicate that equivalent physical effects are operative.

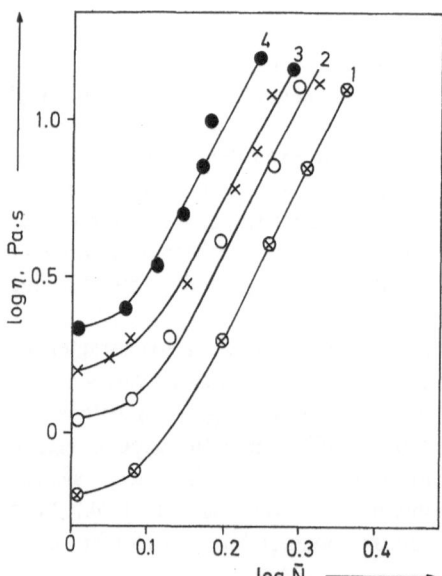

Fig. 7. Dependence of viscosity on the degree of polymerization in the formation of a polyurethane from macrodiisocyanate and 3,3'-dichloro-4,4'-diaminodiphenylmethane [43] at 60 °C (1); 70 °C (2); 80 °C (3); and 90 °C (4)

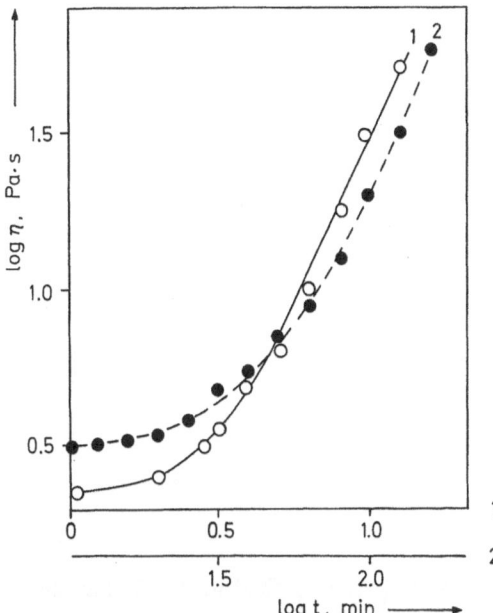

Fig. 8. Changes of viscosity in the process of polyurethane formation: 1 — bifunctional curing agent [43]; 2 — polyfunctional curing agent [41]

The formulas discussed above assume that there exists a point at which $\eta \to \infty$, since $\bar{M}w \to \infty$ where gelation threshold is achieved. Since the molecular mass is a function of conversion the expression for $\eta(\alpha)$ dependence (Eq. (7)) has be suggested employing the following boundary condition $\alpha = 0$, $\eta/\eta_0 = 1$, and $\alpha = \alpha^* \eta/\eta_0 \to \infty$:

$$\eta/\eta_0 = \left(\frac{\alpha^*}{\alpha^* - \alpha}\right)^{A + B\alpha} \tag{7}$$

where α is conversion, α^* is its value at gel point, A and B are the constants equal to 3.5 and -2, respectively.

Equation (7) is close to the expressions of exponential form obtained on the basis of percolation theory in order to describe rheological characteristics in the closest neighborhood of a gel point [51, 52]:

$$\eta \sim (\alpha^* - \alpha)^{-S} \tag{8}$$

where S is the constant whose theoretical value is 0.7 ± 0.007 [51].

Application of Eq. (8) is, generally speaking, limited to a very naroow range of α as the gel point is approached. If we assume that in this narrow interval $\alpha \sim t$, we may pass to the time dependence of the viscosity:

$$\eta \sim (t^* - t)^{-S} \tag{9}$$

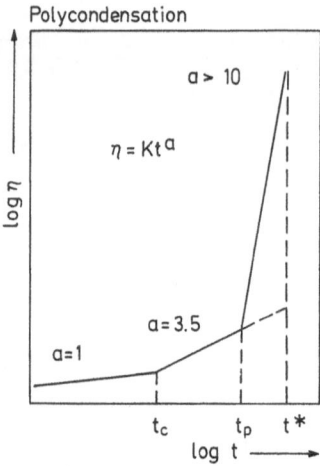

Fig. 9. Viscosity variation in different curing stages [53]

which is equivalent in form to Eq (8). An experiment cited in Ref. 52 has shown that in the time interval $10^3 < \Delta t/t^* < 5 \times 10^{-2}$, Eq (9) is well suited for describing the kinetics of viscosity growth, and the value of S is close to the theoretically expected value.

Each of the equations discussed above describes a certein stage of network formation in reactive systems. In general, the viscosity growth is determined by a set of physico-chemical phenomena accompanying the chemical reaction and change of molecular structure. The main stages of network formation are reffected in the viscosity profile during the process of three dimensional polycondensation. Dependences of viscosity changes in log-log plot are approximated by linear sections each of which corresponds to a certain stage of the gelation process [53]. As is seen from Fig. 9, the exponent of the $\eta(t)$ dependence increases in the initial stages of network formation from ~ 1 ($t \leq t_c$) to ~ 3.5 ($t > t_c$), which is caused by an increase of intermolecular interactions in the reactive system with increasing molecular mass of the resulting polymer. Such a dependence is obvious, since in polycondensation processes a time dependence of viscosity is adequate to a dependence of viscosity on molecular mass [54–56] and the reaction time t_c describes the attainment of the critical molecular mass [57].

If the curing process is ideally homogeneous, such law for viscosity variation should be conserved up to the termination of network formation [21]. In rheokinetic investigations of polymerization [58–62] and linear polycondensation [48, 54, 55] it was shown that the viscosity changes to deep conversions follow exactly the same power law provided that gel-effect and diffusion limitations are absent. A hypothetical viscosity variation according to this law are represented in Fig. 9 by a dashed line. In some real curing processes at a certain critical time t_p, the viscosity increases along with the rise of the exponent up to very large values. The ratio between the times t_c and t_p depends on the structure of reacting components and composition of the system. For instance, increasing functionality of the curing

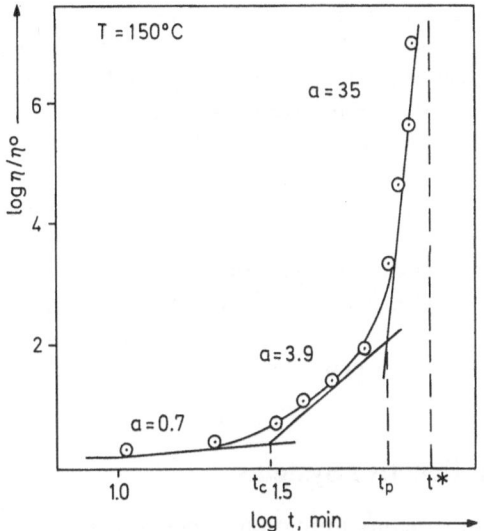

Fig. 10. Viscosity increase in curing of an epoxy resin by a diamine [63]

agent of epoxy oligomers results in a decrease of the linear section of η vs. t curve [32].

The appearance of one more section on the time dependence of viscosity can hardly be explained by variation of a molecular mass as a result of the reactions since such high values of the exponent of the η (M) dependence are unknown.

Characteristic experimental data are shown in Fig. 10 for curing of a particular epoxy resing — diamine system [63]. On the viscosity curve, three sections with different exponents are well distinguished, which are separated by two characteristic points, t_c and t_p.

An attempt to explain an avalanche-like viscosity growth as a physical gel point is approached, were made on the basis of a conception of equalization of the curing temperature and the glass transition temperature T_g of a reactive system [64–67], since it is well known that T_g grows in the course of the process [68, 69]. Starting from this assumption, Gillham has developed a model which supposes that viscosity is determined by two factors, namely, an increase in molecular mass and growth of the glass transition temperature. The effect of the glass transition temperature on viscosity of a reactive system is estimated according to the well-known WLF equation. In this case, a general formula for calculating the viscosity of a curing oligomer has the form [64]:

$$\ln \eta = \ln \eta_\infty + \ln \bar{M}_w + E/RT - \frac{C_1(T - T_0)}{C_2 + T - T_0}, \tag{10}$$

where C_1 and C_2 are the constants of WLF equation, T_0 is the reference temperature. It is possible to calculate viscosity by Eq. (10) if the following relationships are determined beforehand: relation between the molecular mass and conversion $\bar{M}^w(\alpha)$; relation between the glass transition temperature and conversion $T_g(\alpha)$; defendence of conversion on time $\alpha(t)$. If all these parameters are known, this formula gives an unamiguous relationship $\eta[\alpha(t)]$. It is obvious that Eq. (10) describes adequately the viscosity changes determined by the distance from the glass transition temperature [70], irrespective of whether the resulting polymer is linear [65] or cured [64]. When gelation threshold is attained above the glass transition temperature, the last term in Eq. (10) becomes zero and does not affect the value of the viscosity of then reactive system even in the nearest neighborhood of the gelpoint.

Since glass transition is a very particular case in curing of reactive systems, and a sharp viscosity growth as $t*$ is approaches is a general case, we must look for the nature of the viscosity growth at point t_p. The steep increase of the molecular weight is the primary reason. Formation of a new phase considered in the next section may be the other reason.

3.2 Microphase Segregation in Reactive Systems

There are two different approaches to structure formation of polymer networks, which can be called "homogeneous" and "heterogeneous". In the first case, it is supposed that the curing material preserves its homogeniety on the supermolecular scale and the formation of a network continues from the start of a reaction and up to the gel point completely according to laws of macromolecule branching. This approach was used and confirmed in many publications cf. e. g. [70–73], by comparing the theoretical predictions with experimental data. As it was stated in Ref. [72], the statistical theory of molecular branching can successfully predict such structural factors as molecular mass distribution, hydrodynamic characteristics of macromolecules in the pregel stage, critical conversion and exponents at the gel point, many molecular parameters (molecular mass distribution, raduis of gyration of sol, content of elastically active chains, cycle range, etc.) beyond the gel point. Electron microscope investigations have shown that in some network polymers (for example in some epoxy resins) one can see globular grains with characteristic size of the order of 20–40 nm [74]. These results were treated as the evidence of supermolecular structure, but such structures can be observed in any amorphous polymer, and depend on the technique of preparation of the samples.

The "heterogeneous" gelation in formation of polymer network occurs via formation of microgel particles [75–79]. It was stated [79] that the appeerence of microgel particles is consequence of a particular reaction mechanism leading to crosslinking. Chain crosslinking copolymerization is one of the examples [82]. However, if such microgel particles are formed, the kinetics of the chemical reaction will change as it depends of the local concentrations of reactive groups.

In the case of heterogeneous network formation, one can sometimes observe two critical gel points, the first one corresponds to the appearances of the "infinite" macromolecules in the form of "microgel" particles of colloid The second macroscopic gel point correlates with the rheological transition from a solution to the state of a gel, when the system as a whole losses the possibility to flow and is converted into a single three-dimension macromolecules.

In some systems one can observe local gel formation also as a result of phase separation, other systems can be very close to homogeneous gelation and no inhomogeniety can be found [80].

Below we shall speak about the gel point in the rheological sense, as a point of the relaxational transition from a fluid to a non-fluid system, but it is necessary to remember that gel particles can appear before this gel point.

In many cases, the rheological behavior of curing system is in a good agreement with the assumption of a homogeneous process. Very impressive examples are presented by Macosco et al. (summarized in [80a] and Nicolais et al. [80 b–d]. The "homogeneous" model is based on the rigorous solution of kinetic equations. This approach (basedon the theory of branching reactions) allows to calculate the changes of the molecular weights of a curing oligomer before the gel point. Such calculations were performed for a regularly cured silicon oligomer [81 a, d]. There is no doubt that the model of the homogeneous reaction completely explains the the correlation between the viscosity growth and the value of M_w at least up to the certain level of viscosity; for example, in the paper [80a] up to the viscosity level of 10 Pa. s. An interesting development of the "homogeneous" approach was proposed in [80e], where it was found that statisfactory results could be achieved if the average longest linear path through the branched molecules was correlated with the viscosity.

In some other cases the peculiarities of viscosity variation $\eta(t)$ as gel point is approached after a time t_p can not be described in terms of homogeneous chain growth and a more general structural concept reflecting the mechanism of formation of some network polymers is required [61]. Figure 11 shows the curves of viscosity growth in curing of an epoxy resing with 3,3'-dichloro-4,4'-diaminodiphenylmethane in the presence of different amounts of chemically inert plasticizer-dibutylphthalate. Special consideration must be given to the following features of viscosity profile in the gelation process for such plasticized systems:

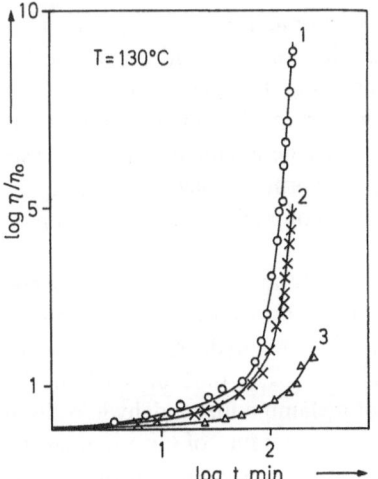

Fig. 11. Changes of viscosity in curing of an epoxy resin by 3,3'-dichloro-4,4'-diaminodiphenylmethane in the presence of dibutyl phthalate: 1) 0% of DBP; 2) 20% of DBP; 3) 40% of DBP

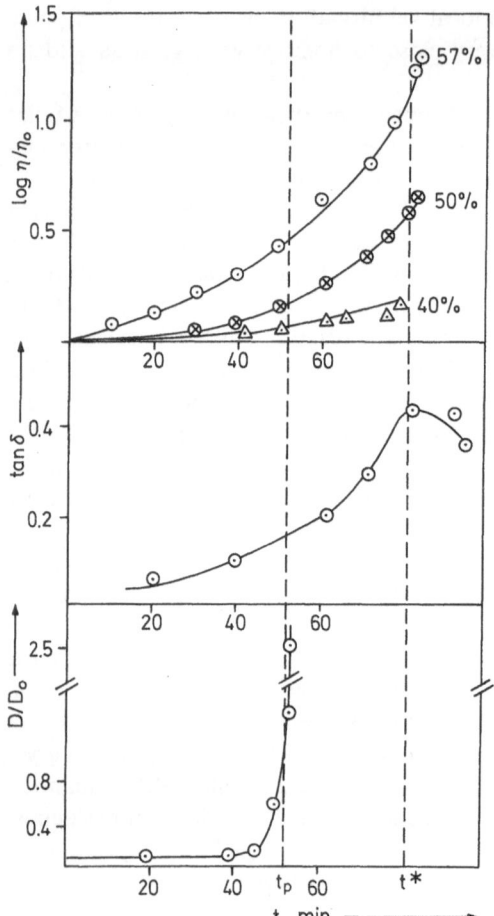

Fig. 12. Dependence of viscosity, loss angle tangent and optical density on time of curing of a melamine-formaldehyde resin in solution and in bulk [81]. T = 80°. Concentration of the solution: 40% (1); 50% (2); 57% (3)

— difference in the values of the viscosity at the moment preceding gelation;
— a very weak interdependence (or practically its complete independence) between the time of reaching gel point and the content of platicizer.

This is a rather unexpected result very important for understanding the mechanism of viscosity variation at gelation. It can be assumed that the *curing process is inhomogeneous*. At a certain stage of curing a new phase may be formed, and fragments of cured structures are the centers of growth of this phase [75–79].

From this point of view, the results given in Fig. 12 are very significant. Here, the relationships are shown between viscosity, light absorption and loss tangent changing in the process of curing of melamine-formaldehyde resin in aqueous solutions of different concentration and in bulk [81]. The time of reaching the gel point does not change with concentration of melamine-formaldehyde resin in solution and coincides with the time of reaching a maximum of tan upon curing of anhydrous resins. Figure 14 shows the temperature dependence of the inverse gelation time, $1/t^*$, in the Arrhenius coordinates. The values of the gelation time determined by the viscometric method upon curing in solution and by a dynamic mechanical method upon curing in bulk are described by a common power relation.

A sharp increase in the light absorption shown in Fig. 12, is observed long before the gel point is reached. This increase is related to segregation of the reactive

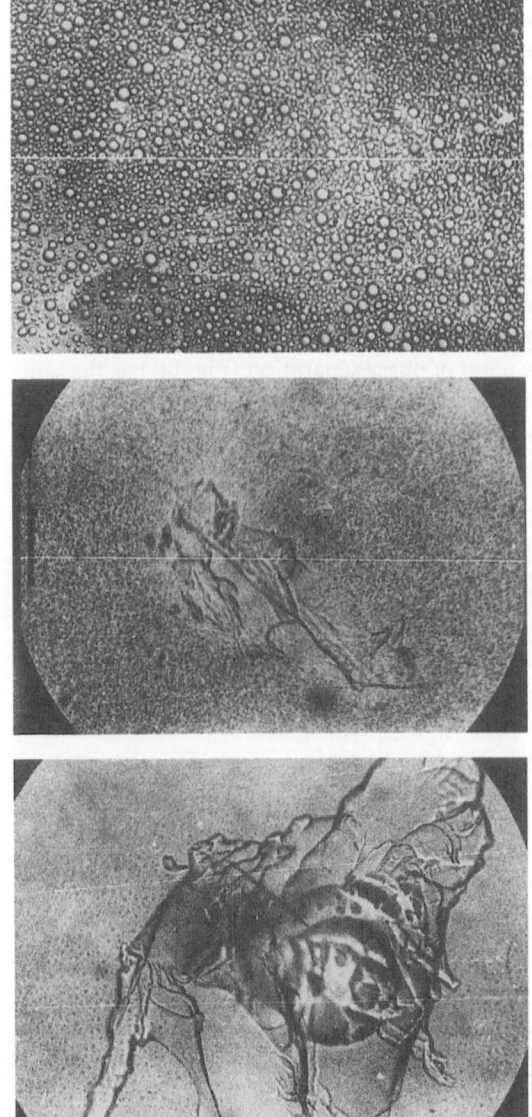

Fig. 13. Photomicrographs of a suspension of microgels at different stages of curing of an epoxy oligomer by a diamine: 160 (1); 220 (2); 280 min (3) [63]

system and appearance in the solution of a new phase consisting of fragments of branched and cured molecules. Such structure formations are known, their existence has long been discussed in the literature and they received the name microgels [71, 79, 82–86]. Investigations have shown that the relation between viscosity and the applied stress corresponds to the moment when phase segregation sets in [87], which is one of the general concepts of rheokinetics of polymerization processes [88].

From these positions different values of viscosity at the moment preceding the breakaway of the reactive system from the working surfaces of a viscometer become understandable and, moreover, the viscosity at this period increases with the concentration of the oligomer in solution. This is due to the fact that before the moment, when a continuous phase is formed from fragments of cured structures, the flow is determined by a dispersion medium whose viscosity depends on the solution concentration.

Information on the structure formed as a result of phase segrgation in the reactive system can be characterized by direct microscopic investigations of the process of curing epoxy oligomer by diamine [63]. However, one must be aware of the nodular structure observed in microphotographs of all crosslinked as well as uncrosslinked amorphous bulk polymers [74]. Figure 13 shows photomicrographs of a suspension of the resulting microgels. The moment of formation of microgels and their subsequent growth up to a complets coalescenc are shown.

Temperature dependence of the point of phase separation (Fig. 14) has the same coefficient as temperature dependence of the gel point t* and the rate constant of a curing reaction (see below). A coincidence of the values of the activation

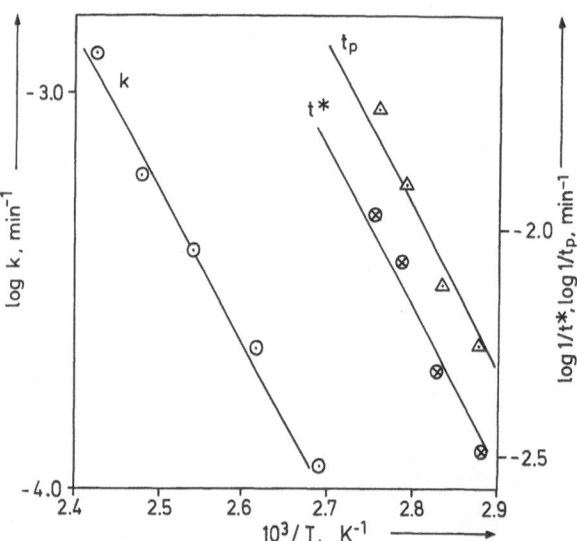

Fig. 14. Temperature dependence of the gelation time (determined by viscometric and dynamic mechanical methods,), the point of phase segregation and rate constant of the reaction [81]

energy of phase segregation gelation and curing indicates that all these phenomena are controlled by the kinetics of the chemical reactions resulting in the formation of a cured product.

Thus, the given results show that before the gel point microgels — cured and branched molecules with dimensions as much as colloidal particles — are formed in curing systems as a result of phase separation. After the onset of phase segregation, t_p, the curing process occurs in a two-phase region. As a rule, phase segregation causes an intensive viscosity growth of a filled system which the reactive mass become as a result an increase in the amount of a "filler", i.e. microgel.

The fact that a viscosity increase after phase segregation (for $t > t_p$) is connected with such mechanism is evidenced by the results of gel chromatographic (GPC) analysis of solfraction in the network formation process of low-molecular siloxane rubbers (Fig. 15). As the reaction proceeds the molecular mass of the sol fraction decreases and so does its viscosity. However, network formation of a number of epoxy resins cured with amines or other curing agents conform the "homogeneous" model without microgel formation [88a].

Figure 16 represents schematically a vatiation of molecular mass and content of sol fraction, calculated on the basis of the branching the ory, in the process of three-dimensional polycondensation of multifunctional compounds [89, 90]. According to these calculations, the weight-average molecular mass of curing oligomer increases monotonically up to the gel point where it diverges. At this same point the gel appears. A real variation of the same characteristics, obtained on the basis of GPC analysis of curing of an epoxy resing with diamine and dibutylphthalate has a different nature (Fig. 17). Molecular mass of the resulting

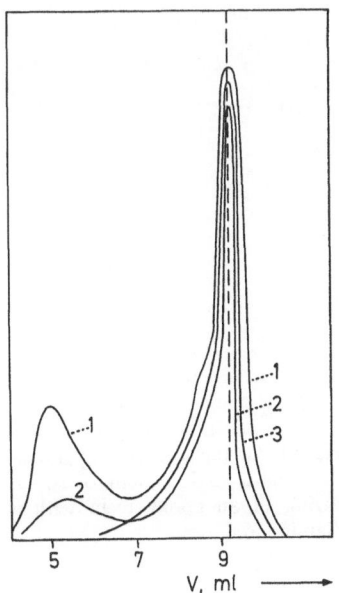

Fig.15. Gel permetion chromatograms of the sol fraction in the network formation from a low-molecular-weight siloxane rubbers. Reaction time: 1 (1); 1.5 (2); 2, 3, 4, 5 and 24 h (3)

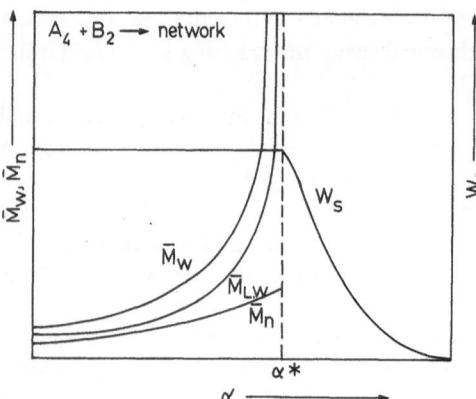

Fig. 16. Results of theoretical calculations of changes of the molecular mass and content of sol in crosslinking polycondensation [89]

polymer varies extremely; moreover a maximum value of the molecular mass is observed at a point of phase segregation, t_p. At this moment and not at the gel point a gel fraction appears in the reactive system although the system us a whole is able to flow. It is interesting to note that a similar picture of extreme variation of molecular mass was observed long ago for a fundamentally different system — polybutadiene — in the vulcanization process [91].

Within the framework of this model, the gel point can be considered as a moment when phase inversion is reached in a reactive system, i.e. when the, fragments of cured structures are transformed from disperse phase into a continuous dispersion

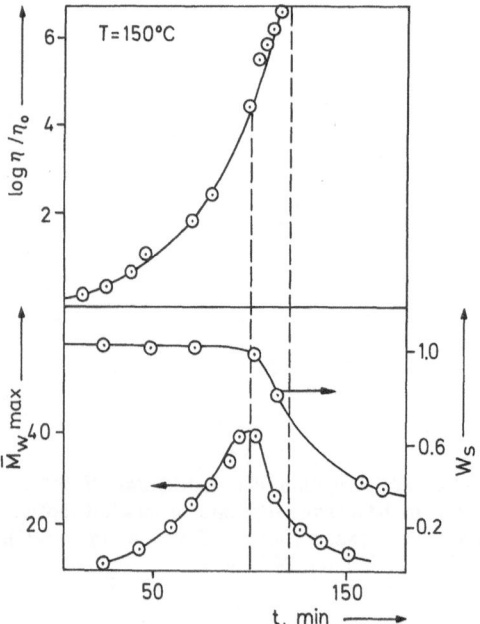

Fig. 17. Changes of viscosity, molecular mass and content of sol in curing of an epoxy resin with a diamine [63]

medium which is connected with a change of relaxation state of the system and transition from viscous to rubbery or glassy state.

Such a reactive system can be considered as a filled composition. This approach makes it possible to apply the well-known concepts of rheology of filled polymers to the description of the $\eta(t)$ dependence.

For filled systems different analytical and empirical expressions are suggested, that describe the relation between the viscosity and concentration of a filler. Expressions of the following form belong to the most widespread expressions [92, 93]:

$$\eta/\eta_0 \sim (1 - \varphi/\varphi_{max})^{-c} \tag{11}$$

where φ_{max} is the value of the ultimate concentration of a filler in the system depending on the method of its packing, c is the constant.

In the special kind of reactive system discussed above, a filler is generated in the course of the reaction. Assuming that $\varphi \sim t$, we can obtain an expression for the kinetics of the viscosity variation in gelation processes quite similar to a scaling formula in Ref. [9]. However, the given approach has a fundamental difference which consists in the fact that the viscosity of the dispersion medium η_0 is not a constant but varies in the course of the process [94]. Then, the viscosity variation with time can be written as:

$$\eta(t)/\eta_0(t) = (1 - t/t^*)^{-b} \tag{12}$$

where $\eta(t)$ is the viscosity of the reactive system, $\eta_0(t)$ is the viscosity of the dispersion medium or matrix, t^* is the time of reaching a gel point, b is the constant.

Experimental verification of Eq. (12) was carried out for a number of quite different systems curing both in a solution and bulk such as epoxy, epoxy silicone and silicone oligomers, melamine-formaldehyde and carbamide resins, derivatives of furan resins [11, 28, 32, 63, 81]. Typical results plotted according to Eq. (12) are given in Fig. 18. After phase segregation ($t > t_p$), the experimental dependences $\eta(t)$ are completely described by formula (12). The viscosity variation of the dispersion medium is a function of molecular mass andis calculated on the basis of Eq. (12).

A position of point t_p is independent of concentration of the reactive oligomer in the system and is determinded by temperature at which the reaction occurs (see Fig. 19) [81]. Figure 19 gives the data showing the variation of t_p/t^* ratio. These results indicate that the higher is the oligomer concentration in the reactive system, the closer is the point of phase segregation to the gel point.

Thus, the presented results show that the type of viscosity variation at gelation is determined, first, by a variation of molecular structure of the oligomer and, second, by a phase separation in the system near the point of relaxation transition, i.e., gelation.

A similar pattern arises when the relaxation state of reactive systems changes as a result of chemical reaction and structure growth without phase separation. This occurs, for example, in the synthesis of oligoimide from 4,4'-diphenylmethane

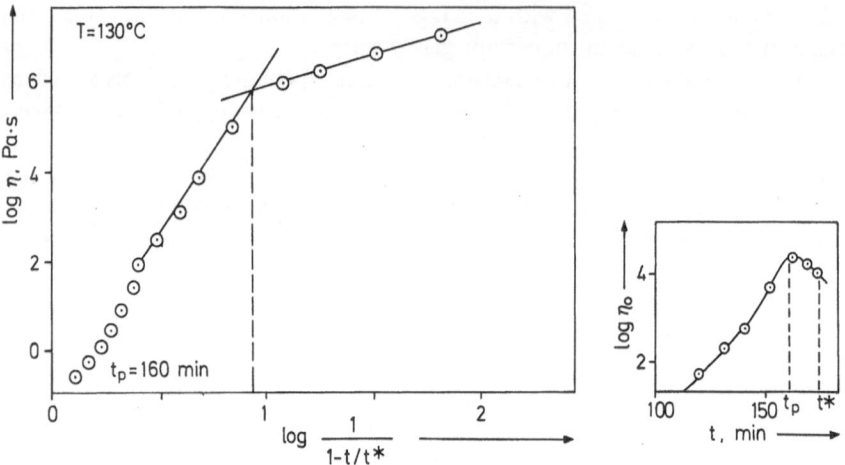

Fig. 18. Time dependence of viscosity in curing of epoxy resin by a diamine according to formula (12) [63]

diisocyanate and dianhydride of 3,3', 4,4'-benzophenone tetracarboxylic acid [95]. The viscosity of the reactive system was measured independently in addition to the concentration of functional groups by IR-spectroscopy. The a number-average molecular mass of oligoimide varied according to Eq. (6). A generalized expression, taking into account the contribution of different factors to the viscosity of a reactive system, can be written as follows [95]:

$$\eta/\eta_0 = \left(\frac{1 + kt}{1 - t/t^*}\right)^a \tag{13}$$

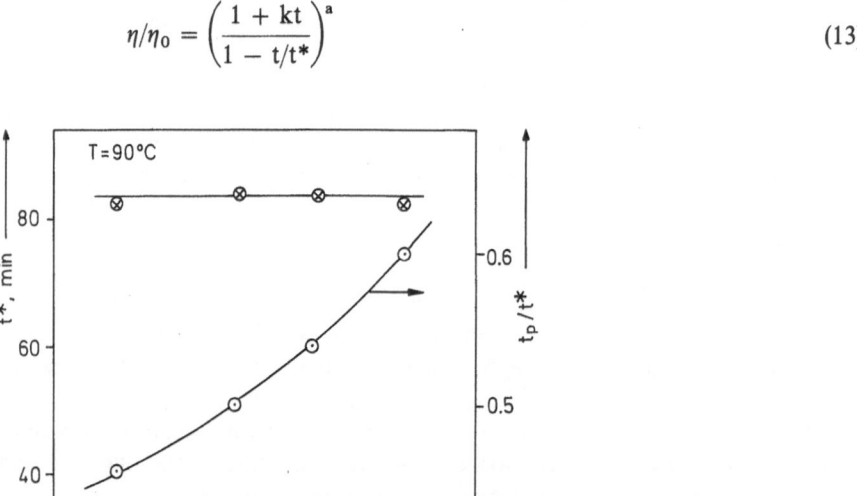

Fig. 19. Position of the gel point and variation of the t_p/t^* ratio in dependence on concentration of melamine formaldehyde resin in solution [81]

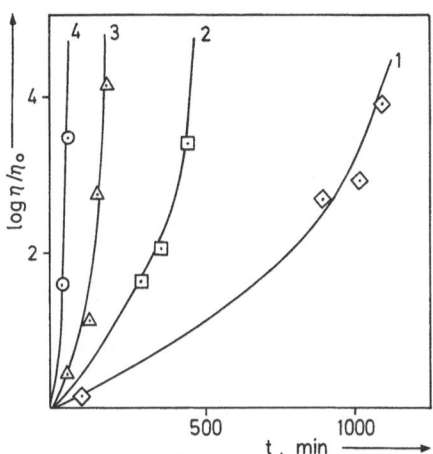

Fig. 20. Experimental (dots) and calculated (solid lines) time dependences of viscosity at different temperatures of oligoimide synthesis (°C): 170 (1); 180 (2); 190 (3); 200 (4) [95]

where k is the rate constant of the reaction, a is the constant. A correspondence between experimetnal and theoretical dependences $\eta(t)$, given in Fig. 20, points to the validity of such a model.

For curing of reactive oligomers, followed by a phase segregation, the last formula becomes more complicated due to a nonmonotonic change of the molecular mass at $t > t_p$. A complete equation for the viscosityariaiation in polycondensation curing processes can be written in the following form [94]:

$$\log \eta/\eta_m = \pm a \log t/t^* + b \log \frac{1}{1 - t/t^*} \tag{14}$$

where a and b are the constants, with $a = 1$ at t t_c, $a = 3.5$ at $t \geq t_c$ and at a point of phase segregation this exponent reverses its sign due to a reduction of the molecular mass; η_m is the constant changing its value at a critical point t_c and at t_p, i. e., in the general case, in contrast to Eq. (13), $a \neq b$.

The first term on the right-hand side of Eq. (14) describes the viscosity change determined by the variation of the molecular mass of a dispersion medium. The sign of the contribution of this term into the system viscosity is determined by the position of a point of phase segregation with respect to a gel point. Figure 21 gives the results of theoretical calculations of the viscosity variation for different specified values of the t_p/t^* ratio and the value of the constant b. These calculations show that different shapes of the $\eta(t)$ dependence may be observed in accordance to the value of the t_p/t^* ratio. A usual shape of these curves with a monotonic viscosity growth is observed only when $t_p/t^* \geq 0.9$, which is typical for the curing in bulk [63]. When t_p/t^* is reduced, for example, by adding a diluent, a plateau shows up on the $\eta(t)$ dependence or even a certain reduction of viscosity is observed.

The curves $\eta(t)$ with a clearly expressed part of reduction in the rate of viscosity growth near the gel point become understandable from this point of view [9, 96, 97]. As a rule, it was considered in this case that the reason for reduction in the

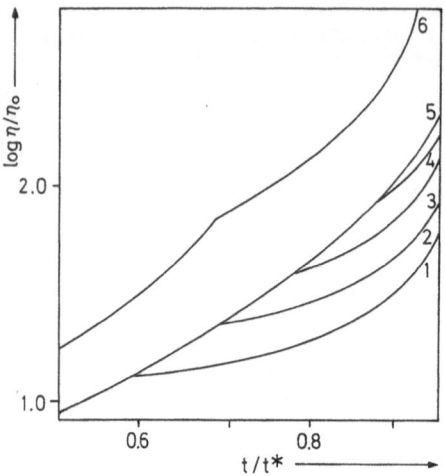

Fig. 21. Different paths of viscosity variations at $b = 1$ and $t_p/t^* = 0.6$ (1); 0.7 (2); 0.8 (3); 0.9 (4); $b = 2$ and $t_p/t^* = 0.7$ (6) [94]

rate of viscosity is the appearance of a slipping effect upon reaching the gelation threshold. However, for certain systems it is possible to experimentally record such a fairly extended part of the $\eta(t)$ dependence. The relationships between the viscosity and tan δ in the process of curing of a silicone oligomer are given in Fig. 22 [98]. During phase segregation (at $t \geq t_p$), the viscosity of the reactive system is somewhat reduced but then it grows up to the gel point. The interdependence of rheological properties and the plateau on the time dependence of the components of the dynamic modulus was observed in the curing epoxy resins by diamines [99]. A similar effect of reduction of the viscosity of a reactive system at the moment of phase segregation was observed in linear free-radical polymerization [62, 100].

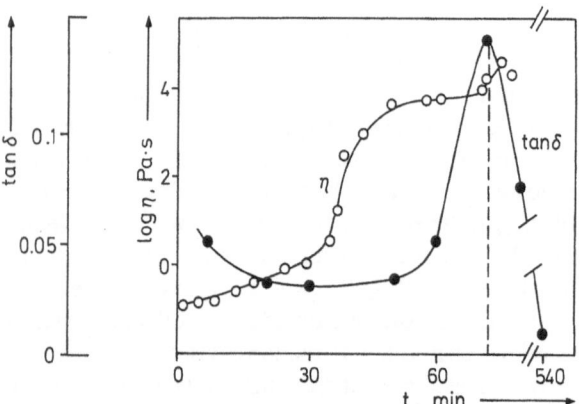

Fig. 22. Changes of viscosity and tan δ in curing of a silicone oligomer at $T = 140$ °C [98]

Phase segregation affects the dependence of the viscosity growth on the applied stress of shear rate [88]. It is possible to see Such an effect is seen experimentally in a sufficiently extended part of flow ability after a microphase segregation; i.e. the ratio t_p/t^* is relatively small, for example, in curing melamine-formaldehyde resins in solution [87]. However, the case of t_p/t^* close to unity, this effect may not always be revealed due to experimental difficulties.

4 Curing After the Gel Point

4.1 Rheology and Other Methods of Investigating the Kinetics of Curing

A characteristic feature of rheological investigations of curing compositions, carried out recently, is the introduction of such parameter as "rheological" degree of conversion β, which takes into account the kinetics of Crosslink formation. In this case, a basic variable is an elastic (or storage) modulus proportional to the crosslinking density, the network bein formed by chemical bonds and physical entaglements. The last statement was verified in special experiments on determining a crosslinking density in curing of low-molecular-weight polydemethylsiloxane rubbers by the method of swelling [101]. A comparison of the kinetics of crosslink formation calculated by a method of swelling and elastic modulus of the curing composition is given in Fig. 23. Such a macroscopic parameter as elastic modulus is actually responsible for the kinetics of crosslink formation [46]. A certain difference in the values of n_c, (20%) is obviously connected with the contribution by physical entanglements [102].

It should be noted that the above results were obtained on endlinked systems [103, 104] yielding model networks. In this case the value of M_c at the end of network formation is close to the number-average MM of the initial sample [105].

Among physicochemical methods for studying curing processes, rheokinetics is the closest to calorimetry, also yielding combined characteristics of the chemical dynamics of the process. Due to this, there was carried out a comparison of quantitative curing parameters obtained by rheological and calorimetric methods.

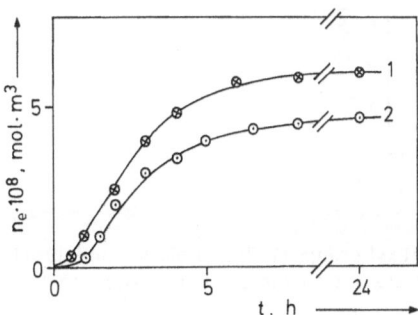

Fig. 23. Changes of crosslinking density n_c in netformation from a low-molecular-weight silicone rubber at 20 °C, calculated from the elastic modulus (1) and equilibrium swelling in toluene (2) [46]

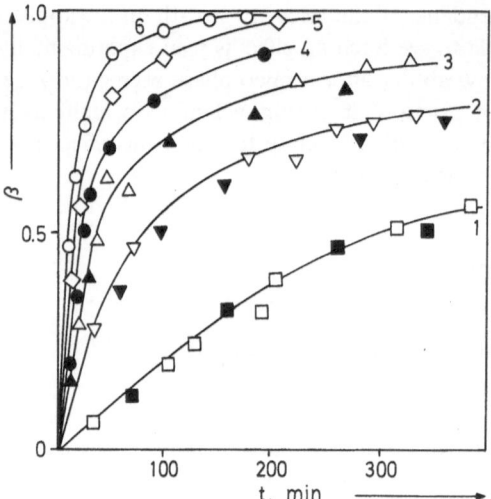

Fig. 24. Time dependence of conversion β (β — open marks, β_c — full marks) of the system: epoxy silicone oligomer — PBTPh (composition 100:7.5) at curing temperatures (°C) of 140 (1), 160 (2), 180 (3), 193 (4), 200 (5), 220 (6) [106, 127]

Figure 24 gives time dependences of "rheological", β_r, and "calorimetric", β_c, degree of conversion of curing reaction of epoxy silicone oligomer by polybutoxytitaniumphosphoroxan. It is clear that β_r and β_c almost coincide under the same conditions [106, 107]. It is possible that the correlation is not universal because, in the general case, the interrelation between rheokinetic and calorimetric parameters of the process depends on thermal effcts of parlicular reactions, since only a part of them is responsible for the formation of crosslinked structures. However, the given data indicate that it is possible to compare kinetic parameters of curing processes calculated on the basis of rheological and calorimetric measurements.

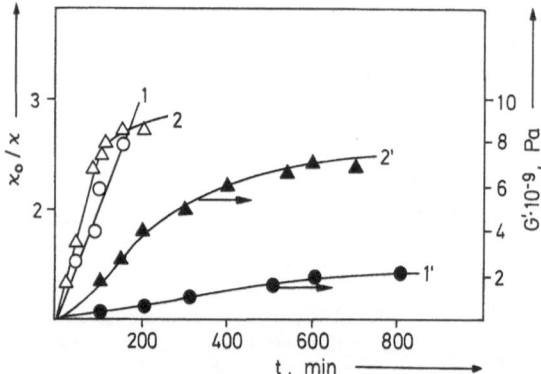

Fig. 25. Changes of the relative content of hydroxyl groups (1, 2) and elastic modulus (1', 2') in the process of curing of polyphenyl methyl siloxane oligomer at temperatures (°C) 170 (1, 1') and 190 (2, 2') [108, 109]

The situation is rather complicated when rheology is compared with experimental methods which measure concentration of functional groups. In this case, no good correlation is observed. Figure 25 gives the time dependences of the concentration of hydroxyl groups and elastic modulus in curing of silicone oligomer [108, 109].

A separate group of experimental methods record the mobility of molecular structures, in particular, a pulse-NMR spectroscopy. This method has shown a high informativeness for investigation of gelling systems [110] and a possibility to directly measure the density of network [111]. Application of a pulse-NMR spectroscopy for studying the kinetics of curing makes it possible to obtain kinetic equations similar to equations based on calorimetric and rheokinetic investigations of the same systems [112].

4.2 Physical Model of Network Formation

Phase segregation and formation of a microheterogeneous structure, if it occurs, affects the kinetics of curing. Polymerization reactions leading to the formation of crosslinked structures are not reactions of isolated macromolecules and, therefore, cannot be considered without taking into account the morphology of the reactive system.

Various experiments show that in some cases the kinetics of curing cannot be described by a phenomenological equation of the n-th order at any reasonable values of n [113, 114]. In a number of cases, the kinetics of curing processes is adequately described by a phenomenological equation which takes into account the self-acceleration effect [115]:

$$\frac{d\beta}{dt} = (k_1 + k_2\beta^m)(1 - \beta)^n \tag{15}$$

where k_1 and k_2 are the rate constants, m and n are the empirical constants.

Equation (15) written above may be of some generality since it contains many constants empirically selected. This fact, however somewhat complicates the application to the analysis of experimental data. Therefore, to describe the rheokinetics of curing of different systems some other notation is often used for the phenomenological equation with a self-acceleration term [47]:

$$\frac{d\beta^*}{dt} = k(1 - \beta^*)(1 + c\beta^*) \tag{16}$$

where β^* is the rheological degree of conversion, t is time and c and k are constants.

The "rheokinetic" degree of conversion means a conversion dependent parameter measured by a rheological method.

We can see a formal similarity of Eqs. (15) and (16), there is, however, a basic difference between them. In Eq. (15), is a true chemical conversion and an autocatalytic overall kinetics is observed it may be due to a specific chemical mechanism, specific diffusion control or physical segregation.

When the time changes are characterized by rheological methods (β^*), the apparent autocatalytic shape of $\beta^*(t)$ may be generated by a non-linear dependence of the rheokinetic parameter β^* on β. For instance, it can be shown that, if β^* is the storage modulus G', G' is proportional to the concentration of elastically active network chain, V_e. For a random trifunctional step polyaddition $G' \propto V_e \alpha \dfrac{2\beta - 1}{\beta}^3$ which clearly gives an S-shaped dependence of G' vs.t. This means that an apparent "self-acceleration" shape of the rheokinetic curves does not have to have anything to do with an autocatalytic chemical process.

An identical mathematical description of the kinetics of curing of reactants different in chemical nature and that obtained on the basis of fundamentally different experimental methods allows us to assume that this apparent "self-acceleration" course of some rheokinetic parameters is common to the processes of formation of materials with a crosslinked structure. It should be emphasized once more that the "self-acceleration" effect must not be identified with the self-catalysis of the reaction of interaction between epoxy monomers and diamines which is studied in detail on model compounds [116, 117]. For each particular curing process the self-acceleration effect is influenced by the mechanism of network formatic, namely, chemical self catalysis [118], the appearance of local inhomogeneities [120], the manifestation of gel effect [78], parallel course of catalytic and noncatalytic reactions [68]. It is probably true that the phenomena listed above may in one form or another show up in specific processes and make their contribution into self-acceleration of a curing reaction.

Phase separation or partial segregation occurring during curing may be an important factor determining the time dependence of rheokinetic parameter. It may be caused first of all by thermodynamic incompatibility or by diffusion controlled fluctuations for some types of reactions. Then, microgel can be formed before the bulk of the system is transformed into a macrogel.

The procedure of analyzing experimental data with the help of Eq. (16) and determining the constants for the kinetic function f (β) of different form is given in Ref. [121]. An integral of Eq. (16) for the reaction of the first order is written in the following form:

$$\ln \frac{1 + c\beta}{1 - \beta} = (1 + c)\,kt \qquad n = 1 \tag{17}$$

or

$$\beta = \frac{\exp\left[(1 + c)\,kt\right] - 1}{\exp\left[(1 + c)\,kt\right] + c} \tag{18}$$

For the reaction of the second order an integral of Eq. (16) has the form:

$$\frac{\beta}{1 - \beta} + \frac{c}{1 + c}\ln \frac{1 + c\beta}{1 - \beta} = (1 + c)\,kt \tag{19}$$

An almost complete agreement between experimental and calculated (Eq. (16)) dependences of conversion on time, $\beta(t)$, is obtained for curing of epoxy [63, 114], epoxy silicone [19], silicone oligomers [20, 109], low-molecular-weight silicone rubbers [46, 105], unsaturated polyester [97, 121], melamine-formaldehyde [122], methylolpolyamide [123] and carbamide resins [122].

By the way of derivation, Eq. (16) must correspond to experimental data beginning from the point of phase segregation and formation of two-phase system, although the results listed above fit actually the whole curing process. This is explained by the fact that phase segregation may occur in the initial stages of the process and the time interval between the beginning of the reaction and point of phase segregation is small with respect to the total duration of the process.

In essence, Eq. (16) describes the formation of a two-phase structure in reactive systems, which takes place according to the mechanism of nucleation and growth under the condition that an increase in concentration of the second phase is determined by the chemical reactions. Crystallization may serve as a physical analogue for such a process. Indeed, in Refs. [124, 125] a new model of crystallization kinetics was developed, which is reduced to a self-acceleration equation similar to Eq. (16).

In the papers quoted above, the authors followed the crystallization kinetics from calorimetric data. From the point of view of rheokinetic methods discussed here, no less interesting is one more example of describing crystallization kinetics given in Fig. 26, where there is a comparison of experimental (obtained in Ref. [126]) and calculated by Eq. (18) time dependences of the relative elastic modulus of crystallization of cis-1,4-polybutadiene. A good agreement between experimental and calculated data implies the possibility of describing the crystallization kinetics, with the help of formulas of the type of Eq. (18).

Thus, a combination of the given results indicates that rheokinetic phenomenology of curing of reactive compounds takes into account the mechanisms of both chemical reactions and phase separation occurring as a result of chemical conversion. As a consequence, a self-acceleration effect exists in the processes of curing of oligomers irrespective of a specific mechanism of the reaction.

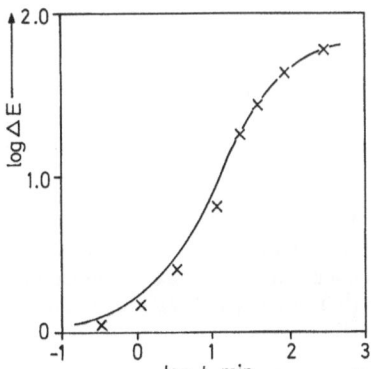

Fig. 26. Changes of the relative elastic modulus in crystallization of cis-1,4-polybutadiene [126]. Dots denote experiment, line — calculation by formula (20)

4.3 Rheokinetic Equations of Curing

An analysis has shown that the characteristic feature of rheokinetics of curing with phase separation is the presence of a self-acceleration effect. Therefore, the rheokinetic equation of curing includes a multiplier, which describes the variations of the reaction rate. However, in some cases interesting deviations from this general rule are observed.

From this viewpoint, of particular interest are the results on curing mechanisms of systems with reactive groups of identical nature but different functionality [19]. Figures 24, 27 present $\beta(t)$ dependences for curing of epoxy silicone oligomers by compounds having identical reactive groups but different functionality – tetra-butoxytitanium (TBT) and polybutoxytitanium phosphoroxane (PBTP). For a curing agent of low functionality, TBT the rheokinetics data are presented in Sect. 4.2. At the same time, for a curing agent of high functionality, PBTP, the $\beta(t)$ dependences do not have a visible S-shaped character and at relatively low temperatures ($<200\,°C$) a curing process is completed for $\beta < 1$.

As regards the visible S-shaped variation of β with t, this fact may be simply related to the way of representation of experimental data, i.e. to that the initial apparent "induction" part of the $\beta(t)$ dependence is too small to be noticeable on the experimental time scale.

The self-inhibition effect [106, 127] means that the conversion may be incomplete when definite restrictions are superimposed on the mobility of the elements of the reactive system [128]. Phenomenological kinetics of such processes may be described by the following equation:

$$\frac{d\beta}{dt} = k(1 - \beta)(1 - \xi\beta) \tag{22}$$

where ξ is a dimensionless parameter, which qualitatively reflects a self-inhibition contribution into the kinetics of the process.

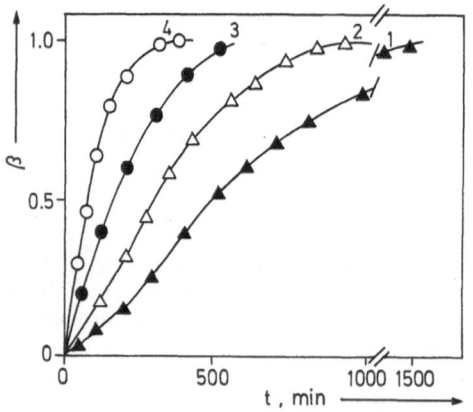

Fig. 27. Time dependence of conversion β for the system: epoxy silicone oligomer–TBT. Temperature (°C): 170 (1), 180 (2), 190 (3), 200 (4) [19]

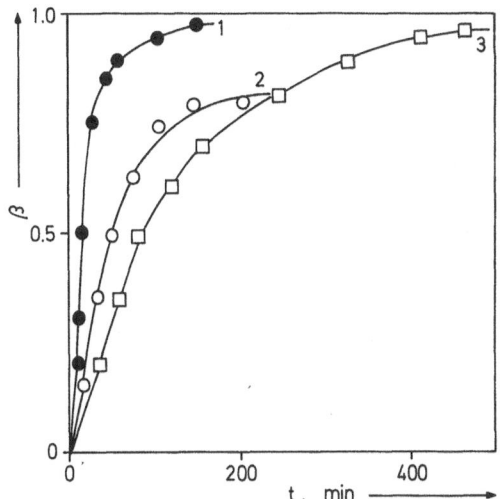

Fig. 28. Time dependence of conversion β for the system: epoxy silicone oligomer–PBTPh (100 : 7.5) (1, 2) and the system: epoxy silicone oligomer–PBTPh + 20% by weight of DBP (3). Temperature (°C): 220 (1); 180 (2, 3)

The constant ξ has a significantly different physical meaning than the quantity C in Eq. (18). The quantity ξ takes into account a mobility drag in the system, which leads to diffusion limitations effective in the course of the reaction. In some cases, these limitations can be removed by a simple increase in temperature or by introducing a plasticizer into a reactive system, i.e., by a change in mobility of reacting molecules. Realization of both possibilities is illustrated in Fig. 28. Increasing temperature (transition from initial curve 1 to 2) or introducing the plasticizer content (transition to curve 3) yields a degeneration of the self-inhibition in curing of an epoxy-silicone oligomer; a certain decrease in the rate after introduction of a plasticizer is obviously caused by a dilution effect. An increase in temperature results in a growth of the parameter ξ up to unity. Obviously, for $\xi = 1$ Eq. (22) is transformed into a second-order kinetic equation. From the physical meaning of the constant ξ as a value of β^{-1} (i.e., the maximum possible degree of curing), it is easy to find for $t \to \infty$ from experimental data the values of the self-inhibition constant. The integral Eq. (22) has the form:

$$\beta = \frac{\exp\left[(1 - \xi)\, kt\right] - 1}{\exp\left[(1 - \xi)\, kt\right] - \xi} \tag{23}$$

or

$$\ln \frac{1 - \xi\beta}{1 - \beta} = (1 - \xi)\, kt . \tag{24}$$

Thus, determining the value of ξ from the condition $\xi = \beta_{\infty}^{-1}$ by plotting β as a function of t in the coordinates $\ln \dfrac{1 - \xi\beta}{1 - \beta} - t$, one can verify the applicability of Eq. (22). Figure 29 represents experimental data in plotted according to Eq. (24)

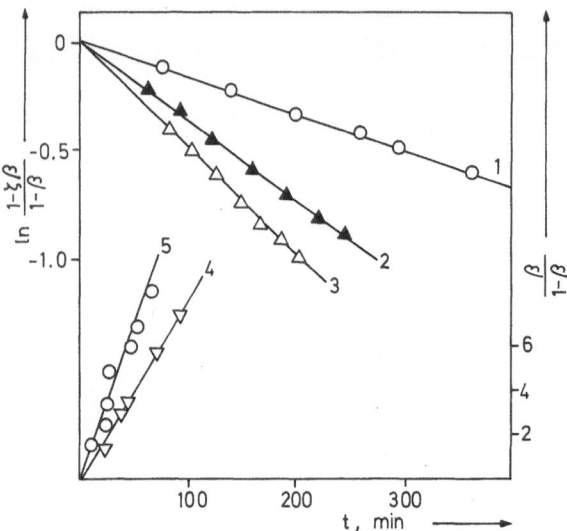

Fig. 29. Time dependence of $\ln (1 - \xi^{\beta})/(1 - \beta)$ (1–3) and $\beta/1 - \beta$ (4, 5) for the system: epoxy silicone oligomer–PBTPh (100 : 7.5) at curing temperatures (°C): 140 (1); 160 (2); 180 (3); 200 (4); 220 (5) [106]

for curing temperatures below 200 °C and according to the second-order kinetics equation $\beta/1 - \beta - t$ for curing temperatures of 200 °C and 220 °C. The fact that linear dependences are obtained in the indicated coordinates supports the validity of Eq. (22) for a low-temperature curing and its applicability to rheokinetic description of such processes [106, 127]. In this case, in a high-temperature region the process goes to the end and, therefore, its kinetics is described by a more simple equation of the second order (i.e., here $\xi = 1$ and $n = 2$ in Eq. (22)).

Equations of such a type (with an incomplete conversion) describe the kinetics of curing with curing agents of high functionality [19]. In such systems, gelation sets in very quickly, so that the stage of microgel formation is "passed over". However, a reduction in functionality of a curing agent results in the appearance of self-acceleration. Moreover, the lower is the functionality, the greater is the value of the self-acceleration constant [19].

The cases just discussed represent extreme cases of self-acceleration and self-inhibition effects in curing processes of reactive compounds, when one of them completely suppresses the other. However, their mutual superposition and simultaneous manifestation is also possible. Figure 30 gives the time dependences of the elastic moduli in the process of curing of melamine-formaldehyde oligomer [122]. The results obtained show that a characteristic incompleteness of the reaction is observed if curing occurs below 130 °C and here the final value of the elastic modulus becomes a function of temperature. Since Eq. (24) is inapplicable for describing the kinetics of curing of a given material it is necessary to write down an equation taking into account a combined action of self-acceleration and

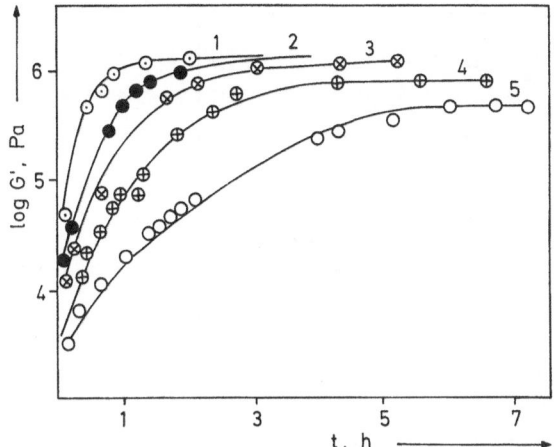

Fig. 30. Changes of the elastic modulus with curing of a melamine formaldehyde resin. T (°C) = 140 (1); 130 (2); 120 (3); 110 (4); and 100 (5) [122]

self-inhibition on phenomenological pattern of curing [121]:

$$\frac{d\beta}{dt} = k(1 - \beta)(1 + c\beta)(1 - \xi\beta) \tag{25}$$

When a curing temperature rises up to 130 °C, $\xi = 1$, and Eq. (25) turns into a second-order kinetic equation with self-acceleration:

$$\frac{d\beta}{dt} = k(1 - \beta)^2(1 + c\beta) \tag{26}$$

Equation (25) can be expressed with boundary condition $\beta = 0$ at $t = 0$ by the following formula:

$$\frac{(1 + c\beta)^{c(1-\xi)}(1 - \xi\beta)^{\xi(1+c)}}{(1 - \beta)^{\xi+c}} = \exp(1 + c)(1 - \xi)(\xi + c)kt \tag{27}$$

or in a form more convenient for calculation:

$$\frac{\ln\left[\dfrac{(1 + c\beta)^{c(1-\xi)}(1 - \xi\beta)^{\xi(1+c)}}{(1 - \beta)^{\xi+c}}\right]}{(1 + c)(1 - \xi)(\xi + c)k} = t \tag{28}$$

Equation (28) makes it possible to compare experimental and calculated dependences $\beta(t)$ as shown in Fig. 31. The given results indicate that Eq. (28) describes adequately the variation of rheological properties of the reactive system in the

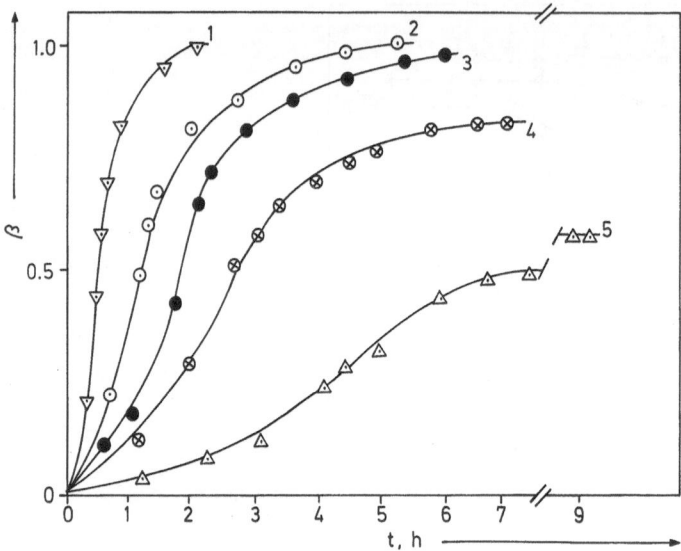

Fig. 31. Time dependence of rheological conversion β in curing of a melamine formaldehyde resin (MFR). Notations as in Fig. 30. Dots denote experiment, lines denote calculation via formula (27) [122]

process of curing of a melamine formaldehyde resin [122]. An increase in the temperature of the process to 130 °C leads to an almost complete degeneration of self-inhibition ($\xi = 1$), and $\beta(t)$ dependence is described by a formula obtained by an second order kinetic equation.

It is interesting to note that Eq. (28) is sufficiently general and describes well the kinetics of curing of fundamentally different systems with phase segregation like, for example, unsaturated polyester resin. This is shown in Fig. 32 where there are compared experimental data of Ref. [129] and curves calculated by Eq. (28).

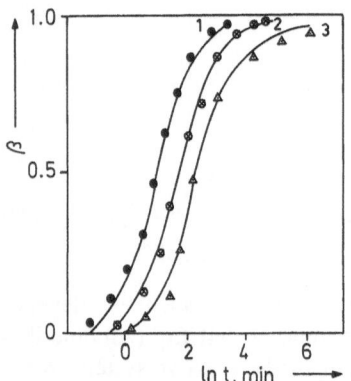

Fig. 32. Experimental (dots) and calculated (Eq. (28), solid lines) time dependences of conversion in semilogarithmic coordinates for the system: polyester resin−benzol peroxide. Curing temperature (°C): 90 (1); 85 (2); 80 (3) [125]

A good agreement between experimental and calculated curves is observed. Similar results were obtained when Eqs. (27) and (28) were used for describing the curing kinetics of reactive compounds different in chemical nature.

Since the physical state of reactive systems is different before gelation and after the gel point, different experimental methods are used — viscosimetry and dynamic mechanical spectroscopy, respectively. Here, a question arises on the generality of the processes proceeding at different stages of curing and on the comparability of results obtained by these two methods.

Equations (24) and (27) must have one common point which corresponds to the gel point. For the condition of reaching the gelation threshold, Eq. (28) can be rewritten as follows:

$$1/t^* = \frac{k(1 + c)(1 - \xi)(\xi + c)}{\ln\left[\dfrac{(1 + c\beta^*)^{c(1 - \xi)}(1 - \xi\beta^*)^{\xi(1 + c)}}{(1 - \beta^*)^{\xi + c}}\right]} \tag{29}$$

where $\beta^* > 0$ is the value of a "rheological" degree of conversion at the moment when a gel point is reached.

The expression (29) points to the existence of proportionality between the reciprocal value of the gelation time and the rate constant. Figure 33 presents the interrelation between $1/t^*$ and the rate constant determined from the rheokinetic equation for the curing of melamine-formaldehyde oligomer. The value of the proportionality factor determined from the data of Fig. 33 coincides with the value calculated from Eq. (29). In particular, for melamine-formaldehyde resin investigated in Ref. [81], Eq. (29) assumes the form:

$$\frac{1}{t^*} = 7 \times 10^{-3}k. \tag{30}$$

The established agreement between t^* and k^* shows that the rate of gelation and the nature of variation of the rheological degree of conversion is governed

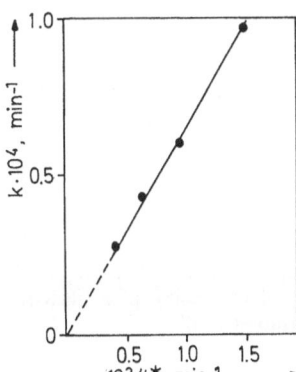

Fig. 33. Interrelation between $1/t^*$ and rate constant of curing [81]

by a single process. In principle, the reaction rate constant can be introduced in explicit form into Eq. (30) and determined from viscometric measurements. For example, the equation that makes it possible to predict the kinetics of viscosity growth in the reaction of epoxy oligomer with aromatic diamine in the neighborhood of the gel point has the following form [63]:

$$\eta/\eta_0 = 1 - \frac{k}{4} t^{-1,2}.$$

5 Rheokinetic Analysis under Optimization of Processing Conditions

A level of operating characteristics of composite materials on the basis of thermosetting binders is determined to a large extent by correctly selected temperature and time at different stages of the process of forming of articles. The problem of selecting optimum conditions may be solved if there are precise criteria determining the length of the respective stage of the production process. Let us discuss this problem taking as an example the production of articles by moulding from preimpregnated materials (prepregs). Empirical approach is based on running over "reasonable" parameters of processing to get the desired physical and mechanical properties of the composite material. At the same time, precise rheological criteria may be suggested related the rheokinetics of curing of compositions, since such criteria are very sensitive to a change in the degree of the processes of gelation and curing.

As an example, Fig. 34 gives the temperature − time diagram for different stages of the process of obtaining glass-reinforced thermoset [130, 131]. Regions I–IV in this diagram actually correspond to different parts of the viscosity curve

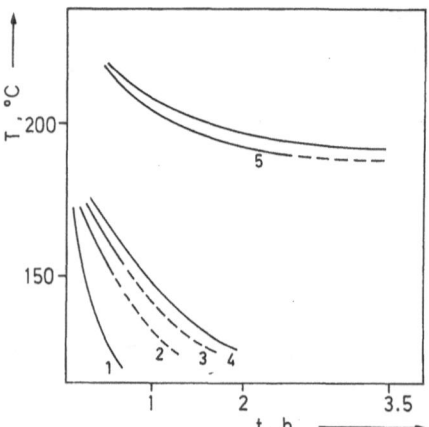

Fig. 34. Temperature-time diagram at different stages of processing of a glass-reinforced thermoset [130]

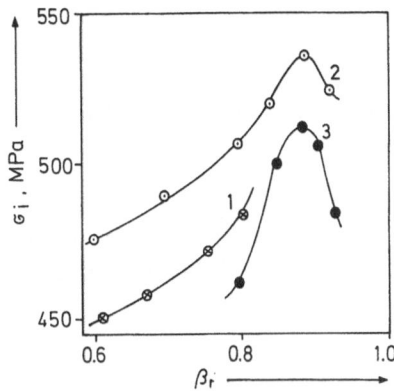

Fig. 35. Dependence of the bending stress of a glass-reinforced thermoset on rheological conversion [131]. T_{cure} (°C) = 180 (1); 190 (2); 200 (3)

given in Fig. 35 and are separated by characteristic points t_c, t_p and t^*. Regions I and II correspond to a homophase stage of the gelation process and are characterized by a relatively low level of the binder viscosity. This corresponds to the optimum conditions for impregnating a filler by a binder (I) and performing of a glass-reinforced thermoset (II), where the filler is reimpregnated under low pressure. The temperature and time of the final forming of glass-reinforced plastic at a given pressure must lie in region III. Region IV includes heterophase stage of gelation and is characterized by an uncontrolled viscosity change up to the gel point. This stage of gelation is "useless" from the point of view of processing, since here the network structure has already been formed and any engineering operations may only deteriorate the quality of the material being formed.

Region V is the region of curing. For the given example, it corresponds to the values of "rheological" degree of conversion 0.86–0.92. Such a selection of optimum values of conversion is explained by the fact that just for such values of β the maximum values of physical and mechanical characteristics of glass-reinforced plastic are reached (Fig. 35).

The example just shown characterizes the application possibilities of rheokinetic analysis of curing of composite material binders from the point of view of optimizing their composition and conditions of curing.

6 Conclusion

The nature of relaxation states of oligomers in the process of curing determines the application of different experimental rheological methods at different stages of the reaction, i.e. viscometry and dynamic mechanical spectroscopy. Rheological properties, viscosity and the dynamic modulus, are very sensitive to the variation of molecular structure of cured oligomers and phase and phase and relaxation transitions in a reactive system, associated with the process of chemical conversion. From this point of view, rheology or rheokinetics, has unique possibilities for studying the process and description of curing in all the variety of manifestations of chemical and physical mechanisms of this complex phenomenon.

Rheokinetic analysis shows that rheokinetic equations suggested to describe changes in viscosity and elastic modulus reflect the characteristics of this process. Therefore, the nature of the viscosity increase up to the gel point is defined not only by a variation of the molecular structure of the material but sometimes also by the effect of formation of a new phase (microgel) from particles of a cured product.

Rheokinetic phenomenology of curing of oligomers is linked with the phase structure and covers a wide circle of thermoset materials. A self-acceleration effect is characteristic for some of them and if molecular mobility is limited conversion may be incomplete.

7 References

1. Gillham JK, Roller MB (1971) Polym Eng Sci 11: 295
2. Babayevsky PG, Gillham JK (1973) J Appl Polym Sci 19: 1065
3. Gillham JK (1979) Polym Eng Sci 19: 319
4. Gillham JK (1979) Polym Eng Sci 19: 676
5. Enns JB, Gillham JK (1983) J Appl Polym Sci 28: 2567
6. Gillham JK (1985) Brit Polym J 17, 2: 224
7. Harrison DJP, Yates WR, Johnson JF (1985) J Macromol Sci Rev Macromol Chem Phys C 25 (4): 481
8. Flory PJ (1953) Principles of polymer chemistry, Ithaca (N4) Cornell University Press
9. Gonzales-Romero VM, Macosko CW (1985) J Rheology, 29 (3) 259
10. Lee DS, Han CD (1987) Polym Eng Sci 27, N 13: 964
11. Kulichikhin SG, Reutov AS, Surova MS, Osipova EV, Malkin AYa (1988) Plastmassy 5: 43
12. Macosko CW (1985) Brit Polym J 17, 2: 239
13. Astakhov PA, Kukichikhin SG, Golubenkova LI, Kozhina VA, Chibisova EI, Chernov YuP, Malkin AYa (1984) Vysokomolek soed 26, B, 11: 864
14. Tung CY, Dynes PY (1982) J Appl Polym Sci 27, 2: 569
15. Winter HH, Chambon FY (1986) J Rheology 30, 2: 367
16. Chambon F, Winter HH (1985) Polymer Bull 13: 499
17. Winter HH (1987) Polym Eng Sci 27, 22: 1698
18. Koton MM, Frenkel SYa, Panov YuN, Bolotnikova LS, Svetlichnyi VM, Shibayev LA, Kulichikhin SG, Krupnova EE, Reutov AS, Ushakova IL (1988) Vysokomolek soed 30, A, 11: 2837
19. Kulichikhin SG, Astakhov PA, Chernov YuP, Kozhina VA, Golubenkova LI, Malkin AYa (1986) Vysokomolek soed 28, A 10: 2115
20. Kulichikhin SG, Chernov YuP, Kozhina VA, Reutov AS, Miroshnikova II, Malkin AYa (1988) Mekhanika kompositnykh materialov 2: 350
21. Vinogradov GV, Malkin AYa (1977) Reologiya polimerov, Khimiya Moscow, p 438 Rheology of polymers (1980) Mir Publishers, Moscow; Springer-Verlag, Berlin Heidelberg New York, p 468
22. Valles EM, Macosko CW (1979) Macromolecules 12: 637
23. Papkov SP Studneobraznoye sostoyanie polimerov (Jelly-Like State of Polymers) Khimiya, Moscow
24. Djabourov M, Maquet J, Thevenau H, Leblond J, Papon P (1985) Brit Polym J 17, 2: 169
25. Danilkin NN, Kanavetz IF (1970) Plastmassy, 1: 29
26. Sokolov AD (1979) Plastmassy, 7: 43
27. Malkin AYa, Merzhanov AG, Frunze TN, Davtyan SP, Kulichikhin SG, Stolin AM, Maizeliya VV, Volkova TF, Shleifman RB, Kotelnikov VA, Kurashev VV (1981) Doklady AN SSSR 258, 2: 402

28. Malkin AYa, Kulichikhin SG, Kozhina VA, Abenova ZD (1986) Vysokomolek soed 28, B, 6: 408
29. Malkin AYa, Beghishev VP (1983) Polymer Proc Eng 1, 1: 83
30. Malkin AYa (1984) Mekhanika kompozitnykh materialov 2: 362
31. Malkin AYa, Shuvalova GI (1985) Vysokomolek soed 27, B, 11: 865
32. Kulichikhin SG, Reutov AYa, Malkin AYa (1988) Vliyanie funktsional'nosti otver-ditelya na polozheniye tochki geleobrazovaniya v protsesse trekhmernoi polikon-densatsii (Effect of functionality of curing agent on the position of a gelpoint in the precess of three-dimensional polycondensation) in: Polikondensatsionnye protsessy i polimery (Polycondensation precesses and polymers) Nal'chik, p 106
33. Castro JM, Macosko CW, Perry SJ (1984) Polym Commun 25: 82
34. Kamal MR, Sourour S (1973) Polym Eng Sci 13: 59
35. White RP (1974) Polymer Eng Sci 14: 50
36. Roller MB (1975) Polymer Eng Sci 15: 406
37. Lee WI, Loos AC, Springer GS (1982) J Composite Matter 16: 510
38. Liska V (1980) Polim-tworz, wielkoczasteczk 25, 6–7: 219
39. Schwesig H, Heimenz C, Milcke W, Menges G (1980) Kautsch and Gummi Kunstst 33: 15
40. Mussati FG, Macosko CW (1973) Polym Eng Sci 13: 236
41. Lipshitz SD, Macosko CW (1976) Polym Eng Sci 16, 12: 803
42. Lipshitz SD, Macosko GW (1977) J Appl Polym Sci 21: 2029
43. Malkin AYa, Beghishev VP, Kulichikhin SG, Kozhina VA (1983) Vysokomolek soed 25, A, 9: 1948
44. Valles FM, Macosko CW, Hickey WJ (1979) Amer Chem Soc, Polymer Prepr 20, 2: 153
45. Valles FM, Macosko CW (1979) 12: 521
46. Naroditskaya EYa, Khodzhaeva ID, Kulichikhin SG, Pozdnyakov VYa, Yunitskii IN, Kireev VV, Malkin AYa (1985) Vysokomolek soed 27, B, 9: 713
47. Malkin AYa, Kulichikhin SG (1985) Reologiya v protsessah obrazovaniya i prevrash-cheniya polimerov (Rheology in the processes of production and transformation of polymers) Khimiya, Moscow, p 240
48. Kulichikhin SG, Kozhina VA, Bolotina LM, Malkin AYa (1982) Vysokomolek soed 24, B, 4: 309
49. Schimazaki A (1968) J Appl Polym Sci 12, 10: 2013
50. Sergeeva LM, Lipatov YuS, Bin'kevich NI (1967) Sintez i fiziko-khimiya poliuretanov (Synthesis ans physicochemistry of polyurethanes), Naukova Dumka, Kiev, 131
51. Stauffer D, Coniglio A, Adam M (1982) Adv Polym Sci 44: 103
52. Adam M, Delsanti M, Durand D, Hild G, Munch JP (1981) Pure Appl Chem 53: 1489
53. Kulichikhin SG Reokinetica protsessov otverzhdeniya epoksidnykh oligomerov (Rheo-kinetics of curing processes of epoxy oligomers) in: Problemy teplo- i massoperenosa v topochnykh ustroistvakh, gasogeneratorakh i khim. reaktorakh (1983) Minsk, p 88
54. Malkin AYa, Kulichikhin SG, Kozhina VA, Bolotina LM (1987) Vysokomolek soed 29, A, 2: 418
55. Bulai AKh, Klyuchnikov VN, Urman YaG, Slonim IYa, Bolotina LM, Kozhina VA, Gol'der MM, Kulichikhin SG, Beghishev VP, Malkin AYa (1987) 28: 1349
56. Kulichikhin SG, Demina GI, Bokareva EZ, Malkin AYa (1980) Plastmassy 1: 57
57. Kulichikhin SG, Kozhina VA, Bolotina LM, Malkin AYa (1981) Proizv i pererab plastmass i sint smol 7: 19
58. Kulichikhin SG, Ivanova SL, Korchagina MA, Malkin AYa (1980) Vysokomolek soed 22, A, 1: 165
59. Malkin AYa, Kulichikhin SG, Ivanova SL, Korchagina MA (1980) Vysokomolek soed 22, A, 1: 165
60. Kulichikhin SG, Malkin AYa (1980) Vysokomolek soed 22, A, 9: 2093
61. Malkin AYa, Kulichikhin SG (1980) Rheology, p 407, Proc 8-th Intern Gongr Rheol, Napoli, Italy, Plenum Press, NY

62. Malkin AYa, Kulichikhin SG, Emel'yanov DN, Smetanina IE, Ryabokon NV (1984) Polymer 25: 778
63. Kulichikhin SG, Nechitailo LG, Gerasimov IG, Kozhina VA, Zaitsev YuS Vysokomolek soed (in press)
64. Enns JB, Gillham JK (1983) J Appl Polym Sci 28, 8: 2567
65. Aronhime MT, Gillham JK (1984) J Appl Polym Sci 29, 6: 2017
66. Lee DS, Han CD (1987) Polym Eng Sci 27, 13: 955
67. Tajima YA, Crozier D (1983) Polym Eng Sci 23: 186
68. Riccardi SS, Adabbo HE, Wiliams JJ (1984) J Appl Polym Sci 29, 8: 2481
69. Apicella A, Nicolais L, Jannon M, Passerini P (1984) J Appl Polym Sci 29, 6: 2083
70. Dusek K (1985) Brit Polym J 17, 2: 185
71. Dusek K (1967) J Polym Sci C 16: 1289
72. Dusek K. (1984) Formation, structure and elastic properties polymer network. Preprints of the International Rubber Conference, Moscow
73. Matejka L, Pokorny S, Dusek K (1985) Macromol Chem 186: 2025
74. Dusek K, Plestil J, Lednicky F, Lunak S (1978) Polymer 19: 393
75. Bobalek E, Moore, Levy S, Lee C (1964) J Appl Polym Sci 8: 625
76. Lipatova TE (1973) Kataliticheskaya polimerizatsiya oligomerov i formirovaniye polimernykh setok (Catalytic polymerization of oligomers and formation of polymer networks) Naukova Dumka, Kiev, p 208
77. Kenon AS, Nielsen LE (1969) J Macromol Chem 3: 275
78. Berlin AA, Korolev GV, Kefeli TYa, Sivergin YuM (1983) Akrilovye oligomery i materialy na ih osnove (Acrylic oligomers and materials on their base) Nauka, Moscow, p 232
79. Lipatov YuS Sovremennye predstavleniya o structure i svoistvakh napolnennykh vulkanizatov (Modern concepts of structure and properties of filled vulcanizers) (1984) Preprinty mezhdunarodnoi konferentsii po kauchuku i rezine, Moscow, Preprint A2
80a. Macosko CW (1985) Brit Polym J 17: 239
80b. Apicella A, Nicolais L (1984) J Polym Sci (1984) v 29, N6, p 2083
80c. Apicella A, Masi P, Nicolais L (1984) Rheol Acta 23: 291
80d. Apicella A, Nicolais L (1985) ACS Polym Prepr 26: 291
80e. Valles EM Macosko CW (1979) Macromolecules 12: 521
81. Kulichikhin SG, Abenova ZD, Bashta NI, Kozhina VA, Blinkova OP, Romanov NM, Matvelashvili GS, Malkin AYa (1988) Vysokomolek soed A 12: 2497
82. Dusek K (1982) In: Haward RN (ed), Developments in polymerization, 3. edn. Applied Science Publishers, London: p 143
83. Boots HMY, Klossterboer JG, Van de Hei GMM, Randey RB (1985) Brit Polym J 17, 2: 219
84. Toussaint F, Cuypers P, D'Hont L (1985) J Coat Techn 57, 728: 71
85. Candau SJ, Ankrim M, Munch JP, Hild G (1985) Brit Polym J 17, 2: 215
86. Galina H, Kolarz BN, Wieczorek PP, Wojczynska M (1985) Brit Polym J 17, 2: 215
87. Malkin AYa, Kulichikhin SG, Abenova ZD, Kozhina VA, Bashta NI, Kus'mina LA, Blinkova OP, Brysin YuP, Romanov NM, Matvelashvili GS (1988) Vysokomolek soed A 11: 2381
88. Kulichikhin SG (1987) Otverzhdeniye reactsionnosposobnykh oligomerov (Curing of reactive oligomers) Review Information. NIITEHIM, Moscow
89. Macosko GW, Miller DR (1976) Macromolecules, 9, 199: 206
90. Miller DR, Valles EM, Macosko CW (1979) Polym Eng Sci 19: 272
91. Kulichikhin VG, Malkin AYa, Vinogradov GV (1970) Vysokomolek soed 12, A, 1: 129
92. Chong JS, Christiansen EB, Bayer AD (1971) J Appl Polym Sci 15, 8: 2007
93. Malkin AYa, Adv Polymer Sci (in press)
94. Kulichikhin SG, Malkin AYa (1989) Vysokomolek soed 30, B, 12: 865
95. Kulichikhin SG, Mikhalin SV, Kotov YuI, Kozhina VA, Shelonina IM, Kachevskii OV, Cherkasov MV, Agapov OA, Vasil'ev VV, Malkin AYa, Matvelashvili GS (1988) Vysokomolek soed 30, A, 4: 707

96. Han CD, Lem KW (1983) J Appl Polym Sci 28, 10: 3155
97. Han CD, Lem KW (1984) J Appl Polym Sci 29, 5: 1879
98. Reutov AS, Kulichikhin SG, Malkin AYa (1989) Plastmassy 10: 41
99. Harran D, Landourd A (1986) J Appl Polym Sci 32: 6043
100. Emel'yanov DN, Golubev AA, Ryabov AV, Belyaev EL (1974) Vysokomolek soed A 16, 11: 2426
101. Flory PJ (1979) Macromolecules 12, 1: 119
102. Charlesworth JM (1988) Polym Eng Sci 28, 4: 229
103. Andrady AL, Llorente MA (1980) J Cem Phys 72, 4: 632
104. Macknight WJ, Chundary D (1983) Polymer Preprints 27, 2: 67
105. Naroditskaya EYa, Kulichikhin SG, Khodzhayeva ID, Pozdnyakov VYa, Yunitskii YuN, Kireev VV, Malkin AYa (1985) Protsessy studneobrazovaniya v polimernykh sistemakh (Processes of gel formation in Polymer systems) SGU Publishers, Saratov, part I, p 89
106. Kulichikhin SG, Astakhov PA, Chernov YuP, Kozhina VA, Golubenkova LI, Malkin AYa (1985) Problemy perenosa v Structuriruyushchihsya zhidkostyah (Transfer problems in Structure-forming liquids) ITMO Publishers, Minsk, AN BSSR, p 63
107. Ponomareva TI, Dzhavadyan EA, Irzhak VI, Rosenberg BA (1988) Mekhanika kompositnykh materialov 2: 347
108. Shuvalova GI, Pshenitsyna VP, Pakhomov VI, Perevertov AS (2983) Plastmassy 7: 57
109. Kulichikhin SG, Shuvalova GI, Kozhina VA, Chernov YuP, Malkin AyA (1986) Vysokomolek soed 28, A, 3: 498
110. Maklakov AI, Skirda VD, Fakullin NF (1987) Samodiffuziya v rastvorakh i rasplavakh polymerov (Self-diffusion in solutions and melts of polymers) KGU Publishers, Kazan, p 224
111. Folland R, Charlesby A (1979) Polymer 20: 211
112. Paci M, Compana F (1985) Polymer 26, 12: 1885
113. Sacher E (1973) Polymer 14, 3: 91
114. Malkin AYa, Kulichikhin SG, Batizat VP, Chernov YuP, Klimova IV, Moskaleva TA (1984) Vysokomolek soed 26 A, 10: 1815
115. Sourour S, Kamal MR (1976) Thermochimica Acta 14: 41
116. Rozenberg BA (1975) Kompozitsionnye polimernye materialy, Kiev, p 39
117. Irzhak VI, Rozenberg BA (1985) Vysokomolek soed 27, A 9: 1795
118. Irzhak VI, Rozenberg BA, Enikolopyan NS (1979) Setchatye polimery: sintez, struktura, svoistva (Network polymers: Synthesis, structure, properties) Nauka, Moscow, p 248
119. Dutta A, Ryan M (1979) J Appl Polym Sci 24, 3: 635
120. Korolev GV, Berlin AA (1962) Vysokomolek soed 4, 11: 1654
121. Kulichikhin GS (1986) Mekhanika kompozitnykh materialov, 6: 1087
122. Kulichikhin SG, Abenova ZD, Bashta NI, Blinkova OP, Matvelashvili GS, Malkin AYa (1988) Plastmassy 4: 57
123. Rodivilova LA, Koroleva LA, Yurchenko NV, Kulichikhin SG, Kozhina VA, Abenova ZD, Chernov YuP, Dubrovskii YuA, Safonova VA (1987) Poliamidnye konstruktsionnye materialy NIITEHIM, Moscow, p 58
124. Malkin AYa, Beghishev VP, Kipin IA (1982) Vysokomolek soed 24, B, 9: 656
125. Malkin AYa, Beghishev VP, Keapin IA, Bolgov SA (1984) Polymer Eng Sci 24: 1396
126. Cesari M et al. (1976) J Polym Sci Polymer Let Ed 14: 107
127. Malkin AYa, Kulichikhin SG, Astakhov PA, Chernov YuP, Kozhina VA, Golubenkova LI (1985) Mekhanika kompozitnyh materialov 5: 878
128. Irzhak VI, Peregudov II, Rozenberg BA, Enikolopyan NS (1982) Dokl AN SSSR 263, 3: 630
129. Han CD, Lem RW (1984) Polym Eng Sci 24, 7: 473
130. Kulichikhin SG, Astakhov PA (1989) Plastmassy 12: 48
131. Astakhov PA, Kulichikhin SG, Kozhina VA, Egorova NV, Chernov YuP, Golubenkova LI, Malkin AYa (1987) Proizv i pererab plastmass i sint smol 2: 9

Editor: K. Dusek

Received July 5, 1990

Author Index Volume 101

Subject Index